Chemistry and Physics of Carbon

VOLUME 27

Chemistry and Physics of Carbon

A Series of Advances

Edited by

LJUBISA R. RADOVIC

Department of Energy and Geo-Environmental Engineering
The Pennsylvania State University
University Park, Pennsylvania

VOLUME 27

CRC Press
Taylor & Francis Group
Boca Raton London New York

CRC Press is an imprint of the
Taylor & Francis Group, an **informa** business

CRC Press
Taylor & Francis Group
6000 Broken Sound Parkway NW, Suite 300
Boca Raton, FL 33487-2742

First issued in paperback 2019

© 2001 by Taylor & Francis Group, LLC
CRC Press is an imprint of Taylor & Francis Group, an Informa business

No claim to original U.S. Government works

ISBN-13: 978-0-8247-0246-5 (hbk)
ISBN-13: 978-0-367-39794-4 (pbk)

Visit the Taylor & Francis Web site at
http://www.taylorandfrancis.com

and the CRC Press Web site at
http://www.crcpress.com

Preface

In this presentation of our 27th volume, I want to briefly recapitulate where we stand with this fascinating "old but new material," as Phil Walker so aptly called carbon in his inaugural preface to this series almost four decades ago. In the intervening period Professors Phil Walker and Peter Thrower, my distinguished predecessors, have assembled authoritative reviews on a remarkable range of materials, including nuclear graphite, activated and molecular-sieving carbons, carbon blacks, cokes, pyrolytic carbons and graphite, doped carbons, carbon fibers and composites, diamonds, filamentous carbon, intercalated graphite, and carbyne. Their applications in a remarkable variety of fields have been analyzed: nuclear reactors, coal gasification, adsorption, separation and pollution control, lubrication and tribology, bioengineering, aerospace engineering, materials reinforcement, metallurgy, electrochemistry, and catalysis, among others. All this using a remarkable variety of chemical and physical characterization methods: optical and electron microscopy, electron spin resonance, infrared spectroscopy, measurements of magnetoresistance and electronic properties, irradiation damage, deformation mechanisms and fracture, chemical kinetics, determination of transport and thermodynamic properties, scattering of light, x-rays, neutrons and electrons. We have also witnessed quite a few shifts in "popularity," from the dominance of nuclear graphite until the 1960s, to the recent "age" of carbon fibers and composites, and now with fullerenes and carbon nanotubes.

Throughout all these years it was essential to have, in the words of Phil Walker from his inaugural preface, a "monograph series of recent advances in carbon research and development and of comprehensive reviews on past achievements in important areas," a series that uses a "truly interdisciplinary approach." In the 21st century, our goals remain the same, reflecting the amazing vigor and fertility of carbon science and the ability of carbon technology to reinvent itself. But the challenges seem to be greater. With the advances in information technology both facilitating and hampering the transfer of knowledge, it will be even more important for scientists and engineers to be able to rely on authoritative, critical, *and* comprehensive reviews. As Mary Anne Fox emphasized in her recent analysis of the future of review articles (*Chem. Rev.*, 2000, 100, 11–12), "the critical contribution of a well written review [is in seeing] connections between published works that would be missed by a computer, particularly if they originate in different subdisciplines." Artificial intelligence and on-line searches are beginning to replace human activities in many areas, but it will be many decades, if ever, before a computer is able to "write" a review of the same caliber as we shall continue to offer our readers in this series.

In Volume 26 we assembled three reviews in which the control of *bulk* properties of carbon was shown to be the key to successful and widely varying applications. The common theme in the present volume is that *surface* properties are crucial in the environmental and electrochemical applications of carbon.

The late Frank Derbyshire and his colleagues set the stage with an eclectic analysis of myriad environmental applications. Frank's sudden death has been a tremendous loss for the carbon community. This chapter will serve as a lasting testimony to how he enriched our science, and our lives: by communicating a firm grasp of the fundamentals and an infectious zeal to appreciate their connectivity and realize their potential. In vintage Derbyshire words, which we shall cherish forever, "we must strive and reach out in our interactions and collaborations to help to translate the positive aspects of carbon science into new technologies."

Dr. Turov and Professor Leboda combine their expertise in nuclear magnetic resonance and adsorption phenomena to propose a new tool for a more incisive analysis of adsorbate–adsorbent interactions. Such an analysis is of critical importance in so many applications where it is becoming increasingly clear that adsorbate–carbon interactions are governed by both pore size and surface chemistry effects. These range from the ubiquitous water adsorption to the design of carbon-coated silicas with tailored ratios of hydrophobic to hydrophilic surface sites.

In assembling this volume, it is only natural that we have included another contribution from Poland, with its long tradition of expertise in the surface properties and behavior of carbon materials (see also Vols. 21 and 22). Drs. Biniak, Świątkowski, and Pakuła have prepared a particularly timely review of carbon electrochemistry, a much needed follow-up on a "call-to-action" chapter by León y León and Radovic in Vol. 24. There is great interest today, and many unanswered questions remain, regarding the virtues of specific carbon materials as electrodes and in electroanalysis, electrosynthesis, electrosorption, and electrocatalysis. It is our hope that the insights offered in this chapter will be particularly helpful in the design of a new generation of batteries and fuel cells, for which a tremendous market has developed over the past few years.

The collaboration of Professors Radovic, Moreno-Castilla, and Rivera-Utrilla has resulted in a comprehensive and unifying in-depth survey of a sorely neglected topic. For an earlier review of liquid-phase adsorption in this series, see the chapter by Zettlemoyer and Narayan in Vol. 2. Here the focus is on activated carbons and aqueous solutions, which are of greatest practical significance, but both organic and inorganic adsorbates are included and contrasted. This has allowed the authors to formulate significant generalizations that translate into recipes for optimizing carbon surfaces and maximizing their uptakes of specific water pollutants. It remains to be seen when and how these advances in our understanding of carbon behavior will result in better carbon products for water treatment, a mature and large-scale technology that Derbyshire and his colleagues have identified as ripe for major (and much needed!) breakthroughs.

As we continue with these efforts to synthesize our knowledge of chemistry and physics of carbon well into the 21st century, I invite your suggestions for reviews to be included in upcoming volumes. Even more important will be your help in identifying potential authors (including yourselves!) who will take the time, and have the expertise, to strike the right balance between comprehensive coverage and critical analysis.

Ljubisa R. Radovic

Contributors to Volume 27

Rodney Andrews Center for Applied Energy Research, University of Kentucky, Lexington, Kentucky

Stanisław Biniak Faculty of Chemistry, Nicolaus Copernicus University, Toruń, Poland

Frank Derbyshire† Center for Applied Energy Research, University of Kentucky, Lexington, Kentucky

Eric A. Grulke Department of Chemical and Materials Engineering, University of Kentucky, Lexington, Kentucky

Marit Jagtoyen Center for Applied Energy Research, University of Kentucky, Lexington, Kentucky

Roman Leboda Faculty of Chemistry, Maria Curie-Sklodowska University, Lublin, Poland

Ignacio Martin-Gullón Center for Applied Energy Research, University of Kentucky, Lexington, Kentucky

---

† Deceased.

Carlos Moreno-Castilla　Department of Inorganic Chemistry, University of Granada, Granada, Spain

Maciej Pakuła　Naval Academy, Gdynia, Poland

Ljubisa R. Radovic　Department of Energy and Geo-Environmental Engineering, The Pennsylvania State University, University Park, Pennsylvania

Apparao Rao　Center for Applied Energy Research, University of Kentucky, Lexington, Kentucky

José Rivera-Utrilla　Department of Inorganic Chemistry, University of Granada, Granada, Spain

Andrzej Świątkowski　Institute of Chemistry, Military Technical Academy, Warsaw, Poland

V. V. Turov　Institute of Surface Chemistry, National Academy of Sciences of Ukraine, Kyiv, Ukraine

Contents of Volume 27

Contents of Other Volumes

Chemistry and Physics of Carbon

VOLUME 27

1

Carbon Materials in Environmental Applications

Frank Derbyshire,† Marit Jagtoyen, Rodney Andrews, Apparao Rao, Ignacio Martin-Gullón, and Eric A. Grulke

University of Kentucky, Lexington, Kentucky

---

† Deceased.

I. INTRODUCTION

The tremendous diversity that is available in the structure and properties of carbon materials underscores their utilization in virtually all branches of science and engineering, ranging from high-technology applications to medicine and heavy industry. Moreover, the list of carbons that are currently available or are under development is impressive and is being continually extended. One of the fastest growing areas is in environmental applications, not the least because carbon is compatible with all forms of life. Enormous interest in carbon materials has emerged over the past several years driven by environmental awareness and regulation and the need to find relatively low-cost solutions for protection and remediation. In this respect, the materials of interest are predominantly activated carbons, although there are areas where nonporous carbons also offer advantages.

This chapter is concerned with technologies that allow carbon materials to be used in environmental applications. In attempting to define the scope of this work, the authors have adopted a relatively broad interpretation and have considered where carbon materials can play a role both directly and indirectly in protecting and improving the quality of the environment and human health.

The first section is mainly concerned with some of the emerging technologies for the production of novel activated carbons that may help to fulfill the increasing demands on performance as regulations on environmental pollution become more stringent. The next two sections deal with applications of conventional and new activated carbons in gas and liquid phase applications. The penultimate section considers the uses of carbons supports for zero-valent metal dehalogenation. The last section briefly describes some of the uses and potential uses of carbons in medical technology and as sensors or detectors for monitoring and control, and to provide means to make more efficient use of energy, thereby lowering the associated emissions.

Over the last two decades, the discovery of new carbon structures in the form of closed cage molecules has demonstrated that carbon science is alive and well and is a productive font for the synthesis of new materials. Advancements in new and more conventional areas of carbon research and development continue at a very fast pace. Now and in the future, there are perhaps more opportunities than ever before for the development of new materials consisting of carbon or containing one or multiple carbon structures.

II. EMERGING TECHNOLOGIES

A. Activated Carbons

Activated carbons are produced with a wide range of properties and physical forms, which leads to their use in numerous applications (Table 1). For example, their high internal surface area and pore volume are pertinent to their being employed as adsorbents, catalysts, or catalyst supports in gas and liquid phase processes for purification and chemical recovery. General information on the manufacture, properties, and applications of conventional activated carbons can be found in *Porosity in Carbons*, edited by John Patrick [1].

Activated carbons are commonly fabricated in the form of fine powders and larger sized granules, pellets, or extrudates (Fig. 1). However, activated carbons can also be produced as fibers, mesocarbon microbeads, foams and aerogels, and in the form of flexible or rigid solids. The form, as well as other properties, helps to determine the suitability for a specific application. Rates of adsorption or reaction can be orders of magnitude higher for activated carbons with narrow dimensions (e.g., powders and fibers with diameters in the range 10 to 50 μm) than for granular carbons (0.5 to 4.0 mm): in the former case, most of the adsorptive surface is readily accessible, and there is much less dependence on intragranular diffusion (Fig. 2).

Powdered activated carbons are used almost exclusively in liquid phase applications, usually in a batch-processing mode. Operation can be flexible because

TABLE 1

Surface area (BET, nitrogen)	500–2500 m^2/g
Total pore volume	0.5–2.6 cm^3/g
Bulk density	<0.1–0.6 g/cm^3
Widely varying hardness and abrasion resistance	
World consumption	~400,000 tonnes/year

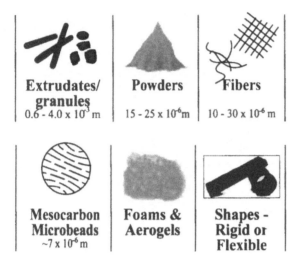

| Extrudates/ granules 0.6 - 4.0 x 10⁻³ m | Powders 15 - 25 x 10⁻⁶ m | Fibers 10 - 30 x 10⁻⁶ m |
| Mesocarbon Microbeads ~7 x 10⁻⁶ m | Foams & Aerogels | Shapes - Rigid or Flexible |

FIG. 1 Forms and properties of activated carbon.

the dosage of activated carbon can be readily adjusted. However, powders present difficulties in handling and are not conducive to use in fixed beds, or to regeneration. In contrast, granular activated carbons are used in both liquid and gas phase applications, lending themselves to continuous or cyclic processing in fixed or moving beds, and are regenerable.

Performance requirements are becoming increasingly more demanding, and there is a growing need for activated carbons with new and improved properties.

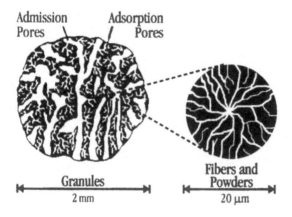

FIG. 2 Schematics showing internal structure of activated carbon forms.

Relatedly, there has been a corresponding increase in research and development, as reflected by the large number of papers that are presented at the various carbon conferences. Advances in activated carbons are expected to emanate from using alternative synthesis routes and precursors, from the development of new forms, and through techniques to modify surface chemistry. Some examples are given in the following sections.

B.   Adsorbents and Catalysts by KOH Activation

In the 1970s, researchers at the AMOCO Corporation, USA, developed a process to produce extremely high surface area carbons (over 3000 m^2/g) by the KOH activation of aromatic precursors such as petroleum coke and coal [2]. The process was commercialized by the Anderson Development Company, USA, in the 1980s and was subsequently licensed and operated on pilot plant scale by the Kansai Coke and Chemicals Co. Ltd., Japan. At this time, only limited quantities of material have been produced. The activated carbon is predominantly microporous, which is responsible for the high surface area, and the total pore volume is exceptionally high: 2.0–2.6 mL/g.

Examination by high-resolution electron microscopy has revealed that the carbons possess a homogeneous cagelike structure, the walls of which consist of one to three graphitelike layers. It is the delicate nature of this structure that may account for the ability of the carbon to swell and accommodate larger molecules than its microporosity would indicate [3]. Despite its extraordinary properties, the cost, low bulk density, and difficulties in handling have presented obstacles to its successful commercialization.

In a variation of this process, activated carbons have been prepared by the reaction of KOH with coals under carefully selected conditions [4,5] to yield catalysts of high activity suitable for the hydrodehalogenation of organic compounds. Such reactions are very much of interest for environmental protection. In experiments to examine the catalytic activity for the hydrodebromination of 1-bromonaphthalene at 350°C and the hydrodehydroxylation of 2-naphthol at 400°C in the presence of a hydrogen donor, 9,10-dihydronaphthalene, it was found that specific KOH-coal catalysts increased the extent of dehydroxylation by about 40% and of hydrodebromination by a factor of about 3 relative to commercial carbons. The conditions used to obtain the maximum selectivity for the catalysts were different in each case, demonstrating that the catalytic properties of these materials can be tailored to a particular application by adjusting the variables used in their preparation (e.g. concentration of KOH, heat treatment profile and gas atmosphere, coal rank, and method of coal cleaning).

C.   Activated Carbon Fibers

There is growing interest in the development and application of activated carbon fibers (ACF), whose unusual properties can be advantageous in certain applications. ACF can be produced with high surface area, and their narrow fiber diameter (usually 10 to 20 μm) offers much faster adsorption, or catalytic reaction, than is possible for granular carbons. Activated carbon fibers are currently produced from carbon fibers made from polyacrylonitrile and from isotropic pitch precursors that are derived from coal tar and petroleum. Activated carbon is also produced from cross-linked phenolic resins [6], and from viscose [7].

Elsewhere, there is considerable interest in exploring alternative routes for the synthesis of ACF and for modifying their properties. Mochida and coworkers at Kyushu University, Japan, have employed ACFs derived from polyacrylonitrile and pitch to study the catalytic oxidation of NO to NO_2 [8] and the oxidation of SO_2 to SO_3 at ambient temperatures: SO_3 is subsequently recovered as sulfuric acid [9,10]. They have found that the catalytic properties of the ACFs can be altered by heat treatment. Economy and coworkers at the University of Illinois, USA, are exploring the production of low-cost alternatives to ACFs by coating cheaper glass fibers with activated phenolic resin [11]. At the University of Kentucky, we have examined the synthesis of fibers and activated fibers from nonconventional isotropic pitches. Whereas most commercial ACFs are microporous, activated carbon fibers produced from shale oil asphaltenes and from some coal liquefaction products are found to possess high mesopore volumes [12,13]. Through the selection of appropriate precursor pitches, it may be possible to produce ACFs with substantially different properties in terms of pore size distribution and surface chemistry, as well as allowing them to be processed more efficiently. In this context, coals present a fertile resource, as heavy coal liquids can be produced relatively cheaply and with a wide range of composition that is determined by the rank and structure of the parent coal and the means by which liquids are produced—pyrolysis, solvent extraction, liquefaction.

D. Monolithic Activated Carbons

Individual fibers and powders present difficulties in handling, containment, and regeneration, and in fixed bed operations they can present an unacceptably high resistance to flow. These problems can be surmounted by their incorporation into forms such as felt, paper, woven and nonwoven fabrics, and rigid monolithic structures. Potential advantages of these forms include facile handling, high permeability, and the possibility of in-situ regeneration. Indeed, regeneration may be accomplished through electrical heating, provided that there is adequate contact between the conducting carbon constituents, thus offering a simple and rapid means of raising the temperature uniformly [11,6,7,14]. In addition, monoliths can be fabricated to a given size and shape, a consequence of which is that completely novel adsorber/reactor designs are possible.

Technology has been developed by the Mega-Carbon Company, USA, to incorporate powders such as high surface area, KOH-activated carbons into robust monoliths (Fig. 3). Shaped carbons are produced from a slip or pourable suspension of the activated carbon powder and a binder, and the mixture is made rigid by mild heating to set the binding agent. The shape and size of the monoliths is determined by the mold, with the advantage that they can be preformed or formed in situ. In this proprietary process, over 80% of the surface area of the free powder is retained in the monolith. The shapes are strong, with crush strengths up to 18 MPa, and they are stable to temperatures in excess of 300°C. The properties can be adjusted by controlling the composition of the binder mix. Prospective applications under investigation are in gas phase adsorption and gas storage.

A collaboration between the University of Kentucky and the Oak Ridge National Laboratory has led to the development of rigid activated carbon fiber composites [15,16]. The composites can be prepared using any type of carbon fiber, although efforts to date have focused on isotropic pitch fibers. The composites are formed by filtering a water slurry of chopped fibers and powdered resin. The filter cake is then dried and bonded by heating to cure the resin, following which the cake can be carbonized and activated in steam or CO_2. Alternately, the cake can be formed using preactivated carbon fibers, when over 90% of the surface area of the free fibers is retained in the composite after curing. Because of their unique architecture (Fig. 4), the composites are strong, are highly permeable to liquids and gases, and can be machined.

It has been found that the composites can be used in gas phase processes such as the separation of CH_4 and CO_2 [16] or the adsorption of volatile organic compounds from air (see later in this chapter). In the liquid phase, column tests have been used to examine the effectiveness of an activated carbon fiber compos-

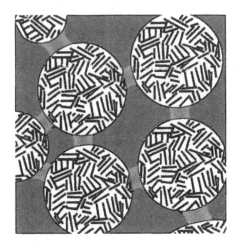

FIG. 3   Activated carbon monoliths (Mega-Carbon Company, USA).

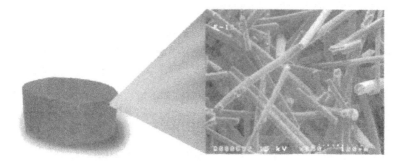

FIG. 4   Structure of rigid activated carbon fiber composite.

ite relative to a commercial granular activated carbon for the adsorption of a common herbicide, sodium pentachlorophenolate or Na-PCP [17]. The breakthrough time was found to be almost ten times longer for the composite than for the equivalent bed of granular carbon. The greater effectiveness of the composite is attributed to its uniform structure, which ensures that the feed is distributed evenly through the column, to the presentation of the adsorbent surface in fiber form, and to the open internal structure, which renders the pore structure readily accessible.

E.   Surface Chemistry

The selectivity of activated carbons for adsorption and catalysis is dependent upon their surface chemistry, as well as upon their pore size distribution. Normally, the adsorptive surface of activated carbons is approximately neutral, such that polar and ionic species are less readily adsorbed than organic molecules. For many applications it would be advantageous to be able to tailor the surface chemistry of activated carbons in order to improve their effectiveness. The approaches that have been taken to modify the type and distribution of surface functional groups have mostly involved the posttreatment of activated carbons or modification of the precursor composition, although the synthesis route and conditions can also be employed to control the properties of the end product. Posttreatment methods include heating in a controlled atmosphere and chemical reaction in the liquid or vapor phase. It has been shown that through appropriate chemical reaction, the surface can be rendered more acidic, basic, polar, or completely neutral [11]. However, chemical treatment can add considerably to the product cost. The chemical composition of the precursor also influences the surface chemistry and offers a potentially lower cost method for adjusting the properties of activated carbons. For example, activated carbon fibers produced from nitrogen-rich isotropic pitches have been found to be very active for the catalytic conversion of SO_2 to sulfuric acid [18].

F. Nonporous Carbons

Outside of the sphere of interest in environmental science and technology that is occupied by activated carbons, the use of other forms of carbon is so broad that there are numerous instances where carbons contribute directly or obliquely to the development, protection, or maintenance of an environmentally friendly society. Examples include diverse applications in the field of medicine, where carbon is attractive, *ceteris paribus*, because of its compatibility with the human body—carbons are chemically and biologically inert; the formation of strong, lightweight, structures that are resistant to chemical attack and can improve the efficiency of energy use; and protection from thermal and acoustic emissions.

The advantages that can be obtained through modifying the properties of concrete by the incorporation of carbon fibers provides an illustration of the diverse applications of carbon materials. The world production of concrete is considerably in excess of 5 billion tons per annum [19]. In certain geographic regions where there is a high incidence of powerful natural phenomena, such as earthquakes, concrete structures are unable to withstand unusually high imposed stresses. Concrete per se is brittle, possessing good strength in compression but having poor tensile strength. Traditionally, improvements in these properties have been effected by reinforcement with steel rods. Much lighter structures, combined with high toughness and increased tensile and flexural strengths, can also be obtained by the incorporation of fibers. In the past, natural fibers, such as asbestos, cellulose, and sisal, were used [16]. Recent work has focused mainly on steel, carbon, and polymer fibers, of which steel is the most common. A recent study compared the relative merits of steel fibers, isotropic carbon fibers (from petroleum pitch), and high-grade polyethylene fibers for concrete reinforcement [20]. Of the three fiber types, the carbon fibers possessed the lowest tensile modulus and strength, but were less expensive than the polyethylene fibers on the basis of either unit mass or unit volume. In spite of the inferior characteristics of the carbon fibers, they conferred the highest tensile strength, indicating that their fiber dispersion and/or fiber–matrix bonding is superior.

It is of note that during the 1989 San Francisco earthquake, certain overhead road supports that were retroactively strengthened by cladding with a relatively thin layer of carbon fiber-reinforced concrete allowed them to survive while conventional concrete pillars failed [20]. Similarly, in Japan, which experiences several hundreds of earthquakes each year of widely ranging intensity, carbon fibers have been extensively adopted for concrete reinforcement. More carbon fiber is employed in the Japanese construction industry than anywhere else [21].

Carbon fibers are also of interest as replacements for asbestos fibers in friction materials, such as brake shoes [22]. This move was initially driven in order to avoid the health hazards associated with asbestos. In comparison to asbestos rein-forced brake formulations, isotropic carbon fibers have been shown to provide much higher friction performance and superior wear resistance, together with many advantages in operating characteristics. These advantages translate to other friction materials, such as clutch plates, and to a much broader range of applica-tions in which carbon fibers are used as reinforcing or filler materials in various matrices. Carbon fibers in the forms of mats, felts, and paper insulation also present viable replacements for asbestos fibers.

III. GAS PHASE APPLICATIONS

A. Removal of Volatile Organic Compounds (VOCs) from Air

With increasing concerns over air quality, the release of volatile organic com-pounds (VOCs) into the environment has become a matter of global concern. In the presence of sunlight, VOCs can combine with NO_x, which can be found in significant concentrations in urban areas as a result of emissions from combustion sources, such as internal combustion engines and power plants, to form ozone. While ozone in the upper atmosphere is beneficial in preventing harmful UV radiation from reaching the earth's surface, at ground level it can cause acute respiratory problems for humans, including decreased lung capacity and impair-ment of the immune defense system. For these reasons, amendments to the Clean Air Act of 1990 were introduced requiring the reduction in emissions of 149 VOCs that are detrimental to air quality [23], and many previously uncontrolled sources are now mandated to reduce VOC emissions [24].

Activated carbons have been used for many years as adsorbents for the re-moval and recovery of organic solvents from air streams emanating from a range of industrial sources. Typically, the activated carbon is in the form of granules that are contained in appropriately sized absorber beds. The beds are usually used in cyclic operation to effect first removal and then recovery of the solvent for recycle. The same principles have been applied for the entrapment and recovery of volatile organics from other sources, such as in the control of fugitive fuel emissions from vehicles. In many situations, traditional granular activated car-bons provide an acceptable level of performance at reasonable cost. However, many of the new and emerging applications present far more demanding condi-tions and have created a need for higher performance adsorbents.

In the field of VOC control, a distinction can be made between processes that involve high concentrations (1–2%) of pollutants in effluent streams, where recovery and recycle is normally the principal objective, and those with low concentrations, typically 10–1000 ppm, where capture and disposal is the preferred route. New regulations limit VOC release to even lower concentrations. In processes where high flow rates are the norm, beds of granular activated carbon (GAC) must be relatively deep in order to provide sufficient contact time for the adequate removal of the adsorbates, thereby incurring a high pressure drop penalty. Over the last ten years several novel forms of activated carbons have been developed that can help to avoid the problems of using deep beds of GAC.

1. Carbon Properties

Although the surface area of activated carbons is an important parameter affecting the performance of these materials, other factors also play significant roles in solvent recovery. A particularly important characteristic is the working capacity, which is the difference between the amount of solvent adsorbed at saturation or equilibrium and the residual amount remaining in the carbon following desorption. Carbons that are highly microporous and have a high equilibrium uptake are not necessarily the best ones for practical use, since narrow micropores can strongly retain the adsorbed phase. There is an optimum pore size distribution for a given application, and that depends primarily on the properties of the adsorbate in question. For example, in automobile canisters containing activated carbon that are used to trap evaporative gasoline emissions, desorption is effected by flowing air at ambient temperature and pressure. To provide a high working capacity under these conditions, the carbon should have significant porosity in the 20–50 Å range [25]. The shape of the adsorption isotherm at different temperatures is also important since this determines the influences of inlet air temperature and adsorptive concentration on the equilibrium adsorption capacity.

The pressure drop over the adsorber bed is controlled by a number of factors including the bed geometry, the size and shape of the activated carbon granules, and fluid density, viscosity, and flow. The energy expenditure associated with overcoming a high pressure drop is undesirable since it can make a considerable contribution to the system operating cost. Hence, a compromise is made between the efficiency of bed use and the process economics. The size of granular carbons can be optimized by using particles that are large enough to give a low pressure drop across the bed, but not so large that there is poor access to the internal surface area: for large pellets the rate of diffusion of organic vapors into the pore structure may be too slow for efficient adsorption at short residence times. In addition to increasing the size of the granules or pellets, the pressure drop can also be reduced by lowering the bed height. However, shallow beds can introduce other problems such as the possibility of channeling and will reduce the time on stream in a cyclic operation.

The mechanical strength of the carbon is another important property. Movement of the granules can occur during cyclic operation causing attrition and creating fines. The generation of fines can contribute to increased pressure drop, adversely affect the flow distribution, and reduce the volume of carbon in the bed. Mechanical strength becomes more important with deep beds, where the carbon at the bottom must support the burden of the bed above.

2. Organic Solvent Recovery

One of the largest gas phase applications for activated carbon is in the recovery of solvents from industrial process effluents. Examples of industries that produce solvent-contaminated air streams are dry cleaning, the manufacture of paints, adhesives and polymers, and printing. In many cases, the solvent concentration is high (of the order of 1–2%). The highly volatile nature of many solvents can create unacceptable problems if they are vented to the atmosphere: they can create health, fire, and explosion hazards as well as pollute the environment. The solvents must therefore be removed before the air streams are vented to the atmosphere. There are economic benefits if the recovered solvent can be reused.

The main technologies available for the removal of solvents and other VOCs are adsorption on activated carbon, thermal or catalytic incineration, biofiltration, or combinations of these methods. Adsorption over activated carbons offers high efficiency for VOC removal at both low and high concentrations, and it is more cost effective at low concentrations than incineration techniques. Incineration offers a method of destroying VOCs with the formation of relatively innocuous products, principally water and CO_2, but it is only viable for effluent streams containing high concentrations of VOCs. At low concentrations the high-energy requirements for treating dilute air streams at high flow rates make this approach prohibitively expensive. Biofiltration is used mainly for air flows with low VOC concentrations, where the low operating cost gives an advantage: at high concentrations ($>$10,000 ppm) the removal efficiency is too low.

In conventional solvent recovery, solvent-laden air is first passed through an adsorber bed of GAC. The solvent is adsorbed on the carbon and clean air is exhausted to the atmosphere. When breakthrough of the solvent occurs (that is, when the solvent concentration in the cleaned gas stream exceeds the limit of acceptability, normally ∼10% of the inlet concentration), the inlet stream is directed to a second adsorber bed, while the first bed is regenerated. An example of a solvent recovery process is shown in Fig. 5. To recover the solvent, it is desorbed by passing a hot low-pressure gas through the bed in a direction countercurrent to the air flow. The most commonly used stripping agent is steam, but hot air or nitrogen are also used. As the flow of steam heats the carbon, the solvent is released and carried away. A typical steam demand is 0.3 kg steam per kg activated carbon [26]. The mixture of solvent and steam is condensed, when the solvent can often be separated and recovered by decanting. If the solvent is soluble in water, distillation is required for separation. The process can be made semicontinuous by the use of multiple carbon beds so that, at any time, one or more beds are being desorbed while others are in the adsorption mode.

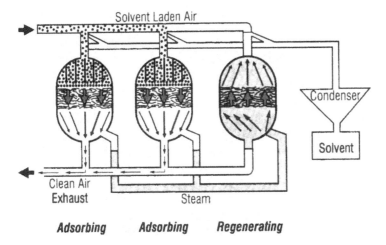

Solvent Laden Air

Condenser

Solvent

Clean Air
Exhaust

Steam

Adsorbing   Adsorbing   Regenerating

FIG. 5   Solvent recovery process using activated carbon beds.

Commercial activated carbon adsorbers have the capacity to treat air volumes from about 50 m³/min to over 15,000 m³/min. A list of solvents that are commonly recovered by activated carbon is given in Table 2 [27,28].

Interactions with the carbon surface can make the recovery of certain solvents, such as ketones and chlorinated hydrocarbons, difficult. Ketones and aldehydes can polymerize, releasing large amounts of heat. When this happens in a part of the bed with poor heat transfer, the temperature can reach the ignition point of the solvent. "Fires always start with hot-spots in parts of the bed where the airflow is reduced due to poor design. The susceptibility of the carbon bed to autoignition can be reduced by removing soluble alkali sodium and potassium salts, that may be present as impurities in the activated carbon and which can function as combustion and gasification catalysts" [29].

Chlorinated hydrocarbons can be hydrolyzed to form highly corrosive hydrogen chloride, which can lead to rapid degradation of the materials of construction

TABLE 2   Solvents Recovered by Activated Carbon Adsorption

toluene	acetone	tetrahydrofuran
heptane	ethyl acetate	white spirit
hexane	methyl ethyl ketone(MEK)	benzene
pentane	naphthalene	xylene
carbon tetrachloride	methylene chloride	petroleum ether

used in the adsorber. These reactions are assisted by the higher temperatures and steam concentrations that are experienced during the desorption cycle. The catalytic effect of the carbon surface is such that even apparently stable solvents such as carbon tetrachloride can be readily hydrolyzed.

3. Evaporative Loss Control Devices (ELCDs)

Automotive emissions make a large contribution to urban and global air pollution. In gasoline fueled vehicles, emissions arise from the by-products of combustion and from evaporation of the fuel itself, Fig. 6. The exhaust from modern internal combustion engines is now cleaned up very effectively by the inclusion of a catalytic converter in the exhaust system, and further improvements have been made via advanced engine design. As older, less efficient vehicles with poor emission control measures are replaced by those conforming to modern standards, interest has inevitably focused upon uncontrolled emissions due to the evaporation of gasoline. To conform with increasingly stringent government air pollution control standards, it has become necessary to fit an activated carbon canister on motor vehicles to prevent the emission of volatile petroleum constituents.

The first Clean Air Act of 1970 required hydrocarbon emissions to be lower than 0.25 grams/km. By 1971 charcoal canisters were installed in US automobiles to trap gasoline vapors. Later amendments to the Clean Air Act state that all US automobiles must have a canister that will cope with both running and refueling losses from 1998. Similar legislation is following in Europe and other parts of the world. The carbon canister, also called an evaporative loss control device (ELCD), is located between the fuel tank and the engine, Fig. 7. Evaporation occurs due to fluctuations in ambient air temperature from night to day. As the temperature rises, it causes the fuel tank to heat and release gasoline vapors.

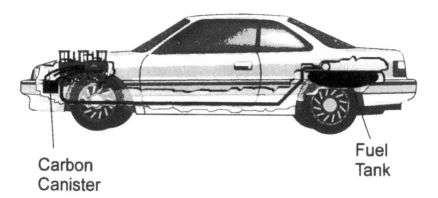

Carbon
Canister

Fuel
Tank

FIG. 6 Evaporative losses from gasoline-powered vehicles.

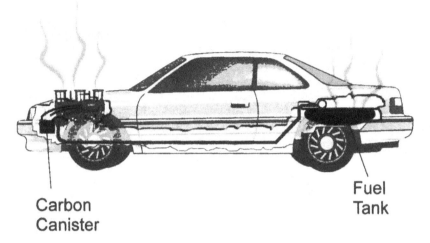

Carbon
Canister

Fuel
Tank

FIG. 7 Evaporative loss control device (ELCD).

Gasoline vapors are also vented from the fuel tank during refueling as they are displaced by liquid gasoline. These vapors are also captured by the ELCD canister. The canister is regenerated by a bypass flow of combustion air when the engine is running. The vapor-laden air is then directed to the inlet manifold. The desorbed gasoline thus forms part of the fuel mixture to the engine, with the secondary benefit of a small but significant increase in fuel efficiency.

4. Removal of VOCs at Low Concentrations

Examples of sources of effluent air containing low concentrations of VOCs are in the vent stacks from flexiographic printing (mixtures of acetates and alcohols), paint booths in automotive assembly plants, and bakeries where the major VOC released is ethanol (~3 kg ethanol/1000 kg dough processed) [30]. Volume flows are typically in the range of 10,000–14,000/hr. Another significant VOC is styrene, a common monomer that is used in the production of a variety of industrial products including the manufacture of fiberglass-reinforced products such as recreational and sports vehicles and car and truck body parts. The consumption of styrene was about 4.0 billion kg in 1989, and reported styrene emissions were around 15 million kg/year in 1990 [30].

New environmental regulations are concerned with controlling low-level VOC emissions that require advanced adsorption technologies. When the concentration of VOCs is in the low ppm range, and the air flow is large, the beds of GAC must be relatively deep in order to provide sufficient contact time for adequate removal of the adsorbates. In turn, this requires large fans that are very energy-demanding to overcome the high pressure drop. In addition, the mass transfer zone (MTZ) in GAC beds is extended under these conditions. The MTZ is the region of the bed between the activated carbon that is already saturated and the point where the gas phase concentration of the adsorptive is at the maximum acceptable limit in the effluent stream, Fig. 8. Breakthrough is reached when the leading edge of the zone advances to the end of the bed. The length of the zone is a measure of the adsorption efficiency of the bed. The longer the zone, the shorter the on-stream time and the smaller is the fraction of the full adsorptive capacity that has been utilized at breakthrough. The rate of adsorption is slow on GAC, particularly at low adsorbate concentrations. Beyond the initial adsorption at the outer layers of the granules, the rate of adsorption is controlled by the slower process of intraparticle diffusion.

New technologies are being developed and commercialized to meet the more stringent demands for VOC removal, and to surmount the problems associated with conventional technology. In one example, a honeycomb structure, made of activated carbon, or a substrate impregnated with activated carbon or zeolite powder, is used in a rotary concentrator to adsorb organic vapors and recover them

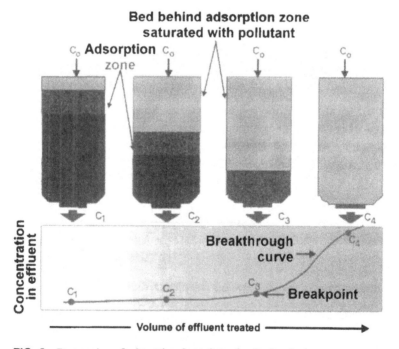

FIG. 8 Progression of adsorption front through adsorber bed.

in concentrated form, when they can be destroyed by thermal or catalytic oxidation.

A schematic of a rotary concentrator is shown in Fig. 9. The wheel rotates slowly (1–3 rotations per hour) with about 90% of its face exposed to the incoming air stream. The remainder of the face is in a regeneration sector where a counterflow of hot air desorbs the VOCs for subsequent incineration. The rotary adsorber increases the VOC concentration by a factor of 100, so there is little need for supplemental fuel in the small oxidizer that can be used to destroy the VOCs. This technology has been used for the removal of styrene emissions in the plastics molding industry. The process is also used with automotive paint booths where modular rotary concentrators and a regenerative thermal oxidizer are used to capture and destroy VOCs.

Activated carbon fibers (ACFs) offer a choice of other carbon forms for VOC removal. As discussed earlier, the narrow diameter of the fibers provides ready access of adsorptive species to the adsorbent surface. The incorporation of ACF into permeable forms such as felt, paper, and rigid monoliths helps to surmount the disadvantages of using loose fibers. Rigid ACF composites have been prepared at the University of Kentucky and examined for their potential for the removal of low concentrations of VOCs [31].

The open internal structure of the composites presents little resistance to the flow of fluids and allows them direct access to the activated fiber surfaces. Consequently, the composites offer a potential solution to the problems of removing low concentrations of VOCs from large volumes of air. In comparative trials, equivalent weights of GAC and an ACF composite were tested for their ability to remove butane at 20 ppm in a flow of nitrogen carrier gas, Fig. 10. The weight uptake of butane at breakthrough was twice as high on the composite as on the GAC. In addition, the structure of the ACF composites means that they are not susceptible to the attrition problems associated with packed granular beds during adsorption/desorption cycling.

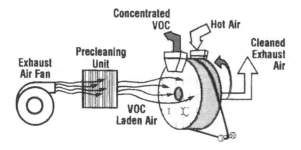

FIG. 9   Schematic of rotary concentrator for VOC recovery.

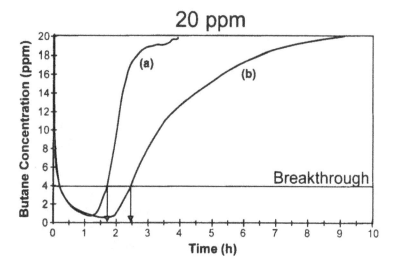

FIG. 10 Butane breakthrough curves for activated carbon beds: (a) fiber composite; (b) granular (20 ppm butane in nitrogen).

B. Purification of Landfill Gas

Landfill sites present sources of air pollution due to the presence of VOCs and to the generation of gases by aerobic or anaerobic microbial digestion of organic wastes. The gases emitted are typically composed of 55% methane and 45% CO_2 with trace amounts of VOCs and other less desirable contaminants. Trace organic compounds have not only been identified in landfill gas but also in the atmosphere downwind from them [32]. The trace contaminants are mainly aromatic hydrocarbons and halogenated aromatic hydrocarbons and include harmful compounds such as toluene, trichloroethylene, and benzene. Many of these compounds are so toxic that strategies to control them have been implemented. Of particular concern are carcinogenic agents such as vinyl chloride [33], for which Canada has a recommended maximum exposure limit of 3 mg/m³. The maximum level recorded by one study performed downwind from a landfill was measured at 2.9 mg/m³. Activated carbon has been found to be an efficient adsorbent for the removal of several of these halogenated hydrocarbons from landfill gas [34].

The significant volumes of gas evolved from large landfill sites and the high proportion of methane that is present makes it a potentially valuable energy resource. This has been recognized by Air Products and Chemicals, Inc., USA, who have developed technology to recover methane at yields of over 90% [35]. In this process (Fig. 11), pressurized landfill gas is passed through a bed of activated carbon to remove the trace impurities. The emergent flow of pure carbon

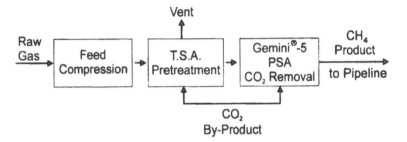

FIG. 11 Landfill gas treatment system.

dioxide and methane is then passed through a second bed of activated carbon maintained at high pressure to selectively adsorb the carbon dioxide and yield the required high-purity methane. The second bed is regenerated off-line by reducing the pressure. The carbon dioxide produced may then be channeled back as a hot gas to regenerate the first bed by stripping impurities, which then are incinerated.

C. Air Conditioning

Increasing public awareness and general concerns over air quality, and a growing incidence of allergic reactions to air pollutants, have generated a demand for improved treatment of the air that is supplied to public spaces. Such environments include airports, hospitals, submarines, office blocks and theaters. Granular activated carbon filters are commonly used in conjunction with air conditioning equipment for the removal of noxious trace contaminants from air that is recycled to populated areas. The recirculation and purification of air has the added economic advantages of reducing heating or refrigeration costs. Other applications include the use of activated carbons in domestic kitchen cooker hoods to adsorb cooking odors and vapors and activated carbon filters incorporated in air purifiers for private homes. High efficiency particulate air (HEPA) filters have been available for some time for the removal of allergens and dust. In more sophisticated air treatment systems, HEPA filters are combined with activated carbon adsorbents for the removal of human and cooking odors.

Smaller occupied spaces are also beginning to receive attention. For example, activated carbon filters have been introduced in automobile passenger cabins to remove odorous contaminants from inlet air and increase passenger comfort. These filters are installed only by a few manufacturers at present: Mercedes, BMW and Porsche in Germany and in the Mercury Mystique and Ford Contour models in the USA [36]. Particulate filters are normally integrated into automobile climate control systems to remove dust and pollens, and these can now be augmented to remove odors that originate in exhaust fumes, rural emissions, and those generated within the car, particularly cigarette smoke and food odors. Cigarette smoke has been found to contain ethene, ethane, acetaldehyde, formaldehyde, NO_x, and isoprene at concentrations from 1000 to 1800 ppm. These contaminants can be reduced to concentrations of 1–2 ppm by the use of an efficient activated carbon filter.

D. Mercury Vapor Adsorption

Mercury is one of a number of toxic heavy metals that occur in trace amounts in fossil fuels, particularly coal, and are also present in waste materials. During the combustion of fuels or wastes in power plants and utility boilers, these metals can be released to the atmosphere unless remedial action is taken. Emissions from municipal waste incinerators can substantially add to the environmental audit of heavy metals, since domestic and industrial waste often contains many sources of heavy metals. Mercury vapor is particularly difficult to capture from combustion gas streams due to its volatility. Some processes under study for the removal of mercury from flue gas streams are based upon the injection of finely ground activated carbon. The efficiency of mercury sorption depends upon the mercury speciation and the gas temperature. The capture of elemental mercury can be enhanced by impregnating the activated carbon with sulfur, with the formation of less volatile mercuric sulfide [37]: this technique has been applied to the removal of mercury from natural gas streams. One of the principal difficulties in removing Hg from flue gas streams is that the extent of adsorption is very low at the temperatures typically encountered, and it is often impractical to consider cooling these large volumes of gas.

E. Protective Filters

The consequences of chemical warfare were first realized in the trenches of World War I when the Germans released chlorine gas on the Allied forces. The Allies quickly developed a means of protection from inhaling the choking gas by the use of gas masks containing granular activated carbon. The continued threat of chemical warfare and the development of highly advanced and toxic chemical and biological weapons has led to a much greater degree of sophistication in the design of adsorbents used in gas masks. Because of the wide range of offensive gases that are potentially available, the activated carbon is required to remove gases by both physical adsorption (e.g., nerve gases) and by chemical adsorption (e.g., hydrogen cyanide, cyanogen chloride, phosphine, and arsine). To accomplish this, the activated carbon is impregnated with a complex mixture of metal compounds that include copper, chromium, silver, and sometimes organic species. These sophisticated carbons are also utilized in filters for underground shelters, and in inlet filters for armored vehicles and other forms of military transport. Threats from percutaneous nerve gases require complete body protection by the use of suits that contain activated carbon in a form that allows rapid and effective adsorption (fine granules or fibers).

The technology used by the military has been adopted by industry to provide protection to workers against hazardous vapors and gases that may be encountered in certain industrial processes. As with the military, this involves individual protective respiratory devices, air treatment filters, and protective clothing. Depending upon the nature of the hazard, the carbons may be impregnated to enhance their ability to remove the toxic species.

F. Flue Gas Cleanup

Appreciable interest has been generated in the use of activated carbons for flue gas cleanup, especially for the removal of SO_x and NO_x: the adsorption of mercury from flue gases was discussed earlier. From the environmental point of view, emissions from the combustion of fossil fuels in power plants and similar industrial processes are major contributors to a lowering of air quality. The flue gases carry traces of SO_2 and NO_x, which can be oxidized and converted to their acid forms in the presence of atmospheric water vapor, and they may also combine with other volatile organics to form ozone and smog. Similarly, low level SO_3 and NO_x emissions from automobiles, while insignificant for individual vehicles, become a large source of pollution when multiplied by the millions of vehicles that are on the roads.

As early as the 1950s [38], after an acid/smog cloud enveloped London, leading to the deaths of thousands from respiratory ailments, measures were introduced to reduce the overall emissions of sulfur compounds and particulates from domestic and industrial combustion. With increasing awareness of the impact on human health and the environment, many other industrialized nations also began to put restrictions on allowable emissions from combustion and other sources. These restrictions have become increasingly more stringent and have encouraged the expansion of research programs in the area of flue gas cleanup, looking towards new materials and processes.

These factors have led to a broad need for technologies capable of reducing both point and distributed sources of SO_x and NO_x. For the control of emissions from power generation fossil fuels, the removal of SO_x has been achieved using wet or dry scrubbing/adsorption processes. These technologies have high energy costs associated with the operation of the scrubbing towers, as well as significant costs associated with raw materials, and materials handling and disposal. The removal of NO_x has been addressed through the introduction of low-NO_x burners (which has created a secondary problem of carryover of carbon into the ash, making the latter unacceptable in many cases for use in cement manufacture) and by selective reduction processes, thermal (SR) and catalytic (SCR), in which reaction with NH_3 reduces the NO_x to nitrogen. The driving force for future technological development has been towards lower cost, more effective technologies. Activated carbons present a possible basis for new approaches to emissions control for both point combustion sources and distributed sources, including automotive emissions.

1. SO$_x$ Removal

The SO_2 in flue gas streams is typically present at concentrations between 500 and 2000 ppm. It can be removed by adsorption on activated carbons, where it can be oxidized to SO_3 and converted to sulfuric acid if oxygen and water are present. The pathway for SO_2 removal depends upon conditions [39], dry or humid. Under dry conditions (with/without O_2),

$$SO_2 + \sigma_v \rightarrow \sigma_{SO_2} \qquad SO_2 \text{ adsorption}$$

$$\sigma_{SO_2} + \frac{1}{2} \sigma_{O_2} \rightarrow \sigma_{SO_3} \qquad \text{oxidation step}$$

Under humid conditions,

$$SO_{2,gas} \rightarrow SO_{2,aq.} \qquad \text{diffusion into water film}$$

$$SO_{2,aq.} + \sigma_v \rightarrow \sigma_{SO_2} \qquad SO_2 \text{ adsorption}$$

$$\sigma_{SO_2} + \frac{1}{2} \sigma_{O_2} \rightarrow \sigma_{SO_3} \qquad SO_2 \text{ oxidation}$$

$$\sigma_{SO_3} + H_2O \rightarrow \sigma_{H2SO4} \qquad \text{reaction to sulfuric acid}$$

$$\sigma_{H2SO4} + H_2SO_4 + \sigma_v \qquad \text{acid desorption}$$

where σ_i = surface site containing component i and σ_v = vacant site.

(a) Dry Adsorption and Oxidation. Under dry conditions, SO_2 can be adsorbed onto activated carbon and then oxidized to SO_3 in the presence of oxygen. In this process, when the adsorptive capacity is reached, the carbon must be regenerated to recover gaseous SO_2 or SO_3. Many factors influence the capacity of activated carbons for SO_x adsorption. Davini [40] found that activated carbons with basic surface groups have a higher capacity to adsorb SO_2 than carbons with predominately acidic functional groups. Heat treatment of activated carbons in an inert atmosphere provides a means to produce more basic functionality by the removal of CO_2 from carboxylic acid groups, leaving more basic C=O groups [41]. This phenomenon was first discovered during investigations of the "active coke" process for desulfurization by Juntgen [39]. Moreno-Castilla et al. [42] support this conclusion and further show that the active sites of interest for this process are in narrow micropores (i.e., those pores accessible to benzene and *n*-hexane).

(b) *Catalytic Conversion to H_2SO_4.*　The presence of both oxygen and humidity enhances SO_2 uptake on activated carbon [43]. The adsorbent capacity is increased by a factor of 2 or 3 in the presence of oxygen and by a factor of 20 to 30 if water is also present [44], Table 3.

When water is present in the gas stream, it reacts with the SO_2 and O_2 to produce sulfuric acid on the carbon surface, and can subsequently desorb. The overall SO_2 adsorption capacity is enhanced due to its solubility in the water film that forms on the carbon surface. Conversely, active sites for SO_2 capture are simultaneously reduced by water coverage. In general, the SO_2 adsorption characteristics of an activated carbon are dependent upon its physical form, the pore structure, the surface area, and the surface chemistry. Similarly, both temperature and contact time also affect the efficiency of the process. The temperature for practical application is usually between ambient and 200°C, with ambient to 50°C being favored due to the decreasing solubility of SO_2 in water at higher temperatures.

The sulfuric acid tends to desorb slowly from the carbon surface, and if the acid occupies all the active sites for SO_2 oxidation and hydration, the carbon becomes inactive for further adsorption of SO_2, as illustrated in Fig. 12. When a gas stream containing SO_2 is passed through a bed of activated carbon, there is complete removal of SO_2 for a period, during which the SO_2 is removed by both physical and chemical adsorption (Zone I). At the breakthrough point, the SO_2 concentration downstream of the adsorber unit begins to rise, as all the physical adsorption sites are occupied and only chemical adsorption and reaction to produce H_2SO_4 are available for SO_2 removal (Zone II). The SO_2 removal may further decrease due to the production of acid, which occupies some of the active sites and can lead to catalyst deactivation (Zone III) as the carbon surface becomes saturated. If there is almost no acid desorption, Zone III hardly exists, and the breakthrough plots go directly from 0 to 100%.

For this process to be viable, the activated carbon must be regenerable, or, preferably, the acid is removed at a rate fast enough to allow a steady state of

TABLE 3

Sample	Adsorbed amount (mmol/g)		
	Only SO_2	5% O_2	5% O_2, 10% H_2O
A	0.10	0.34	2.3
FE-100-600			
B	0.18	0.56	3.8
FE-200-800			
C	0.13	0.26	2.7
FE-300-800			

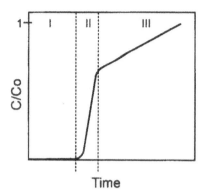

FIG. 12   Typical breakthrough plot for the SO_2 adsorption over activated carbon in the presence of oxygen and humidity.

SO_2 removal. Regeneration can be accomplished in two ways: (a) One can continuously or cyclically flush the activated carbon with water to remove the sulfuric acid as a by-product. Either trickle bed operation [45,46] or a series of cyclically loaded and purged beds may be employed. (b) One can heat the activated carbon in an inert atmosphere to decompose the sulfuric acid into SO_2 and water and obtain a concentrated stream of gas phase SO_2. Thermal regeneration may lead to the loss of carbon as CO and CO_2.

In case (b), the efficacy of the activated carbon to adsorb SO_2 may change with the number of regeneration cycles. The reaction to cycling may be positive or negative, the outcome being dependent upon the original pore structure of the activated carbon. In general, the activity of microporous carbons with relatively low surface area remains constant or increases with each regeneration (due to further activation in the thermal regeneration cycle), while for carbons with wide and well developed porosities, the activity can decrease with each regeneration (as microporosity is widened and enlarged). The final regeneration temperature also influences activity. Adsorbed H_2SO_4 is desorbed completely at temperatures above 380°C, decomposing to SO_2 and H_2O while donating an oxygen group to the surface. Low-temperature regeneration (<380°C) does not produce the same type of active sites as those originally present, and the total adsorption capacity is not fully recovered. Regeneration must take place at higher temperatures in order to decompose those oxygen groups.

Most research and development has focused on the use of granular or extruded activated carbons, and all commercial processes to date use this technology. However, within the last several years there has been appreciable research to examine the application of activated carbon fibers. Activated carbon fibers (ACFs) have properties that are relevant to this application due to the inherently narrow distribution of micropores, and direct access to the adsorbent surface, which minimizes diffusional limitations. The fibrous form also allows more facile removal of sulfuric acid, allowing for the continuous removal of SO_2 and the production of H_2SO_4 [44,47,48].

Activated carbon fibers made from various precursors have been investigated (i.e., polyacrylonitrile or PAN, coal tar pitch, petroleum pitch, and oil shale tars) and have all exhibited high activity for SO_2 conversion [47]. It has also been shown that heat treatment of the fibers can increase the catalytic activity, the extent of change being dependent upon the type of fiber, and the heat treatment temperature and atmosphere [48].

Figure 13 shows breakthrough plots for a PAN-based ACF sample using a simulated flue stream containing 1000 ppm of SO_2, 5% O_2, and 10% H_2O at different operating temperatures. At 30°C, 100% removal is attained. The steady-state removal activity decreases as the temperature is increased. At 100°C, there is no production of sulfuric acid, which suggests that the existence of a thin water film over the fiber surface is crucial to the mechanism of H_2SO_4 extraction. A second effect of increasing the operating temperature is to decrease the solubility of SO_2 in water, which results in a lower uptake from the gas phase and a lower concentration of dissolved SO_2 in the water film on the surface of the fiber [49].

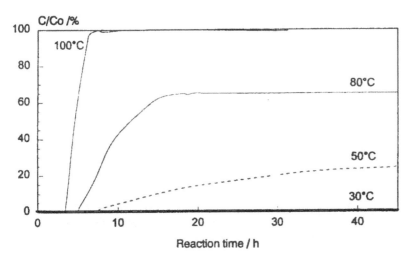

FIG. 13 Breakthrough plots for PAN-based ACF at various temperatures (1000 ppm SO_2, 5% O_2, and 10% H_2O).

2.   NO$_x$ Removal

Nitrogen oxides, NO, NO$_2$, and N$_2$O, are formed in the combustion of fossil fuels. Typical flue gas concentrations range from 400 to 1500 ppm. Activated carbon can be used as a reducing agent, as a catalyst, or as an adsorbent for the removal of these compounds [50].

Carbons can react with nitrogen oxides to reduce them to N$_2$, with a corresponding loss of carbon as CO or CO$_2$. Depending on the particular carbon, this reaction begins at about 600°C and is most effective at temperatures around 700°C. Illan-Gomez et al. [51] report a good correlation between NO reduction and the specific surface area of the carbon. In addition, it has been found that the presence of O$_2$ greatly increases the activity for the reduction of NO$_x$. However, concomitant carbon gasification leads to carbon consumption, making the process less attractive.

The removal of nitrogen oxides can be also be accomplished by the reaction of reducing gases such as ammonia, carbon monoxide, and hydrogen over activated carbon. Here, the carbon acts as a catalyst rather than as a reactant. Most researchers have used ammonia as the reducing gas in an analogue to the process of selective catalytic reduction (SCR) of NO with ammonia over a catalyst such as V-TiO$_2$ at temperatures in the range of 200–350°C [52]. (Because NO comprises the majority of the nitrogen oxides present, most research has focused on this compound.) SCR over activated carbon occurs at temperatures ranging from 100 to 150°C. The reactions are

Anoxic condition:

$$6NO + 4NH_3 \rightarrow 5N_2 + 6H_2O$$

Oxygen present:

$$4NO + 4NH_3 + O_2 \rightarrow 4N_2 + 6H_2O$$

The reactivity is higher over activated carbons that contain basic nitrogen in their structure [53,54], and there seems to be no correlation between the BET specific surface area and the catalytic activity [55]. In addition, it has been shown that the adsorption of NO is enhanced when surface oxygen is present and that the oxidative posttreatment of the carbon with sulfuric acid produces a more active catalyst due to a resultant increase in acid surface functionalities including carboxyl and carbonyl groups [55]. Figure 14 shows the effect of gas phase oxygen concentration and temperature on the SCR of NO. The presence of oxygen increases the activity by a factor of about 50. On the other hand, the presence of water vapor inhibits the reaction, and a small amount of moisture has a large adverse influence. For example, at 5% humidity the activity decreased by 25% compared to that in a dry atmosphere, while at 10 or 20% humidity there was little further change [39].

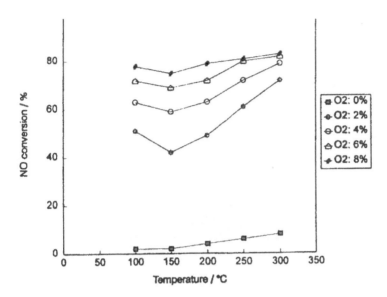

FIG. 14 Effects of oxygen concentration and temperature on NO conversion during SCR over activated carbon. (Adapted from Ref. 55.)

The conversions of NO and NO_2 have been studied as separate processes. The SCR of NO_2/NH_3 in the absence of oxygen is faster than the conversion of NO in 4% O_2. This suggests that the limiting step in the process of NO removal is the oxidation of NO on the carbon surface, followed by the subsequent reduction of NO_2, which is more reactive.

(a) Metal/Carbon Catalysts for NO_x Removal.   Several research groups have reported that metals loaded onto activated carbon can reduce the temperature required for NO_x reduction by SCR in the presence of oxygen to as low as 300°C versus 600–700°C for the uncatalyzed reaction. A significant enhancement of the reaction is seen with copper present as the metal component [56], Fig. 15. On the other hand, this reaction does not occur to any significant extent when oxygen is absent, regardless of the metal. Illan-Gomez et al. [57] tested the catalytic effects of K, Ca, Cr, Fe, Co, Ni, and Cu on activated carbon. All these metals reduced the reaction temperature, with K, Co, and Fe being most effective at low temperatures.

3.   De-SO_x and De-NO_x Combined Processes

(a) Large-Scale Process for Combined SO_2 and NO_x Removal.   It has been shown that activated carbons can be used to remove both SO_2 and NO_x. In the case of SO_2, regeneration of the activated carbon is necessary if water is not

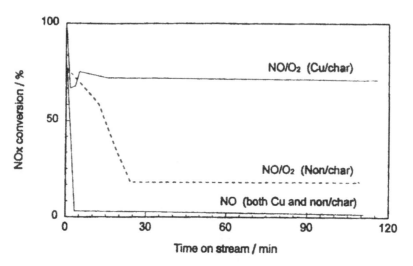

FIG. 15 Influence of copper on catalyst activity for NO reduction by activated carbon at 300°C. (Adapted from Ref. 56.)

present in sufficient amounts to form and remove sulfuric acid. On the other hand, NO_x can be reduced by reaction with NH_3, where the activated carbon acts as a catalyst. Several processes have been proposed for the simultaneous removal of these compounds. A number of aspects must be taken into account, as the processes may interfere with each other. The SO_2 removal process must be carried out in the presence of humidity, while NO_x reduction is partially inhibited by water. If ammonia is present in a flue gas stream containing SO_2, the SO_2 will react to form $(NH_4)_2SO_4$, which may deposit on the surface of the activated carbon, causing deactivation. Therefore, in a combined process for removing SO_2 and NO_x, sulfur dioxide should be removed prior to the injection of NH_3.

(b) Bergbau-Forschung Process.   In the 1960's, Bergbau-Forschung, now Deutsche Montan Teknologie (DMT), developed a combined SO_2 and NO_x removal process using a low surface area activated coke. Originally, this process was developed to remove only SO_2 and consisted of an adsorber where a bed of activated carbon moving downwards was contacted with a flue gas stream entering at the top of the column and exiting SO_2-free at the bottom. This process operated at temperatures in the range 100–200°C. The activated coke was continuously regenerated in a separate unit by heating to 600–700°C, to recover a gas stream containing about 30% SO_2, which was used to produce sulfur, liquefied SO_2, or H_2SO_4, Fig. 16.

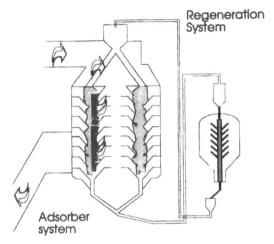

FIG. 16   Schematic of Bergbau-Forschung process for SO_2 removal. (Adapted from Ref. 56a.)

4. Commercial Processes

(a) Mitsui Process. In the 1980s, the Burgbau-Forschung process was expanded to remove NO_x. The adsorber was divided into two vertical sections. Activated coke flows downwards through both sections and the flue gas is passed upwards countercurrently. Most of the SO_2 is removed in the lower section prior to the injection of NH_3 when the gas enters the upper section, Fig. 17. This process has been tested satisfactorily in a 350 MW coal power plant at Luenen, Germany, removing 99% of SO_2 and over 70% NO_x.

The two-stage process was licensed by Mitsui Mining Company (MMC) in Japan in 1982, and by 1993 a modified form of the process was installed in four commercial plants in Japan and Germany [58]. The granular carbon or activated coke used in this process has a surface area of initially 150 to 250 m^2/g, which is much lower than that of commercial activated carbons. It is produced from a bituminous coal and a pitch binder. Low surface area carbons have been found to be the most effective in this process; they are cheaper than high surface area activated carbons, they retain their SO_2 adsorption capacity more efficiently on repeated cycling, and their relatively low porosity contributes to strength and abrasion resistance.

In the MMC combined process for SO_2 adsorption, NO_x removal, and air toxics adsorption, the granular activated coke flows downwards through the two adsorbers, countercurrent to the flow of flue gas. A booster fan is required in the flue gas stream to overcome the pressure drop across the beds. Oxides of nitrogen are

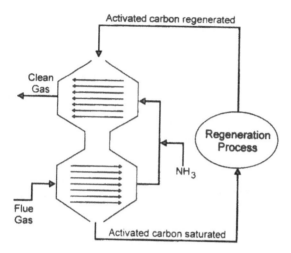

FIG. 17 Schematic of Mitsui process for de-SO$_x$ (lower adsorber) and de-NO$_x$ (upper adsorber). (Adapted from Ref. 58.)

reduced by reaction with NH$_3$ in the upper bed. If SO$_2$ is present, some ammonia is consumed in the formation of ammonium salts, resulting in the inefficient use of ammonia and possible blockages due to salt deposition. Hence NO$_x$ reduction is preferably conducted after substantial SO$_2$ removal, which occurs in the lower, where air toxics are also adsorbed. The adsorbed SO$_x$ is mainly present on the activated coke as sulfuric acid, and to some extent as ammonium salts.

The activated coke is discharged from the bottom of the lower adsorber for regeneration and by-product recovery, before recycling to the upper adsorber. Regeneration takes place by heating to decompose the adsorbed species where the principal effects are some loss of carbon as CO$_2$, which further activates the coke by increasing its surface area, and ammonium salts (if present) are decomposed to liberate ammonia, SO$_2$, nitrogen, and water, simultaneously producing nitrogen-containing surface groups. The combined process can be conducted over the temperature range 100–200°C, with lower requirements for additional heat input than technologies employing metal oxide catalysts for SCR. The removal of NO$_x$ is preferably conducted at the higher end of the temperature range, while SO$_2$ (and SO$_3$) adsorption is more efficient at lower temperatures, such that conditions in the two beds would ideally be separately controlled.

This process has not gained widespread use due to the large capital investment required and its high operating costs. Power is consumed in overcoming the pressure drop across the adsorbers, and fresh activated coke must be constantly added to make up for the losses of carbon on regeneration and attrition in the moving beds. The relevance of this technology to power generation may also diminish as more efficient coal-fired plants, based on integrated gasification combined-cycle (IGCC), come into commercial use, using other advanced gas cleanup technologies.

G. Gas Storage on Activated Carbons

Activated carbon has been considered as a medium for the storage of fuel gases, notably natural gas and hydrogen. Potential advantages are that the gases can be stored in a densified, adsorbed form at moderate pressures, thus allowing the use of relatively lightweight containment systems and reducing the hazards associated with high-pressure storage and transport.

1. Natural Gas as Transportation Fuel

Natural gas systems are being developed to provide alternative transportation fuels to traditional liquid petroleum fuels for both strategic and environmental reasons. Instability in the world's petroleum markets and increasing dependence on foreign crude sources has stimulated interest in using domestic natural gas reserves for vehicular transport. Environmentally, natural gas has many advantages over gasoline. It is composed mostly of methane, with lesser amounts of ethane, propane, and butane, contains few or no contaminants, and burns cleanly and efficiently. Hence emissions of SO_x, NO_x, and CO are minimal, and the high hydrogen content (25 wt% for pure CH_4) means that the CO_2 emissions per unit of energy produced are much lower than for liquid fuels. Compressed natural gas (CNG) and liquefied petroleum gas (LPG), which consists of higher condensable paraffins, are already used to a significant extent in countries such as Italy and the Netherlands.

Interest in the US focuses mainly on fuelling fleet vehicles, such as taxis and commercial transport: the absence of any appreciable infrastructure effectively excludes privately owned automobiles. There were more than 333,000 alternative fuel vehicles (AFVs) in the United States in 1995, of which three quarters were vehicles designed to operate on LPG, primarily propane [59]. The use of AFVs is expected to continue to grow at a rate of about 7.6% per year. Natural gas fueled vehicles make up two-thirds of the non-LPG AFVs in use, or approximately 55,000 vehicles as of 1995.

The major shortcoming of natural gas as a fuel is the relatively low calorific value per unit volume (energy density) compared to liquid fuels. One solution is to liquefy the gas by compression, yielding liquefied natural gas (LNG), which has a comparable energy density to that of liquid fuels, Table 4. In practice, this is too costly and too dangerous for use in a vehicle fleet [60]. Compression to a lesser degree, without liquefaction, produces an increase in energy density in proportion to the applied pressure. However, to obtain an adequate energy density

TABLE 4

Fuel	Energy density MJ/L	Pressure MPa
Diesel	37	—
Gasoline	32	—
Methanol	16	—
Liquid propane	29	—
LNG	23	—
CNG	0.8	2
CNG	5.6	10
CNG	9.7	20
ANG	3.8–6	3.5

for use as a vehicle fuel, compression to ~25 MPa is required. There are associated disadvantages due to the weight and bulk of the containment vessel that is needed, the associated safety risks, and the costs associated with multistage compression cycles. In contrast, natural gas adsorbed onto activated carbon at a pressure of 3.5 MPa has a significant energy density, and the development of adsorbent carbons with controlled properties could allow this technology to compete with compressed natural gas (CNG).

2. Activated Carbons for Natural Gas Storage

Traditionally, adsorptive capacity is measured on a mass basis. Since the size of any natural gas storage system for transportation fuel will be subject to space limitations, the amount of stored gas must be considered on a volumetric basis. In considering the use of activated carbon as a gas storage medium, Quinn and MacDonald [61] have suggested that a vessel packed with an adsorbent be viewed as if it were composed of four parts:

1. Interstitial voids
2. Macropore volume
3. Volume occupied by the adsorbent
4. Micropore volume

Of these four regions, methane adsorption only occurs in the micropores of an activated carbon. At ambient temperatures, there is negligible methane adsorption in the macropores, and the macropore volume can be considered as void space: methane is present in these voids only at the density of the compressed gas. Hence, the volume occupied by macropores and interstitial voids must be minimized for optimum storage capacity, consistent with the need to allow sufficient pathways for the transport of methane through the adsorbent mass. The micropores are preferably of uniform diameter and of a dimension that provides high storage density without being so narrow that there is limited gas release upon desorption.

The relative distribution of space within a storage vessel is shown in Fig. 18 for vessels filled with a traditional GAC, a "super activated" powdered carbon, and a compressed carbon monolith produced from a polymer precursor [61]. The monolith has the advantage of greater micropore volume, resulting in a higher storage volume of methane. Monoliths also present advantages in allowing their formation into different geometries and being resistant to abrasion.

The heat of adsorption of methane onto activated carbon is about 15–18 kJ mol^{-1} [62]. On the scale envisioned for an adsorbed natural gas (ANG) fuel system, this enthalpy change could present significant problems of thermal management during the cyclic operations of vessel filling and discharge. The exothermic heat released upon vessel filling can result in a temperature rise of as much as 140°C [60]. The temperature excursion can be restricted by controlling the rate of filling, although this can be a practical inconvenience, and a means of heat removal would be more desirable. Conversely, upon discharge, heat must be supplied to counteract the endotherm and permit an acceptable rate of fuel delivery. These factors must be considered in the design of the storage vessel, with large surface-to-volume ratios being favored for more rapid thermal transport. Monolithic carbons offer the advantage of possessing generally superior heat transfer characteristics.

3. Hydrogen Storage

Similar considerations to those described for the natural gas have led to the development of interest in the adsorbent-enhanced storage of hydrogen. Again, acti-

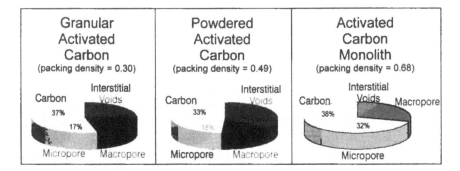

FIG. 18 Distribution of volume in adsorbent-filled gas storage vessel. (Adapted from Ref. 61.)

vated carbons are attractive candidate materials. Amankwah et al. [63] have investigated the storage of hydrogen on activated carbon at refrigeration temperatures (150–165 K). Using "super-activated" carbon (AX-31M, Anderson Development Company), the authors report a storage value of 17 kg H_2/m^3 (or ~90 g H_2/kg carbon) at 54 atm and 150 K. This value is higher than is obtained for pressurized hydrogen (11 kg H_2/m^3) but lower than the 36.5 kg H_2/m^3 for a $FeTiH_2$ metal hydride system. On a weight basis (systems with Kevlar storage cylinders), the adsorbed hydrogen system is more efficient: 5.13 kg H_2/kg (54 atm, 150 K) versus 1.40 kg H_2/kg for the metal hydride system. These authors also analyzed the cost of various hydrogen storage methods and found adsorbed hydrogen storage systems to be the least expensive ($7/$10^6$BTU) compared to pressurized hydrogen gas ($11/$10^6$BTU), liquid hydrogen ($18/10^6BTU), or metal hydride systems ($13/$10^6$BTU).

The unique properties of carbon nanotubes include adsorbent characteristics that offer the prospect of yet more efficient hydrogen storage. Dillon et al. [64] used temperature programmed desorption to determine the hydrogen storage capacity of single-wall nanotubes. The authors predict a storage capacity of ~50 kg H_2/m^3 at ambient temperature and pressures for nanotubes with 20 Å diameters.

IV. WATER TREATMENT

A. Introduction

The largest liquid-phase application of activated carbons is in water treatment, where the demand for activated carbon has increased rapidly over the last few years: the average annual rate of increase is about 6.6% in the USA [65]. It is important at the outset to note that there are various areas of water treatment, which are distinguished principally by the objectives of the treatment process.

In the production of potable water supplies, the primary aims are to produce water to an acceptable quality for human consumption. Potable water treatment can be broadly divided into different categories, according to the scale of operation. Municipal water treatment involves the large-scale treatment of water from rivers, lakes, and reservoirs for distribution to urban communities. Groundwater treatment is usually conducted on a smaller scale and often provides a drinking water supply to individual households or small communities. Possibly the fastest growing area in potable water treatment is in the point-of-use treatment of drinking water, which can involve the following: the batch treatment of tap water; in-line water treatment for single and multiple dwellings, and commercial buildings; and portable water treatment units for use in remote areas. In each of these applications, the original role of activated carbon was in a "polishing" step to remove species that adversely affect taste and odor. While this is still a requirement, a general deterioration in the quality of water supplies has created a situation where activated carbon is no longer used simply for aesthetic reasons but also to remove contaminants that are considered to be hazardous to health. Taking this a step further, activated carbon is employed specifically in the remedial treatment of groundwater to remove contaminants that have infiltrated from various sources.

The water employed in industrial processes such as heat exchangers, cooling towers, and steam generators is obtained from the most convenient local sources. This water must be treated to remove species that can give rise to corrosion, the formation of deposits, and fouling. Here too, adsorption on activated carbon is often an integral part of the treatment process.

A third important area of water treatment where activated carbons are also employed is in the treatment of effluent or waste waters that derive from a broad range of manufacturing plants. Before this water can be discharged, it must be rendered free of toxic substances that can pollute natural water sources; and the biochemical oxygen demand that it would place on such systems must be limited.

B. Properties of Activated Carbons for Water Treatment

Use of granular activated carbon (GAC) is considered to be the best currently available technology for removing low-solubility contaminants such as disinfection by-products (usually from chlorination) that include trihalomethanes (THM), detergents, pesticides, herbicides, polyaromatic hydrocarbons, and some trace metals. The amendments to the Safe Drinking Water Act state that other treatment technologies must be at least as effective as GAC [66].

Most of the granular activated carbons used in adsorber beds are produced from bituminous coal because their hardness, abrasion resistance, relatively high density, and pore size distributions render them suitable to withstand operating conditions and to adsorb the small organic molecules often present in drinking water.

Powdered activated carbon (PAC) is used for the same purposes as GAC. The main difference between PAC and GAC is the smaller particle size of the powders (typically around 44 μm versus 0.6–4.0 mm for granules), which allows faster rates of adsorption [67]. However, because of the greater difficulty in handling, fine powders cannot be used in fixed-bed operations without incurring a high pressure drop, and there are associated problems in regenerating them for reuse. Hence, they are invariably used as disposable additives.

Powdered activated carbons offer the advantage of low cost compared to granules in terms of both purchase price and capital expenditure (investment in adsorber units, pumps, etc). The cost of PAC is about $1.00/kg versus $2/kg for GAC [19]. For a system treating 4 million/day the cost is about $0.03/1,000 L [67]. A wider range of impurity removal levels can be attained with powdered carbon, where the dose of carbon per batch can be adjusted, depending on the type and concentration of the contaminants [68]. GAC is normally used in continuous flow deep beds and is advantageous when variations in adsorption conditions occur, such as large spikes in contamination levels. Although granular carbons are more expensive, they may be more cost effective if the usage rate is high, since they can be regenerated and reused [69]. Services have emerged over the last few years to increase the convenience of dealing with spent GAC and its regeneration. Granular carbon can be supplied in a module that serves as an adsorber bed and as a shipping container to facilitate transportation to a regeneration facility [70]. Regeneration typically involves heat treatment in steam to temperatures from 200 to 800°C to desorb organics and other contaminants.

The adsorptive capacity is the most important property of GAC relevant to water treatment. The value of this parameter determines the quantity of water that can be treated per unit mass or volume of carbon and hence provides information for sizing equipment. In general, the adsorptive capacity is not simply related to the total surface area or pore volume. The surface area is normally determined by the Brunauer, Emmett, and Teller (BET) method [71] using nitrogen adsorption at 77 K, whereas the important parameter in water treatment applications is the total wetted surface area, which may be quite different. The pore size distribution also affects the efficiency and selectivity of adsorption, i.e., the distribution of pores among micro (<2 nm diameter), meso (2–50 nm) and macropores (>50 nm). A consideration of the dimensions of some pollutants shows that activated carbons can feasibly be used to remove many of the impurities occurring in water: small organic molecules with low solubility have sizes in the range 0.6 to 0.8 nm and can be adsorbed in micropores; larger compounds such as color molecules and humic acids have dimensions around 1.5–3.0 nm that will favor their adsorption in mesopores. The distribution of pores in activated carbons can vary significantly depending upon the precursor material. For example, anthracite and some bituminous coals yield carbons with a high proportion of micropores, lower rank bituminous coals give a broad pore size distribution, while lignite and wood precursors produce mesoporous carbons, Fig. 19.

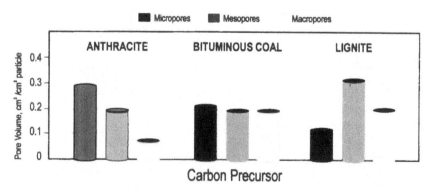

FIG. 19 Dependence of pore size distribution of activated carbons on the precursor.

In addition to pore size distribution, the surface chemistry of the activated carbon can have an important influence on the adsorption of certain compounds. As the adsorptive surface of most activated carbons is hydrophobic, they are best suited for the removal of neutral organic molecules, while polar and ionic compounds show much less affinity for adsorption. For the adsorption of polar compounds such as phenol, research has shown that the carbon surface chemistry is more relevant than the total available adsorption capacity or surface area [72–74]. It has been found that the presence of acidic surface oxides, whose concentration can be increased by oxygen adsorption or chemical treatment, leads to a decrease in adsorptive capacity for compounds such as phenols and increases the base adsorption capacity [75].

Isotherms are normally developed to evaluate the capacity of the carbon for the adsorption of different contaminants. Data are obtained in batch tests, which determine the equilibrium relationship between the compound adsorbed on the carbon and that remaining in solution. The isotherms are used as screening tools to determine which carbon is suitable for a given application. Batch equilibrium tests are often complemented by dynamic column studies to determine system size requirements, contact time, and carbon usage rates [19]. Other parameters that are used to characterize activated carbons for water treatment include phenol number, an index of the ability to remove taste and odor, and molasses number, which correlates with the ability to adsorb higher molecular weight substances. However, these parameters still do not reflect performance in service, and they can only be considered as guidelines.

To design and implement systems for water treatment, it is not only the adsorption characteristics of the activated carbon that must be considered but also the effect that the carbon may have on the practical operation of the unit. In this context, the pressure drop generated across a GAC bed is one of the most important factors, as it also is for gas phase applications. The particle size distribution should be optimized to attain an acceptable pressure drop commensurate with the desired rate of adsorption. The carbon attrition resistance is another important parameter. Part of the operating cost of adsorbers is due to the loss of carbon fines during transport, handling, and regeneration.

C. Bacterial Colonization

It has long been known that, under appropriate conditions and especially in the liquid phase, synergistic associations can develop between microbiological systems and activated carbons or other support media (e.g., in trickling bed filters for aerobic water treatment). In liquid phase applications, bacterial colonization of activated carbon can occur quite readily [76–79]. For example, the adsorptive capacities of activated carbon beds used in water treatment are often greatly enhanced by the presence of microorganisms, and the useful filter life is extended beyond that expected for a process of purely physical adsorption. Essentially, the adsorption of biodegradable compounds on activated carbon provides a convenient and enriched substrate for bacterial nutrition. Bacteria can become resident on the external surfaces of activated carbon granules and occupy the void spaces between them. Bacteria and viruses typically range in size from 0.4 to 20 μm and 10 to 300 nm, respectively, the dimensions being dependent upon their shape. Hence, viruses and some bacteria can also be accommodated in macropores inside the carbon matrix. For this reason, complete disinfection can be difficult, since microorganisms in such locations are not readily accessed by disinfectants, and the carbon offers a protective environment against attack.

The presence of microorganisms is generally advantageous in water treatment, and systems have been developed that deliberately attempt to intensify biological reaction rates, such as biologically based fluidized beds that use GAC as the fluidizing medium [77]. There are also drawbacks to GAC colonization [76]. For instance, excessive bacterial growth can reduce the flow through fixed bed adsorbers and raise the pressure drop; microorganisms can also enter into the effluent stream by periodic sloughing and by the attrition of carbon granules when fine particles can act as a carrier. Conditions may be created that are favorable to the buildup of pathogens. The presence of microorganisms may also inhibit the adsorption of other compounds as they occupy active sites on the carbon surface like other organic molecules.

In some applications, such as food processing and the purification of water for beverage industries or for domestic drinking water, bacterial colonization is neither acceptable nor is it easily avoidable. Bacterial growth in beds can be prevented by steam injection at temperatures higher than 80°C for prolonged periods, which does not always present a practical solution. In water treatment processes, secondary disinfection posttreatment with activated carbon provides protection against the entry of living bacteria into the water supply system. Domestic water filters are also affected by bacterial colonization. In these cases, a chemical that can inhibit bacterial growth may be added to the carbon. The most common additive is silver, which exhibits bactericidal properties at high concentrations [80].

D. Potable Water Treatment Processes

1. Municipal Water Treatment

Powdered activated carbons have been used for more than 50 years to treat public water supplies for the removal of naturally occurring organic compounds that adversely affect taste and odor. Granular activated carbons (GAC) are now used widely in municipal and groundwater treatment plants.

In municipal water treatment, the adsorption of contaminants on activated carbon is the final polishing step. It is essential to pretreat the water to remove debris and particulates and other inorganic contaminants before the organics are

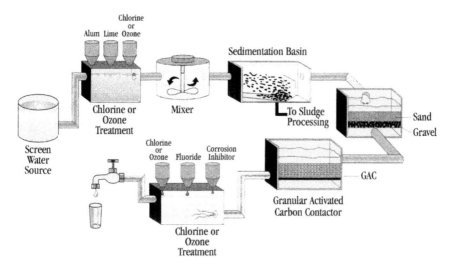

FIG. 20   Schematic of water treatment system.

removed in an activated carbon unit. A general schematic of a chlorine- or ozone-based water treatment system is shown in Fig. 20. The initial screening removes leaves and other large debris that can clog the system. This is followed by a disinfection process where the water is treated with either chlorine or ozone. Coagulants such as lime are added at this stage to aid in water clarification. The coagulants cause particulates to cluster and settle at the bottom of the sedimentation basin. Residual particles are removed in the sand filters. A granular activated carbon bed is placed either as a layer in combination with the sand filter or in a separate tank. The effluent water is then postdisinfected with small amounts of chlorine prior to distribution.

The current trend towards decreasing quality of water supplies is of great concern because of the implications to public health. In recent years synthetic organics have been appearing in public water supplies in ever-increasing concentrations. Of most concern are volatile organic compounds (VOCs), pesticides, and disinfection by-products that are among the most frequently found contaminants in groundwaters. Each of these groups of compounds is believed to be detrimental to public health. Pesticide ingestion has been associated with health problems that include cancer and nervous system disorders [81]. Organic disinfection by-products (DBPs) are formed when disinfectants that are used to sterilize water supplies react with natural organic matter. For example, the reaction of chlorine with humic acids in groundwaters can produce chlorophenols and halomethanes. The DBPs are potentially toxic substances, and some may promote cancer in humans [82].

In GAC adsorbers, carbon is lost through a number of mechanisms, making it necessary periodically to complement the bed inventory with makeup carbon to maintain bed volume and treatment efficiency levels. Some attrition occurs during everyday operation through movement of the granules and as a result of operations such as bed loading, pre- and postregeneration handling, and back-flushing of the bed to remove fine solids. Microporous activated carbons also tend to lose porosity due to the blockage of pores by inorganic matter that accumulates during adsorption.

2. Regeneration

A significant operational cost for treatment plants utilizing granular carbons lies in adsorbent regeneration. Depleted carbons are taken out of the operational stream and subjected to thermal processing in order to remove and/or destroy the adsorbed compounds from the carbon surface and to reactivate the carbon for subsequent reuse. The long-term economics of carbon regeneration can often be the single most important parameter in determining overall treatment plant design and operational procedures.

For treatment operations requiring high carbon tonnage and/or frequent bed regeneration, it may be economically beneficial to carry out carbon regeneration in a thermal processing unit maintained on site. While there is a need to deal with the secondary waste streams produced during regeneration, there is an advantage in the rapid and controlled turnaround of the carbon bed material.

Smaller plants utilizing carbon-based treatment units may have neither the bed volume nor the usage rates that would justify the construction or operation of an on-site treatment facility. These sites are better served by contractual services whereby spent bed material is collected and sent to a third-party-operated regeneration facility. The benefits of employing off-site regeneration can include decreased plant energy and maintenance costs associated with the operation of a regeneration facility, as well as reducing the inventory of waste streams produced on-site.

E. Groundwater Treatment

Groundwaters may need to be treated either to render them suitable for drinking or to remove pollutants. Groundwater treatment is becoming more common because of problems similar to those encountered in municipal water treatment, namely that water sources are becoming contaminated.

Approximately 50% of US drinking water supplies are drawn from groundwater, and of these about 25% have been detected with VOCs above the Maximum Contaminant Level [73]. Sources of contamination include agriculture and industry, hazardous waste disposal, underground storage tanks, and accidental spills. The organic contaminants of particular concern are chlorinated aliphatic and aromatic solvents. These compounds have little affinity for soil and pass quickly into the aquifer. They are also resistant to degradation and can persist for long periods of time [73]. It is not uncommon to encounter high concentrations of pesticides and chemicals in the water table as deep as 30–40 m below the ground surface [83]. Remedial treatment is necessary to remove the major contaminants before they infiltrate other natural water supplies and adversely affect aquatic life as well as municipal water treatment systems. Approaches to substantially removing chlorinated hydrocarbons involve liquid phase or gas phase adsorption on activated carbons, and chemical reaction.

The type of activated carbon used for the treatment of groundwater depends upon the other organic compounds present. For example, humic acids will often compete with other contaminants for adsorption sites. Hence, for each situation, isotherms should be determined, and the efficiency of contaminant removal should be evaluated on a TOC basis (total organic carbon). As noted, the compounds most prevalent in groundwaters are chlorinated organics. Their removal by adsorption on GAC is the most effective treatment in liquid or gas phase processes [66].

1. Liquid Phase Adsorption

A typical unit for contaminant removal by liquid phase adsorption on activated carbon is shown in Fig. 21. It consists of a downflow fixed bed adsorber, to which water is fed under gravity or pressure. Typically, the vessel is about 10 ft in diameter and contains about 20,000 lb of activated carbon. Adsorbers of this size are routinely used, as they are convenient for transporting GAC to and from the site on a single trailer. The adsorbers can treat up to 350 gpm at a contact time of 15 minutes.

In contaminated waters, usually there is a mixture of compounds. The treatment system uses two or more vessels filled with activated carbon and connected in series or parallel. At any time, one (or more) vessel(s) will be in an adsorption cycle while the other one(s) is regenerated. To illustrate the carbon usage rate, for an inlet contaminant concentration of 200 mg/L, 8000 gallons of water can be processed per lb of GAC, to give an acceptable effluent concentration of below 5 mg/L.

2. Gas Phase Adsorption

In some cases, it is more economical to use air stripping to remove VOCs from water and then use a GAC bed to purify the airstream. GAC has a higher capacity for VOC adsorption in the gas phase than in the liquid phase. Also, gas phase adsorption occurs more rapidly, so that shallower beds can be used [84].

Air stripping is achieved by pumping groundwater to a packed column. The packing materials are used to provide large void volumes and a high surface area. The water flows downwards under gravity countercurrent to a stream of air. The

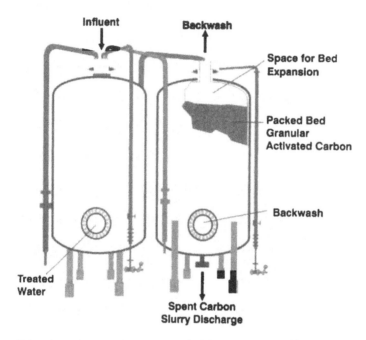

FIG. 21   Schematic of system for liquid phase water purification.

removal efficiency of organic compounds is typically in the range 80–95%. The effluent air is then run through an activated carbon bed to purify the airstream and recover the stripped organics.

The regeneration of activated carbon beds used in gas phase adsorption requires less severe conditions than for liquid phase processes. Regeneration can be conducted in situ by stripping with steam. Newer and more efficient systems use regeneration by hot inert gas, nominally at 350°F, to recover a greater portion of contaminants with their subsequent recovery. This is particularly attractive if the disposal of condensate from steam regeneration becomes a problem [73].

F.   Domestic Water Treatment Units

The sale of home treatment units for tap water purification has increased greatly over the last few years. Although the market is not large in terms of the quantity of activated carbon consumed, it is significant because of the high prices it commands. Most units are fitted to a faucet and are made with GAC as the adsorbent. However, recently there has been a change in this trend: GAC has been replaced with monolithic carbon blocks containing powders or granules held together by a binder [85]. The main advantage of the monoliths lies in that the small diameter of the powders minimizes mass transfer limitations and allows more rapid rates of adsorption than can be achieved with GAC. Because of their efficient adsorption capabilities, many of the new filters can remove contaminants that GAC is not able to remove, including microorganisms such as the cysts cryptosporidium and giardia. Outbreaks of cryptosporidiosis in Milwaukee and Las Vegas in 1993 and 1994, respectively, resulted in several deaths [85]. The outbreak in Milwaukee caused illness in 403,000 people, and many who were immunocompromised died. The probability of an outbreak of cryptosporidiosis is higher than for bacteria and viral pathogens because of the resistance of the cysts to the disinfectants used in drinking water treatment plants [86]. One way of removing the cryptosporidium from water is to use filters with aperture sizes less than 2 µm. Several of the new monolithic activated carbon filters on the market can remove 95–99% of the cryptosporidium from drinking water.

In addition to faucet units for water treatment, activated carbon filters are used with water jugs and in portable water treatment systems for use in remote areas. Of the home water treatment units, there are point-of-entry units that are placed in line and treat water supplies to whole households and commercial buildings. Both point-of-use and point-of-entry filters often contain a combination of activated carbon for the removal of organics and cryptosporidium and ion exchange resins for the removal of metals and inorganic compounds.

V. CARBON SUPPORTS FOR ZERO-VALENT METAL DEHALOGENATION

The treatment of halogenated hydrocarbons by zero-valent metals, mainly iron, tin, and zinc, offers several attractive features for the remediation of contaminated waters: relatively high reaction rates, and the low-cost and noncontaminating nature of the metal. These reactions can be enhanced by the use of a carbon support to increase the area of metal available for reaction.

A. Background: Zero-Valent Metal Dehalogenation

The emphasis of most research in this area has been on the use of iron [87–93], or a modified iron, such as palladium-plated iron granules [90]. Alternatively, relatively fast reaction rates have been obtained when using very fine divided Zn or Sn particles in an inert atmosphere [94].

While the dehalogenation of chlorinated species is often referred to as zero-valent metal catalysis, it is not strictly a catalytic but rather an electrochemical corrosion process. The metal is consumed by reaction, leading to the formation of metal salts and dechlorinated by-products.

If M is the metal,

(1) $M^0 \rightarrow M^{n+} + n\ e^-$

(2) $R\text{-}Cl + e^- \rightarrow R + Cl^-$

(3) $M^{n+} + n\ Cl^- \rightarrow M(Cl)_n$

The zero-valent metal reaction ranges from 5 to 15 orders of magnitude faster than those observed with natural abiotic processes. For example, using a reactive bed containing iron filings, about 90% of the trichloroethylene (TCE) found in groundwater at the Canadian Forces Base, Borden, Ontario, site could be removed [92]. It was concluded that the reaction rate was independent of TCE concentration. In addition, it was found that granular iron was an excellent catalyst promoting the decomposition of various chlorinated species, including 13 out of 14 halogenated methanes, ethanes, and ethylenes [92].

In a pilot scale study of TCE dechlorination using iron filings, several factors were found to control the process [93]. The alkalinity of the groundwater and the concentration of dissolved oxygen affect the life of the catalyst bed: the presence of carbonate- or oxide-forming species in the water can lead to the formation of an inert layer on the iron surface. This layer greatly reduces the overall reaction rate and shortens bed life [87,91,93,94]. A tenfold increase in aqueous alkalinity was found to reduce the reaction rate by a factor of 3.

Metal surface area is another rate controlling factor. Large metal pellets are ideal in terms of handling and process operation, but reaction rates are low due to the small amount of surface area available. In contrast, fine metal particles offer a means of substantially increasing the reaction rate, but they introduce problems of containment, handling, and high resistance to flow in packed beds. Cryoparticle zinc with a surface area of $>65\ m^2/g$ reduced the concentration of CCl_4 in water by over 90% in 3 hours, while granular zinc, having a surface area $<1\ m^2/g$, achieved a reduction of only 25% [87,16]. A similar correlation between surface area and dechlorination ability was seen in studies using Sn.

Of the most commonly studied zero-valent metals, Fe, Zn, and Sn, Fe is the cheapest, is easily obtained as scrap, and is considered to be the most environmentally friendly. However, Zn and Sn have been shown to be more reactive and have a higher tolerance for oxygen before forming an inert oxide layer.

B.   Carbon-Supported Zinc Catalysts

An obvious solution to the surface area limitation is to distribute the selected metal over a high surface area support. However, as metal is consumed in the process, a conventional metal support system would require frequent replacement. A more attractive proposition is offered by the availability of permeable monolithic carbon fiber composites, which were discussed earlier in this chapter. The ability to make these composites electrically conducting allows metals to be deposited on the fiber surfaces by electroplating. The advantages of this system are that it can produce a catalyst with high surface area and low resistance to flow, and it could be regenerated in situ.

In recent investigations, such a composite was plated with a thin layer of zinc and examined for its effectiveness for the dechlorination of TCE [95]. A composite measuring 10 cm in length and 2.5 cm in diameter was prepared by established methods [16] and heat treated to 800°C in nitrogen to attain an electrical conductivity of less than 200 ohm^{-1} cm^{-1}. It was then plated with zinc in a zinc chloride bath using a low current density (Fig. 22).

The plated composite was sectioned and examined for elemental composition. This analysis showed that the zinc coating on the fiber surface was uniform throughout the length and over the cross-section of the composite. The plating process was simple to operate and control, which suggests that the zinc layer could be regenerated in situ on a spent composite by the application of current while flowing zinc chloride solution.

In activity tests, sections of the Zn composite were used as agitation paddles in a batch reactor and compared to granular zinc, which was loaded into a similar vessel with an inert agitator (Fig. 23). The reactors were charged with 1 L of water and sparged with nitrogen to remove dissolved oxygen. TCE was then added to a concentration of 1000 ppm. A chloride selective electrode was used to monitor chloride ion formation as a function of reaction time. First order kinetics were assumed, and rate constants were calculated per mass of Zn.

The granular zinc had a measured surface area of 0.0022 m^2/g, while that of the composite was calculated to be 0.06 m^2 Zn/g composite. The reaction rate with the composite was found to be an order of magnitude higher than with the granular zinc; 1×10^{-5} min^{-1}g^{-1} for granules versus 2.5×10^{-4} min^{-1}g^{-1} for

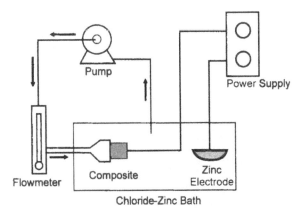

FIG. 22 Diagram of composite plating apparatus.

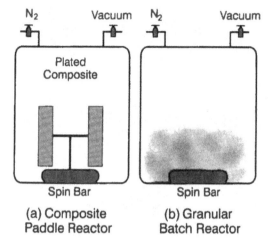

FIG. 23 Schematic of reactors used to evaluate catalytic removal of trichloroethylene. (a) Zn-plated carbon–fiber composite; (b) Zn granules.

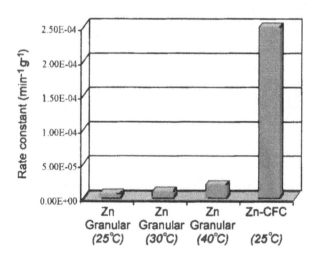

FIG. 24 Reaction rates of Zn granules and Zn-plated carbon–fiber composites for removal of trichloroethylene from water.

the composite (Fig. 24). The rate correlates directly with the metal surface area, indicating that this parameter controls the reaction rate. Zinc plating was found to have no measurable effect upon the pressure drop through the composite. The combination of high permeability and activity of the zinc-plated composite suggests that it is a viable candidate for use in a fixed-bed flow system.

Locating a metal source upstream of the composite and supplying an electric current would allow the surface area of metal available for reaction to be replenished as the metal is depleted during the reaction. A practical scheme would be the simultaneous operation of two beds, in which a fresh bed would be on stream while the depleted bed would be regenerated by back flushing with plating solution while metal is coated onto the carbon support. The metal chloride found during this process could be recovered upstream by ion exchange or pH-shift mediated precipitation.

VI. OTHER APPLICATIONS

A. Detectors and Measuring Devices

Carbon-based sensors are finding extensive use in various fields that include the electronics, transportation, environmental science, and medicine. With the discovery of fullerenes in the past decade [96], then of nanotubes, and subsequently of synthesis routes [97] for fullerenes and carbon nanotubes [98,99], novel devices such as transistors based on thin fullerene films [100] and carbon nanotubes [101] are beginning to emerge. Due to a remarkable combination of the physical and chemical properties of these and other structures such as diamond and diamondlike carbons, it is predicted that a new generation of carbon-based integrated circuits and power devices will become a reality in the near future [102]. It is also predicted that carbon-based technology will have a significant impact on the automobile and aircraft industries. The availability of strong, lightweight materials can lead to the improved design and manufacture of efficient transportation systems for use on land, in air, and in space. With respect to mass transportation, benefits will include low energy consumption and smart devices to optimize vehicle performance to reduce emissions. Below, a few examples of emerging carbon-based environmentally friendly devices are presented.

1. Diamondlike Carbon and Hard Carbon-Based Sensors

Sensors that are based upon diamond technology include thermistors, pressure and flow sensors, radiation detectors, and surface acoustic wave devices [103]. The relative ease of depositing prepatterned, dielectrically isolated insulating and semiconducting (boron-doped p type) diamond films has made polycrystalline diamond-based sensors low-cost alternatives to those based on conventional semiconductors. Diamondlike carbon and diamond films synthesized by chemical vapor deposition (or similar methods) are found to exhibit properties similar to those of crystalline diamond, e.g., extreme hardness, high thermal conductivity, good optical transparency, wide band gap, high piezoresistivity (i.e., variation in resistance with mechanical strain), radiation immunity, and chemical inertness. At least two types of diamond-based sensors have already been put to practical use: a resistance thermometer or thermistor [104] and a diamond-based force sensor test chip [103]. In the case of boron-doped diamond thermistors, the temperature-dependent electrical resistance varies logarithmically with temperature, and the device exhibits a stable response from −80 to 900°C [104]. Figure 25 shows a diamond-based force sensor (pressure, acceleration, etc.) chip that is capable of operating in a harsh environment. This chip comprises a pressure-sensing diaphragm, a prototype field effect transistor (FET) structure, freestanding diamond beam, and a Hall sensor structure [104].

With the advent of faster aircraft, the US air force is interested in materials with high damage threshold for use as antireflection coatings on forward-facing components such as infrared transmitting windows and domes. Hard carbon coatings are used because of their extreme durability and transparency in the infrared. They can be obtained via the dissociation of methane on substrates such as germanium [104]. Such hard carbon coatings have been tested by the aircraft industry for optical transmission as well as resistance to damage due to rain and sand impact. The performance of hard carbon coatings was found to be highly depen-

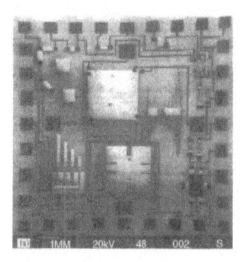

FIG. 25 A diamond-based force chip composed of a pressure-sensing diaphragm, an FET, a diamond beam, and a Hall sensor. (Adapted from Ref. 104.)

dent on the speed of the aircraft. A quarterwave thick hard carbon coating on Ge tested successfully only at low aircraft speeds (~30 m/s) [105]. At higher speeds (~200 m/s), the damage due to rain and sand impact was severe and resulted in unacceptably large optical transmission losses. This damage stems from the fact that hard carbon coatings generally have high compressive stress and can readily deteriorate upon impact, subsequently leading to damage of the substrate. Hard carbon coatings with low stress have been developed by the Hughes Aircraft Corporation. These carbons have been found to possess two to three times better resistance to impact damage. However, the gain in durability has so far been achieved at the expense of low transmission. The poor quality in transmission can be overcome by depositing amorphous silicon to fabricate a multilayer stack of Si and hard carbon [105]. Amorphous silicon films are deposited by plasma-assisted chemical vapor deposition in which pure silane gas is dissociated.

2. Carbon-Based Nanoprobes

Electron transfer plays a central role in scanning electron transmission microscopy (STM) and is accomplished through the overlap of the electronic wave functions of the sample and the STM probe tip. If a tip and a conducting sample are separated by only a few angstroms, their electronic wave functions overlap. Upon the application of a bias voltage, a tunneling current flows between the tip and the sample in the direction dictated by the sign of the applied voltage. It has been shown theoretically, and observed experimentally, that the adsorption of atoms or molecules on a STM tip changes the tip structure and the electronic states of the tip [106–108]. Thus a chemically modified STM tip can be "tailored" to discriminate certain species (e.g., toxins) through specific chemical interaction between that species and the tip. Nakagawa et al. [109] have developed a novel atomic force microscope (AFM) in which sensor molecules can be immobilized by a chemical adsorption technique in order to measure specific intermolecular forces between sensor molecules on the tip and on the surface of the sample. AFM tips that were chemically modified with octadecyltrichlorosilane or perfluorotrichlorosilane were more sensitive than unmodified tips for detecting the adhesive force on alkyltrichlorosilane monolayers on silicon substrates.

Fullerene C_{60} adsorbed onto STM tips has been reported to enhance atomic resolution images of highly oriented pyrolytic graphite [110]. Recently, Dai et al. [111] have demonstrated that multiwalled carbon nanotubes (MWNTs) attached to the silicon cantilever of a conventional atomic force microscope (AFM) can be used as well-defined tips with exceptionally high resistance to damage from "tip crashes." The MWNT were attached by first coating the bottom 1–2 mm section of the silicon tip with an acrylic adhesive by inserting them into an adhesive-coated carbon tape. The tip was then brought into contact with the side of a bundle of 5–10 MWNTs while being directly observed through an optical microscope using dark-field illumination. Once attached, the MWNT bundle was pulled free from its connections with other MWNTs, leaving a single 5 nm diameter MWNT extending alone for the final 250 nm of the tip (Fig. 26).

Because of the small diameters of MWNTs (~5–20 nm) and their inherent slenderness, images can be made of sharp recesses in surfaces, allowing much better definition of surface topography. Furthermore, since the MWNTs are electrically conductive [112], they may also be used as probes for the scanning tunneling microscope. An atomic-scale resolution STM image of a freshly cleaved 1T-TaS$_2$ surface [111] obtained using a MWNT tip is shown in Fig. 27. Slender tips were also formed by attaching a single "rope" of single-walled carbon nanotubes (SWNTs) to the side of a MWNT probe tip, similar to the one shown in Fig. 26. It has been demonstrated that a chemical derivatization of the end-cap of a single (10,10) nanotube tip would serve as the "ultimate" nanoprobe [113].

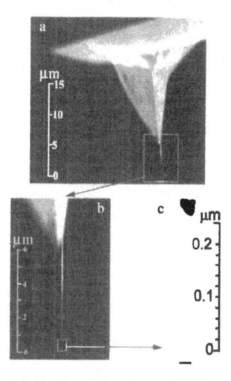

FIG. 26 A single 5 nm diameter MWNT extending from the AFM tip. (Adapted from Ref. 111.)

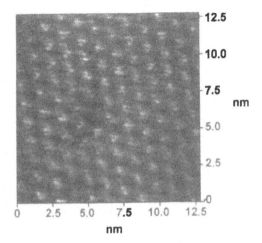

FIG. 27 An atomic-scale resolution STM image of a freshly cleaved 1T-TaS$_2$ surface obtained using an MWNT tip.

3. Carbon Nanotube Transistor

Microelectronic devices are very much of interest to the electronic and computer industries. The possibility of using individual molecules as functional electronic devices is at the heart of most miniaturizing strategies. Progress in this field has been fraught with difficulties, since attaching electrical leads to molecules is a challenging task. Recently, Dekker and coworkers [101] succeeded in fabricating and characterizing a TUBEFET (single carbon nanotube field-effect transistor), which is a hybrid of molecular and solid-state materials and can function at room temperature. Figure 28 shows an atomic force microscope image of a TUBEFET in which a SWNT lies across two predeposited platinum contacts on a SiO$_2$ over-coated silicon substrate. The Si substrate serves as the gate electrode, while the two platinum contacts act as the source and drain electrodes. The ON/OFF state required of a transistor is achieved by applying a gate voltage to the conducting substrate to move charge carriers onto the nanotube. A negative bias applied to the substrate induces hole carriers in the initially nonconducting nanotube and switches it into the ON state. This behavior is analogous to a conventional p-type metal-oxide FET.

4. QCM Sensors Coated with Lecithin and Activated Carbon

Quartz crystal microbalance (QCM) sensors that are coated with lecithin and activated carbon can be used to detect environmental pollutants with high recognition ability. The pollutants are detected from a measurable change in crystal oscillator frequency, which is caused by a small increase in mass deposited on

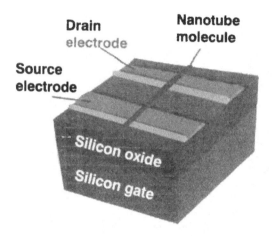

FIG. 28   An AFM image of a TUBEFET. (Adapted from Ref. 101.)

the crystal. The nature of the environmental pollutants can be inferred by comparing the normalized patterns of their resonant frequency shift and the relative response intensities [114]. Figure 29 depicts the response of two sensors to propanol in a controlled experiment: (1) lecithin with activated carbon coated QCM and (2) lecithin coated QCM [114]. The increase in the frequency shift intensity is proportional to the amount of propanol injected into the system. The superior response exhibited by lecithin with the added activated carbon was attributed to the increased surface area of the activated carbon [114]. While the QCM sensors are quite versatile, one of their major drawbacks is that they may not be chemically stable when exposed to a broad spectrum of pollutants. It was found that the response of a lecithin and activated carbon coated QCM to various concentrations of hexane leads to a decrease in the frequency shift intensity, implying a net mass loss on the QCM sensor.

5.   Carbon Paste Electrodes for Biosensors

Composite carbon paste electrodes prepared by mixing carbon paste with biocomponents (enzymes, tissues, etc.) is a recent interesting development in the field of electrochemical biosensors. The biological entities show a strong decrease in activity when incorporated in the carbon paste matrix due to their confinement in a hydrophobic environment. A new approach to the preparation of electrochemical biosensors based on a Sepharose-carbon paste electrode has been reported by Gasparini et al. [115]. As another example of a modified carbon electrode, Chen et al. [116]. developed a sensitive and selective detection scheme for dipeptides based on carbon electrodes. The modified electrode was found to be more selective for α-dipeptides over β- and γ-dipeptides as well as amino

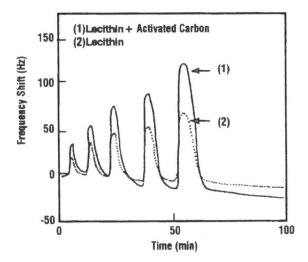

FIG. 29 Response of a lecithin with activated carbon coated QCM (trace 1) and a lecithin coated QCM (trace 2) sensor for propanol. (Adapted from Ref. 114.)

acids at pH 9.8, whereas it was selective for all dipeptides over amino acids at pH 8.0.

Lyne and O'Neil [117] reported the *in vivo* detection of dopamine using stearate-modified carbon-Nujol paste electrodes. Prior to their work, the detection of dopamine by voltammetric techniques was hindered primarily due to the coexisting ascorbic acid in the extracellular fluid of the mammalian brain. Ascorbic acid oxidizes at electric potentials similar to that of dopamine on many electrode materials. These authors found that the use of stearate-modified carbon-Nujol paste electrodes retards the electro-oxidation of anionic species (such as ascorbate) to such an extent that the cationic dopamine species could be detected in their presence.

B. Medical Applications of Carbon

Carbon, especially activated carbon in the form of charcoal, has been used in medical and health applications for centuries. Early Egyptian writings around 1550 BC document the use of charcoal for treating putrefying wounds and problems in the gastric intestinal tract. In these two applications, charcoal functions as an adsorbent to remove odors and to provide an effective treatment against bacteria and toxic materials. Hippocrates (460 BC) and Pliny record the use of charcoal for treating epilepsy, chlorosis, and anthrax. Anthropologists excavating Phoenician shipwrecks have found drinking water stored in charred wooden barrels, presumably affording a method for purifying potable water and extending its storage. Hindu documents of similar age also describe charcoal filters for water treatment. More recently, in the late eighteenth century, Lowitz and Kehl each reviewed the use of chars for controlling odors from a range of medical conditions.

Medical applications of activated carbon adsorbents and purifying aids continue today. Examples of the many applications in current use include the use of activated carbons to adsorb bacterial toxins in the GI tract and in dialysis equipment for the purification of blood.

1. Adsorption Applications

Activated carbons are used for the adsorption of specific chemical species and also for wide classes of compounds. Odor adsorption applications include deodorizing gaskets for colostomy patients [118]. In haematology, activated carbons have been developed to adsorb IgG antibodies [119] and toxic agents from plasma [120]. In the development of medical equipment, activated carbons with broad specificity are used for the purification of liquids and gases. For example, activated carbon filters continue to be used as a part of the water purification systems in haemodialysis [121], and as microfilters for blood transfusions [122]. Some activated carbons are encapsulated for easier use as implants in the body [123,124]. In some cases, chemicals can be sorbed onto activated carbon for release in the body to serve as drug delivery systems, such as gold-coated carbon implants for treating arthritis [125].

2. Structural Applications

The advent of high-strength carbon-based fibers has led to a number of structural applications in medical treatment. The specifications for implanted materials required that they offer acceptable long-term mechanical properties and surfaces that are biocompatible [126]. Surface compatibility affects immediate acceptance, while the long-term mechanical performance is determined by the bulk properties of the implant.

Carbon fibers, which are relatively inert under physiological conditions, are selected particularly to enhance the mechanical properties of various biomedical materials, such as their incorporation in bone cement [127]. Metal implants for total hip joint replacements do not match the mechanical properties of human bone, and epoxy-graphite implants may have better properties [128].

The foreign body response to carbon fibers is not chronic unless they fragment. One major problem has been the accumulation of wear debris near the acetabular components [129]. However, when used as ligament replacements, carbon fibers appear to induce a type of fibrosis in which collagen fibers align with the fibers. This suggests that carbon fibers can be used as a template for the formation of fibrous tissue [130] and indeed tendon fibroblast cells have been grown on a variety of fibers, including carbon. The morphologies of the cells and their orientations are affected by the fiber diameter and by the surface defects [131]. Other workers have found that carbon fibers can cause continuous irritation of collagenous tissue when used as ligaments. The histology of implanted ligaments after one year showed a high density of histiofibroblasts produced by the carbon fiber irritation, and deficiencies in collagen density. The tensile strength of the implants was due entirely to the carbon fibers [132].

Three types of ligament replacements have been tried: permanent prostheses that stay in the body, inductive prostheses that allow collagen to grow around them, acting as a scaffold for a new tendon, and resorbable prostheses that act as a scaffold and then degenerate [133]. Carbon fibers could be used as components for the first two types of ligament replacements. Braided carbon fiber ligaments have been used to replace the anterior cruciate ligament in animal knees. The stiffness and elasticity of the replacements was between 50 to 100% of the natural material [134]. The growth of tendon fibroblasts on three-dimensional carbon fiber matrices is different from that on standard culture plates, suggesting that there are differences in the mechanisms controlling cell growth on such structures [135]. Table 5 provides examples of tissue scaffold, ligament, dental implant, medical implant, composite implant, and other implant devices.

3. Electrical/Conducting Applications

Carbon fibers formed into filaments provide very compatible biomaterials. Carbon does not corrode in the body, nor does it generate a foreign body response. Further, its high electrical conductivity suggests that a variety of electrode and conductor applications are possible. For example, filamentary carbon cathodes have been used to stimulate the growth of bone and soft tissues [136], when adjusting the level of current can be used to influence the growth.

Microelectrodes made of carbon have been used for in vivo neuroelectrochemistry to measure catecholamines in neural tissues [137]. A major technical issue is distinguishing one compound from another using redox potential as a basis and differential pulse voltammetry as the method.

Carbon fibers are flexible enough to permit implantation in a number of tissues, including the wall of the intestine [138]. However, sharp angles must be avoided, since the fibers will break and their small diameter makes them difficult to handle. Table 6 gives examples of conducting devices in medical applications.

4. Future Biomaterials and Medical Implants Needs: Potential for Utilizing Advanced Carbons

A 1995 workshop sponsored by NIH reviewed directions and needs in biomaterials and medical implants. The workshops developed recommendations for research priorities in several topic areas that are summarized here. Research in basic design principles for medical implants and biomaterials should include building

Derbyshire et al.

TABLE 5 Examples of Carbon-Based Structural Devices for Medical Applications

Application	U.S. Pat. No.	Year	Authors
Tissue scaffolds			
Bioabsorbable tissue scaffold	4,512,038	1985	Alexander, Parsons, Strauchler, and Weiss
	4,411,027	1983	Alexander, Parsons, Strauchler, and Weiss
	4,329,743	1982	Alexander, Parsons, Strauchler, and Weiss
Medical putty for tissue augmentation	4,595,713	1986	St. John
Ligaments			
Prosthetic ligament	4,932,972	1990	Dunn, Lewis, Sander, Davidson, Beals, and Gill
	4,790,850	1988	Dunn, Lewis, Sander, Davidson, Beals, and Gill
	4,731,084	1988	Dunn, Lewis, Sander, Treharne
Dental implants			
Root canal implant	4,525,147	1985	Pitz, Jacob
Dental implants	4,722,688	1988	Lonca
Implants			
Porous fibrous implant	3,992,725	1976	Homsy
Porous implant of PTFE and carbon fibers	4,129,470	1978	Homsy
Implant material for living body hard tissue	5,171,492	1992	Kawakubo, Ohtani
Other structural applications			
Tri-leaflet all carbon heart valve	5,207,707	1993	Gourley
Carbon coated tubular endoprosthesis	5,163,958	1992	Pinchuk
Bio-absorbable copolymer	4,643,734	1987	Lin

Description	Number	Year	Inventors
Device for atraumatic access to the blood circuit	4,654,033	1987	Lapeyre, Slonina
Wig-like cool cap	5,218,977	1993	Takahashi
Biocompatible carbon prosthetic devices	3,952,334	1976	Bokros, Horsley
Composite implants			
Composite using bone bioactive glass	5,468,544	1995	Marcolongo, Ducheyne, Ko, LaCourse
	5,645,934	1997	Marcolongo, Ducheyne
Composite orthopedic implant	5,181,930	1993	Dumbleton, Lin, Stark, Crippen
Composite implant prosthesis	4,902,297	1990	Devanathan
Manufacturing method for composite implant prostheses	4,978,360	1990	Devanathan
Reinforced fiber bone replacement implant with treated surfaces	4,714,467	1978	Lechner, Heissler, Scheer, Siebels, Asherl
Composite femoral implant with increased neck strength	5,522,904	1996	Moran, Salzstein, Daniel, Cairns, Smith
Composite orthopedic implant	5,443,513	1995	Moumene, Lin, Stark
Surgical prosthetic implant for vertebrae	5,192,327	1993	Brantigan
Implant for intervertebral spinal fusion	5,425,772	1995	Brantigan
Surgical implant design	4,851,005	1989	Hunt, Mundell, Strover
Prosthetic device: cured fiber reinforced triazine resin	4,356,571	1982	Esper, Gohl
Coatable implantable medical device	5,609,629	1997	Fearnot, Kozma, Ragheb, Voorhees
Plastic knee femoral implants	5,358,529	1994	Davidson
Vertebral prosthesis	5,306,310	1994	Siebels
Composite femoral implant having increased neck strength	5,163,962	1992	Salzstein, Toombes
Surgical implant	4,813,967	1989	Renard, Chareire

TABLE 6 Examples of Carbon-Based Conducting Devices

Application	U.S. Patent No.	Year	Authors
Leads for integrated defibrillation/ sensing electrodes	5,534,022	1994	Hoffmann and Bush
Improved electrode structures	5,143,089	1992	Alt
Difibrillation electrodes	5,411,527	1995	Alt
Implantable carbon electrode	4,281,668	1981	Richter, Weidlich
Composite defibrillation electrode	5,683,444	1997	Huntlev, Zvtkovicz, Geiger
Defibrillation lead employing electrodes fabricated from woven carbon fibers	5,336,254	1994	Brennen, Williams, Gabler

biological structure and function into biomaterials, developing strategies for synthesis and methodologies to generate new materials and coatings, emphasizing reliability analysis, failure analysis, and corrosion, and expanding current tools for understanding and characterizing materials. Research on biocompatibility should tap into developments of contemporary receptor and cell biology, improve understanding of geometry, morphology, porosity, and biomechanics, assess biological responses, develop nanocharacterization of implants and materials, and develop methods for controlling biodegradation. Research on commercial implant materials should develop new biostability data, new materials, and new evaluation and measurement systems.

VII. CONCLUSIONS

The information that has been presented in this chapter on various applications of carbon materials in environmental technologies illustrates the tremendous significance and scope for the use of carbons and carbon-containing materials. It has been shown how different forms of carbon can serve as adsorbents and catalysts for gas and liquid phase processes, how they can function as detectors and measuring devices, and how they can be employed in a variety of ways in medical technology. Even within these defined areas, the advances that are being made in the research and development, across a dimensional scale of carbon structures and materials that ranges from nanometers to meters, are so fecund that it is difficult to create a text that is fully comprehensive and authoritative. Over the time that it has taken to prepare this chapter, a considerable volume of additional material has been made publicly available on new technologies that address environmental issues. Hence the authors have given here an overview or snapshot that is indicative of the current state and extent of the field and of its future potential, as a source of reference to those who wish to delve further.

Perhaps the most pertinent insight that is conveyed by this survey is the extent of the carbon world. As noted in the introduction, it reaches into virtually all branches of science and engineering and involves many other disciplines. It sends a message to what has traditionally been a rather small community of researchers that, collectively and individually, we must strive and reach out in our interactions and collaborations to help to translate the positive aspects of carbon science into new technologies.

ACKNOWLEDGMENTS

The authors wish to express their gratitude to Ms. Marybeth McAlister, ·who provided the overall management and coordination of this project, and without whose help and prompting it would never have been completed. Thanks are also due to Ms. Kathie Sauer for original graphic design of the figures, Dr. Christopher Lafferty for additional contributions, and Mr. Terry Rantell for assistance with editing and preparation.

Finally, the authors wish to thank the editor, Dr. Ljubisa Radovic, for his patience during this manuscript preparation.

REFERENCES

1. J. W. Patrick, ed., *Porosity in Carbons*, London, Edward Arnold, 1995.
2. A. N. Wennerberg and T. M. O'Grady, U.S. Pat. 4,082,694, 1978.
3. H. Marsh, D. Yan, T. O'Grady, and A. Wennerberg, *Carbon 22*, 603–611, 1984.
4. M. Farcasiu, S. Petrosius, and E. Ladner, *Journal of Catalysis* 146, 313–316, 1994.
5. P. B. Kaufman, E. Ladner, and M. Farcasiu, Carbon catalysts from coal for environmentally relevant reactions, American Chemical Society, Division of Fuel Chemistry Preprints, New Orleans, LA, 1996, pp. 447–450.
6. M. Lordgooei, K. Carmichael, T. Kelly, M. Rood, and S. Larson, Separation and concentration of volatile organic contaminants by activated carbon cloth for cryogenic recovery, American Chemical Society, Division of Fuel Chemistry Preprints, New Orleans, LA, 1996, pp. 369–373.
7. P. Le Cloirec, C. Brasquet, and E. Subrenat, Adsorption onto fibrous activated carbon—applications to water and air treatments, American Chemical Society, Division of Fuel Chemistry Preprints, New Orleans, LA, 1996, pp. 379–384.
8. I. Mochida, S. Kisamori, M. Hironaka, S. Kawano, Y. Matsumura, and M. Yoshikawa, *Energy Fuels 8*, 1341–1344, 1994.
9. S. Kisamori, K. Kuroda, S. Kawano, I. Mochida, Y. Matsumura, and M. Yoshikawa, *Energy Fuels 8*, 1337–1340, 1994.
10. I. Mochida, K. Kuroda, Y. Kawabuchi, S. Kawano, A. Yasutake, M. Yoshikawa,

and Y. Matsumura, Oxidation of SO_2 into recoverable aq.H_2SO_4 over pitch based active carbon fibers, American Chemical Society, Division of Fuel Chemistry Preprints, New Orleans, LA, 1996, pp. 335–338.

11. J. Economy, M. Daley, and C. Mangun, Activated carbon fibers—past, present, and future, American Chemical Society, Division of Fuel Chemistry Preprints, New Orleans, LA, 1996, pp. 321–325.

12. Y. Q. Fei, F. Derbyshire, M. Jagtoyen, and I. Mochida, Advantages of producing carbon fibers and activated carbon fibers from shale oils, *Proceedings of the Eastern Oil Shale Symposium* Lexington, KY, Nov. 16–19, 1993, pp. 38–45.

13. G. M. Kimber, A. Vego, T. D. Rantell, C. Fowler, A. Johnson, and F. Derbyshire, Synthesis of isotropic carbon fibers from coal extracts, *Proceedings of the Pittsburgh Coal Conference*, Pittsburgh, PA, 1996, pp. 553–558.

14. G. M. Kimber and A. Johnson, Enhanced desorption from activated carbon composites via electrical heating, *Proceedings of Carbon '96*, Newcastle upon Tyne, UK, 1996, pp. 52–53.

15. M. Jagtoyen, F. Derbyshire, N. Brubaker, Y. Fei, G. Kimber, M. Matheny, and T. Burchell, Carbon fiber composite molecular sieves for gas separation, *Materials Research Society Symposium Proceedings*, 1994, pp. 344, 77–81.

16. G. Kimber, M. Jagtoyen, Y. Fei, and F. Derbyshire, *Gas Sep. Purif. 10*(2), 131–136, 1996.

17. F. Derbyshire, M. Jagtoyen, C. Lafferty, and G. Kimber, Adsorption of herbicides using activated carbons, American Chemical Society, Division of Fuel Chemistry Preprints, New Orleans, LA, 1996, pp. 472–475.

18. Y. Q. Fei, Y. N. Sun, E. Givens, and F. Derbyshire, Continuous removal of sulfur oxides at ambient temperature using activated carbon fibers and particulates, American Chemical Society, Division of Fuel Chemistry Preprints, Chicago, IL, 1995, pp. 1051–1055.

19. M. H. Stenzel and W. J. Merz, *Environmental Progress 8*, 257–264, 1989.

20. D. Chung, *Carbon Fiber Composites*, Boston, Butterworth-Heinemann, 1994, pp. 18–20.

21. K. Okuda, Applications of carbon fibers to construction materials in Japan, *Tanso 155*, 426–9, 1992.

22. W. P. Hettinger, J. Newman, R. Krock, P. Richard, and D. Boyer, *CARBOFLEX and AEROCARB—Ashland's New Low Cost Carbon Fiber and Carbonizing Products for Future Brake Applications*, SAE Earthmoving Industry Conference, Peoria, IL, USA, 8th April 1986.

23. S. Falcone, *Pollution Engineering*, May, 34–37, 1993.

24. K. R. Woodside, *Proceedings of the Air and Waste Management Association Emerging Solutions to VOC and Air Toxics Control*, San Diego, CA, February 26–28, 1997, pp. 109–123.

25. R. S. William, *The Use of Activated Carbon for Evaporative Emission Control, Proceedings, 1995, Workshop on Adsorbent Carbon*, University of Kentucky, Lexington, KY, July 12–14, 1995, pp. 15.

26. C. S. Parmele, W. L. O'Connell, and H. S. Basdekis, *Chemical Engineering 86*(28), 58–70, 1979.

27. M. J. Ruhl, *Chemical Engineering Progress*, July, 37–41, 1993.

28. D. W. Oakes, *Journal of Coated Fabrics 16*, 171–189, 1987.
29. R. G. Jenkins, S. P. Nandi, and P. L. Walker, *Fuel 52*, 288–293, 1973.
30. A. Cupta, *Proceedings for Emerging Solutions to VOC Air Toxics Control*, San Diego, CA, Feb. 26–28, 1997, pp. 286–304.
31. M. Jagtoyen and F. Derbyshire, *Novel Activated Carbon Fiber Composites for Removal of VOC's, CD ROM Proceedings, Emerging Solutions to VOC and Air Toxics Control*, March 4–6, Florida, 1998.
32. L. Stiegler, M. Stallard, R. Lang, and G. Tchobanoglous, *43rd Purdue Industrial Waste Conference Proceedings*, Lewis Publishers, Chelsea, MI, 1989, 27, pp. 212–219.
33. M. D. Corbridge, A. G. Ng, and G. B. DeBrou, *Air and Waste Management Assn. 88th Annual Meeting*, San Antonio, Texas, 1995, v. 10.
34. M. Schafer, H. J. Schroter, and G. Peschel, *Chemical Engineering Technology 14*, 59–64, 1991.
35. T. Golden, *Energeia 1*(3), 1989.
36. J. W. Cobes and D. T. Doughty, *Fluid/Particle Separation Journal 8*, 84, 1995.
37. P. L. Walker, *Carbon 10*(4), 369–382, 1972.
38. *Clean Air Act 1956*, United Kingdom.
39. H. Juntgen and H. Kuhl, in P. A. Thrower, ed. *Chemistry and Physics of Carbon*, New York, Marcel Dekker, 1989, v. 22.
40. P. Davini, *Fuel 68*, 145–148, 1989.
41. J. A. Menendez, J. Phillips, B. Xia, and L. R. Radovic, *Languir 12*, 4404–4410, 1996.
42. C. F. Moreno-Castilla, E. Carrasco-Marin, E. Utrera-Hidalgo, and J. Rivera-Utrilla, *Langmuir 9*, 1378, 1993.
43. J. Zawadzki, *Carbon 25*, 495–502, 1987.
44. S. Kisamori, I. Mochida, and H. Fujitsu, *Langmuir 10*, 1241–1245, 1994.
45. S. K. Gangwal, G. B. Howe, J. J. Spivey, P. L. Silveston, R. R. Hudgins, and J. G. Metzinger, *Environmental Progress 12*, 128, 1993.
46. P. M. Haure, R. R. Hudgins, and P. L. Silveston. *AIChE Journal 35*(9), 1437–1444, 1989.
47. Y. Fei, Y. N. Sun, E. Givens, and F. Derbyshire, *ACS FCD 40*(4), 1051, 1995.
48. R. Andrews, E. Raymundo-Pinero, and F. Derbyshire, *Proc. Carbon '98*, Strasbourg, France, 1998, pp. 275–276.
49. M. Hartman and R. W. Coughlin, *Chemical Engineering Science 27*, 867–880, 1972.
50. A. M. Rubel and J. M. Stencel, *Fuel 76*(6), 521–526, 1997.
51. M. J. Illan-Gomez, A. Linares-Solano, C. Salinas-Martinez de Lecea, and J. M. Calo, *Energy Fuels 7*, 146–154, 1993.
52. K. Tsuji and I. Shiraishi, *Fuel 76*(6), 549–553, 1997.
53. H. Jüngten, *Fuel 65*, 1436–1446, 1986.
54. H. Jüngten, E. Richterand, and H. Kühl, *Fuel 67*, 775–780, 1988.
55. A. N. Ahmed, R. Baldwin, F. Derbyshire, B. McEnaney, and J. Stencel, *Fuel 72*, 287–292, 1993.
56. H. Yamashita, A. Tomita, H. Yamada, T. Kyotani, and L. R. Radovic, *Energy Fuels 7*, 85–89, 1993.

56a. K. Knoblauch, E. Richter, and H. Juntgen, *Fuel 60*,832–838, 1981.
57. M. J. Illan-Gomez, A. Linares-Solano, L. R. Radovic, and C. Salinas-Martinez de Lecea, *Energy Fuels 9*, 97–103, 1995.
58. K. Tsuji and I. Shiraishi, *Fuel 76*(6), 555–560, 1997.
59. *Alternatives to Traditional Transportation Fuels 1995*. Vol. 1. United States Department of Energy, Energy Information Administration, DOE/EIA-0585(95), December, 1996.
60. N. D. Parkyns and D. F. Quinn, in J. W. Patrick, ed., *Porosity in Carbons*, London, Edward Arnold, 1995, pp. 291–325.
61. D. F. Quinn and J. A. MacDonald, *Carbon 30*(7), 1097–1103, 1992.
62. J. Alcañiz-Monge, M. A. de la Casa-Lillo, D. Cazorla-Amorós, and A. Linares-Solano, *Carbon 35*(2), 291–297, 1997.
63. K. A. G. Amankwah, J. S. Noh, and J. A. Schwarz, *J. Hydrogen Energy 14*(7), 437–447, 1989.
64. A. C. Dillon, K. M. Jones, T. A. Bekkedahl, C. H. Kiang, D. S. Bethune, and M. J. Heben, *Nature 386*, 377–379, 1997.
65. Frost & Sullivan Co., Chemical Online.com at http://news.chemicalonline.com, Ian Lisk and Nick Basta, eds., August 1998.
66. J. A. Goodrich, B. W. Dykins, Jr., and Robert Clark, *J. Environmental Quality 20*(7), 707–716m, 1991.
67. I. N. Najm, V. L. Snoeyink, M. T. Suidan, C. H. Lee, and Y. Richard, *J. AWWA 82*, 1:65–72, Jan. 1990.
68. F. S. Baker, *Kirk-Othmer Encyclopedia of Chemical Technology*, 4th ed., v. 4, John Wiley, 1992.
69. R. A. Hutchins, *Chemical Engineering 87*(4), 101–110, 1980.
70. S. Irving-Monshaw, *Chemical Engineering*, 43–46, Feb. 1990.
71. S. Brunauer, P. H. Emmett, and E. Teller, *J. Am. Chem. Soc. 60*, 309–319, 1938.
72. O. P. Mahajan, C. Moreno-Castilla, and P. L. Walker, Jr., *Separation Science and Technology 15*, 1733–1752, 1980.
73. P. C. Millett, *J. NEWWA*, 141–148, June 1995.
74. A. S. Goldfarb, G. A. Vogel, and D. E. Lundquist, *Waste Management 14*, 145–152, 1994.
75. R. D. Vidic, M. T. Suidan, G. A. Sorial, and R. C. Brenner, *Water Environment Research 65*, 156–61, 1993.
76. J. Wildman and F. J. Derbyshire, *Fuel 60*, 655–661, 1991.
77. P. M. Sutton and P. N. Mishra, *Water Science Technology 29*(10–11, Biofilm Reactors), 309, 1994.
78. K. P. Olmstead and W. J. Weber, Jr., *Chem. Eng. Commun. 108*, 113–125, 1991.
79. B. Haist-Gulde, G. Baldauf, and J. J. Brauch, in J. Hrubec, ed., *Handbook of Environmental Chemistry*, Berlin, Springer-Verlag, 1995, pp. 103–28.
80. V. G. Collins and D. C. Popma, Water reclamation and conservation in a closed system, *Proceedings of Ecological Technology Space-Earth-Sea Symposium*, 1st, Hampton, VA, 1966, pp. 165–95.
81. H. Bouwer, *Civil Engineering 59*(7), 60–63, 1989.
82. P. P. T. Chen and G. B. Rest, *Public Works*, 36–38, Jan. 1996.

83. M. Abbaszadegan, M. N. Hasan, G. P. Gerba, P. F. Roessler, B. R. Wilson, R. Kuennen, and E. V. Dellen, *Wat. Res. 31*, 574–582, 1997.

84. M. H. Stenzel and W. J. Metz, *Environmental Progress 8*(4), 257–264, 1989.

85. J. T. Lisle and J. B. Rose, Cryptosporidium contamination of water in the USA and UK: a mini-review, Aqua, 44, 103–117, 1995.

86. R. S. Carter, W. H. Stiebel, P. J. Nalasco, and D. L. Pardieck, *Water Quality Res. J. 30*, 469–491, 1995.

87. T. M. Sivavec, D. P. Horney, and S. S. Baghel, *Emerging Technologies in Hazardous Waste Management VII, Special Symposium*, Industrial and Engineering Chemistry Division, American Chemical Society, Atlanta, Georgia, Sept. 17–20, 1995.

88. L. Liang and J. D. Goodlaxson, *Emerging Technologies in Hazardous Waste. Management VII, Special Symposium*, Industrial and Engineering Chemistry Division, American Chemical Society, Atlanta, Georgia, Sept. 17–20, 1995.

89. R. G. Orth and D. E. McKenzie, *Emerging Technologies in Hazardous Waste Management VII, Special Symposium*, Industrial and Engineering Chemistry Division, American Chemical Society, Atlanta, Georgia, Sept. 17–20, 1995.

90. N. Korte, C. Grittini, R. Muftikian, and Q. Fernando, *Emerging Technologies in Hazardous Waste Management VII, Special Symposium*, Industrial and Engineering Chemistry Division, American Chemical Society, Atlanta, Georgia, Sept. 17–20, 1995.

91. A. Agrawal, L. Liang, and P. G. Tratnyek, *Emerging Technologies in Hazardous Waste Management VII, Special Symposium*, Industrial and Engineering Chemistry Division, American Chemical Society, Atlanta, Georgia, Sept. 17–20, 1995.

92. S. F. O'Hannesin, R. W. Gillham, and J. L. Vogan, *Emerging Technologies in Hazardous Waste Management VII, Special Symposium*, Industrial and Engineering Chemistry Division, American Chemical Society, Atlanta, Georgia, Sept. 17–20, 1995.

93. P. D. MacKenzie, S. S. Baghel, G. R. Eykholt, and R. P. Homey, *Emerging Technologies in Hazardous Waste Management VII, Special Symposium*, Industrial and Engineering Chemistry Division, American Chemical Society, Atlanta, Georgia, Sept. 17–20, 1995.

94. T. Boronina, K. Klabunde, and G. Segeev, *Environmental Sci. Tech. 29*, 1511–1517, 1995.

95. R. Andrews, B. Spears, and E. Grulke, The use of carbon fiber composite as substrate for zero valent metal dechlorination system, *ACS Preprints, Div. Fuel Chem. 41*, 385–388, 1996.

96. H. W. Kroto, J. R. Heath, S. C. O'Brien, R. F. Curl, and R. E. Smalley, *Nature 318*, 162–165, 1985.

97. M. S. Dresselhaus, G. Dresselhaus, and P. C. Eklund, *Science of Fullerenes and Carbon Nanotubes*, New York, Academic Press, 1996.

98. A. Thess, R. Lee, P. Nikolaev, H. Dai, P. Petit, J. Robert, C. Xu, Y. H. Lee, S. G. Kim, A. G. Rinzler, D. T. Colbert, G. E. Scuseria, D. Tománek, J. E. Fischer, and R. E. Smalley, *Science 273*, 483–487, 1996.

99. C. Journet, W. K. Maser, P. Bernier, A. Loiseau, M. Lamy de la Chapelle, S. Lefrant, P. Daniard, R. Lee, and J. E. Fischer, *Nature 388*, 756–758, 1997.

100. R. C. Haddon, A. S. Perel, R. C. Morris, T. T. M. Palstra, A. F. Hebard, and R. M. Fleming, *Appl. Phys. Lett. 67*, 121–123, 1995.
101. S. J. Tans, A. R. M. Verschueren, and C. Dekker, *Nature: 393*, 49–52, 1998.
102. L. S. Pan and D. R. Kania, *Diamond—Electronic Properties and Applications*, Kluwer Academic Publishers, 1994.
103. M. Aslam, I. Taher, M. A. Tamor, T. J. Potter, and R. C. Elder, *Transducers, Proceedings of the International Conference on Solid-State Sensors and Transducers*, Yokohama, Japan, June 7–10, 1993.
104. N. Fujimori and H. Nakahata, *New Diamond 2*, 98, 1990.
105. W. Hasan and S. H. Propst, *Proceedings of SPIE—The International Society for Optical Engineering 2253*, 228–235, 1994.
106. P. Sautet, J. C. Dunphy, D. F. Ogletree, C. Joachim, and M. Salmeron, *Surface Science 315*, 127–142, 1994.
107. S. Rousset, S. Gauthier, O. Siboulet, W. Sacks, M. Belin, and J. Klein, *Phys. Rev. Lett. 63*, 1265–1268, 1989.
108. L. Ruan, F. Basenbacher, I. Stensgaard, and E. Laegsgaard, *Phys. Rev. Lett. 70*, 4079–4082, 1993.
109. T. Nakagawa, K. Ogawa, and T. Kurumizawa, *J. Vac. Sci. Technol. B 12*, 2215, 1994.
110. J. Resh, D. Sarkar, J. Kulik, J. Brueck, A. Ignatiev, and N. J. Halas, *Surface Science 316*, L1061–L1067, 1994.
111. H. Dai, J. H. Hafner, A. G. Rinzler, D. T. Colbert, and R. E. Smalley, *Nature 384*, 147–150, 1996.
112. T. W. Ebbesen, H. J. Lezec, H. Hiusa, J. W. Bennett, J. F. Ghaemi, and T. Thio, *Nature 382*, 54–56, 1996.
113. S. S. Wong, E. Joselerich, A. T. Woolley, C. L. Cheung, and C. M. Lieber, *Nature 384*, 52–55 (1998).
114. S. M. Chang, Y. H. Kim, J. M. Kim, Y. K. Chang, and J. D. Kim, *Mol. Cryst. Liq. Cryst. 267*, 405–410, 1995.
115. R. Gasparini, M. Scarpa, F. Vianello, B. Mondovi, and A. Rigo, *Analytica Chimica Acta 294*, 299–304, 1994.
116. J. Chen, E. Vinski, K. Colizza, and S. G. Weber, *J. Chromat. A. 705*, 171–184, 1995.
117. P. D. Lyne and R. D. O'Neill, *Anal. Chem. 62*, 2347–2351, 1990.
118. E. V. Savrikoc, L. E. Frumin, T. S. Odaryuk, P. V. Tsar'kov, V. A. Sadovnichii, and P. V. Eropkin, Deodorizing gaskets for colostomy patients, Biomed. Eng. *21*(4), 140–142, 1987.
119. H. Klinkmann and E. Behm, Adsorption of immunologically relevant molecules— its present and future, *Biomater. Artif. Cells Artif. Organs. VII International Symposium on Hemoperfusion, Kiev, USSR 15*(1), 41–58, 1987.
120. D. M. McPhillips, T. A. Armer, and D. R. Owen, *J. Biomed. Mater. Res. 17*(6), 993–1002, 1983.
121. D. A. Luehmann, Water purification for hemodialysis, *Med. Instrum. 20*, 74–79, 1986.
122. N. A. Belyakov, A. L. Belkin, A. S. Vladyka, K. Y. Gurevich, V. G. Nikolaev, A. R. Ox'mak, V. V. Petrash, I. P. Serzhantu, and N. N. Shchegrinov, *Biomed. Eng. 18*, 130–133, 1984.

123. Y. Mori, S. Nagaoka, H. Tanzawa, Y. Kikuchi, Y. Yamada, M. Hagiwara, and Y. Idezuki, *J. Biomed. Mater. Res. 16*, 17–30, 1982.
124. C. J. Holloway, K. Hartsick, G. Brunner, and I. Gaeger, *Biomater. Med. Dev. Artif. Organs 9*, 167–180, 1981.
125. E. J. Jacob, U.S. Pat. 4,606,354, Aug. 19, 1986.
126. G. W. Hastings, *Polymer Spec. Polym. 26*(9), 1331–1335, 1984.
127. S. P. Saha, Mechanical properties of machine-mixed carbon fiber reinforced bone cement, *ASME, Appl. Mechanics Div. 21*, 57–60, 1985.
128. S. L. Iyer, T. Jayasekaran, C. F. J. Blunck, and R. Selvam, *ISA Trans. 23*, 471–494, 1984.
129. L. S. Stern, M. T. Manley, and J. Parr, Particle size distribution of wear debris from polyethylene and carbon-reinforced acetabular components, Trans.-Second World Congress on Biomaterials, Washington, DC, 1984, v. 7, p. 66.
130. K. J. J. Tayton, *J. Medical Eng. Technol. 7*(6), 271–272, 1983.
131. J. L. Ricci, A. G. Gona, H. Alexander, and J. R. Parsons, *J. Biomed. Mat. Res. 18*, 1074–1087, 1984.
132. D. G. Mendes, M. Iusim, D. Angel, A. Rotem, D. Mordehovich, M. Roffman, S. Lieberson, and J. Boss. *J. Biomed. Mat. Res. 20*, 699–708, 1986.
133. S. Nelson, *Chemistry in Britain 23*, 1152–1153, 1987.
134. L. Claes, C. Burri, and R. Neugebauer, *Adv. Bioengineering, ASME 52*, 1984.
135. J. L. Ricci, A. G. Gona, and H. Alexander, *J. Biomed. Mat. Res. 2595*, 651–666, 1991.
136. M. Zimmerman, J. R. Parsons, H. Alexander, and A. B. Weiss, *J. Biomed. Mater. Res. 18*, 927–938, 1984.
137. C. A. Marsden, M. P. Brazell, and N. T. Maidment, *J. Biomed. Eng. Electrochem. 6*, 184–186, 1983.
138. A. J. M. Starrenburg and G. C. Burger, *IEEE Trans. Biomed. Eng. BME 29*, 352–355, 1982.

PATENT REFERENCES

H. Alexander, J. R. Parsons, I. D. Strauchler, and A. B. Weiss, U.S. Pat. 4,411,027, Oct. 25, 1983, assigned to University of Medicine and Dentistry of New Jersey.
H. Alexander, J. R. Parsons, I. D. Strauchler, and A. B. Weiss, U.S. Pat. 4,512,038, Apr. 23, 1985, assigned to University of Medicine and Dentistry of New Jersey.
E. Alt, U.S. Pat. 5,143,089, Sep. 1, 1992, no assignee.
E. Alt, U.S. Pat. 5,411,527, May 2, 1995, Intermedics, Inc.
J. W. Brantigan, U.S. Pat. 5,192,327, Mar. 9, 1993, no assignee.
J. W. Brantigan, U.S. Pat. 5,425,772, June 20, 1995, no assignee.
K. R. Brennen, T. M. Williams, and R. A. Gabler, U.S. Pat. 5,336,254, Aug. 9, 1994, assigned to Medtronic, Inc.
J. A. Davidson, U.S. Pat. 5,358,529, Oct. 25, 1994, assigned to Smith & Nephew Richards Inc.
T. N. C. Devanathan, U.S. Pat. 4,902,297, Feb. 20, 1990, assigned to Zimmer, Inc.
T. N. C. Devanathan, U.S. Pat. 4,978,360, Dec. 18, 1990, assigned to Zimmer, Inc.

J. H. Dumbleton, R. Y. Lin, C. F. Stark, and T. E. Crippen, U.S. Pat. 5,181,930, Jan. 26, 1993, Pfizer Hospital Products Group, Inc.

R. I. Dunn, D. H. Lewis, T. W. Sander, and R. W. Treharne, III, U.S. Pat. 4,731,084, Mar. 15, 1988, assigned to Richards Medical Company.

R. I. Dunn, D. H. Lewis, J. A. Davidson, N. B. Beals, and Y. L. Gill, U.S. Pat. 4,790,850, Dec. 13, 1988, assigned to Richards Medical Company.

R. I. Dunn, D. H. Lewis, J. A. Davidson, N. B. Beals, and Y. L. Gill, U.S. Pat. 4,932,972, June 12, 1990, assigned to Richards Medical Company.

F. Esper and W. Gohl, U.S. Pat. 4,356,517, Nov. 2, 1982, assigned to Robert Bosch GmbH.

N. E. Fearnot, T. G. Kozma, A. O. Ragheb, and W. D. Voorhees, U.S. Pat. 5,609,629, Mar. 11, 1997, assigned to MED Institute, Inc.

D. A. Hoffmann and M. E. Bush, U.S. Pat. 5,534,022, July 9, 1996, assigned to Ventritex, Inc.

C. A. Homsy, U.S. Pat. 3,992,725, Nov. 23, 1976, no assignees.

C. A. Homsy, U.S. Pat. 4,129,470, Dec. 12, 1978, no assignees.

M. S. Hunt, P. J. Mundell, and A. E. Strover, U.S. Pat. 4,851,005, July 25, 1989, assigned to South African Invention Development Corporation.

S. Huntley, D. J. Zytkovicz, and M. Geiger, U.S. Pat. 5,683,444, Nov. 4, 1997, no assignee.

T. Kawakubo and S. Ohtani, U.S. Pat. 5,171,492, Dec. 15, 1992, assigned to Mitsubishi Pencil Co., Ltd.

D. Lapeyre and J. P. Slonina, U.S. Pat. 4,654,033, Mar. 31, 1987, assigned to Biomasys.

F. Lechner, H. Heissler, W. Scheer, W. Siebels and R. Ascherl, U.S. Patent 4,714,467, Dec. 22, 1987, assigned to MAN Technologie GmbH.

S. Lin, U.S. Pat. 4,643,734, Feb. 17, 1987, assigned to Hexcel Corporation.

P. Lonca, U.S. Pat. 4,722,688, Feb. 2, 1988, no assignee.

M. S. Marcolongo, P. Ducheyne, F. Ko, and W. LaCourse, U.S. Pat. 5,468,544, Nov. 21, 1995, assigned to the Trustees of the University of Pennsylvania.

M. S. Marcolongo and P. Ducheyne, U.S. Pat. 5,645,934, July 8, 1997, assigned to the Trustees of the University of Pennsylvania.

J. M. Moran, R. A. Salzstein, I. M. Daniel, D. S. Cairns, and D. B. Smith, U.S. Pat. 5,522,904, June 4, 1996, assigned to Hercules Incorporated.

M. Moumene, R. Y. Lin, and C. F. Stark, U.S. Pat. 5,443,513, Aug. 22, 1995, assigned to Howmedica Inc.

R. J. Pitz and E. J. Jacob, U.S. Pat. 4,525,147, June 25, 1985, no assignee.

P. Renard and J. L. Chareire, U.S. Pat. 4,813,967, Mar. 21, 1989, assigned to Société Nationale Industrielle Aérospatiale.

G. Richter and E. Weidlich, U.S. Pat. 4,281,668, Aug. 4, 1981, assigned to Siemens Aktiengesellschaft.

R. A. Salzstein and G. R. Toombes, U.S. Pat. 5,163,962, Nov. 17, 1992, assigned to BHC. Laboratories, Inc.

E. V. Savrikoc, L. E. Frumin, T. S. Odaryuk, P. V. Tsar'kov, and K. St. John, U.S. Pat. 4,595,713, June 17, 1986, assigned to Hexcel Corporation.

W. Siebels, U.S. Pat. 5,306,310, Apr. 26, 1994, assigned to MAN Ceramics GmbH.

M. Takahashi, U.S. Pat. 5,218,977, June 15, 1993, no assignee.

2

¹H NMR Spectroscopy of Adsorbed Molecules and Free Surface Energy of Carbon Adsorbents

V. V. Turov

Institute of Surface Chemistry, National Academy of Sciences of Ukraine, Kyiv, Ukraine

Roman Leboda

Maria Curie-Sklodowska University, Lublin, Poland

I. INTRODUCTION

Methods of nuclear magnetic resonance (NMR) spectroscopy have found much use for researches into the structure of adsorbents and their interaction with various classes of organic and inorganic compounds. Especially great successes have been attained in applying procedures of solid-state NMR with magic angle spinning (MAS NMR). The application of these procedures to numerous solids made it possible to record separately the signals of atoms in various chemical environments and to collect valuable information on the structure of these substances [1–5]. In the case of adsorbents on whose surface there are chemisorbed organic molecules, the MAS NMR method enables one not only to establish the presence of substances adsorbed on the surface but also to make their identification [6]. However, application of solid-state NMR to carbonaceous adsorbents is complicated by the fact that their carbon atoms may be constituents of a large number of structures of various types with sp^2- and sp^3-hybridized orbitals. The ^{13}C chemical shifts of these substances are in the range from 70 to 200 ppm, which results in a large number of lines in the NMR spectra. Besides, graphitic clusters forming the base of many materials are characterized by an anisotropic broadening of NMR signals caused by a strong temperature dependence of their diamagnetic susceptibility [7]. Therefore, when recording ^{13}C MAS NMR spectra for carbonaceous adsorbents, if often turns out that we can succeed in observing only one broad signal whose maximum position is determined by the concentration ratio of sp^2- and sp^3-hybridized carbon atoms.

The overwhelming majority of experimental results that up to the present have been achieved by the NMR method are related to researches into organic substances and water adsorbed on carbon surfaces. Here we can distinguish three main directions of studies: measurements of chemical shifts for adsorbed molecules, determination of their dynamic characteristics (mobility, diffusion coefficient), and investigations of thick layers of a liquid adsorbed on the surface. Each of these directions permits one to characterize certain properties of a material. Thus the chemical shift value makes it possible to evaluate the coordination positions of molecules on carbon surfaces [8–13]. The very first experiments with recording ^1H and ^{19}F NMR spectra for toluene and benzene adsorbed on graphitized carbon blacks (Spheron, Graphon, Carbopack) have already shown [8] that such a spectrum may have two signals shifted into a region of strong magnetic fields. The authors of this paper made the assumption that the main causes of the phenomenon observed were the influences of the local magnetic anisotropy of the fused system of benzene rings forming the carbon surface and adsorption on sites of various types. The chemical shift value calculated on the basis of local magnetic field strengths proved to be close to that acquired by experiment. Thus it has been shown that spectral characteristics of signals of adsorbed molecules can be used for identification of the adsorption site structure. These inferences were corroborated later [9–13], but the range of the adsorbents studied remained limited to several types of activated carbons and graphitized carbon blacks.

The dynamic characteristics of adsorbed molecules can be determined in terms of temperature dependences of relaxation times [14–16] and by measurements of self-diffusion coefficients applying the pulsed-gradient spin-echo method [17–20]. Both methods enable one to estimate the mobility of molecules in adsorbent pores and the rotational mobility of separate molecular groups. The methods are based on the fact that the nuclear spin relaxation time of a molecule depends on the feasibility for adsorbed molecules to move in adsorbent pores. The lower the molecule's mobility, the more effective is the interaction between nuclear magnetic dipoles of adsorbed molecules and the shorter is the nuclear spin relaxation time. The results of measuring relaxation times at various temperatures may form the basis for calculations of activation characteristics of molecular motions of adsorbed molecules in an adsorption layer. These characteristics are of utmost importance for application of adsorbents as catalyst carriers. They determine the diffusion of reagent molecules towards the active sites of a catalyst and the rate of removal of reaction products. Sometimes the data on the temperature dependence of a diffusion coefficient allow one to ascertain subtle mechanisms of filling of micropores in activated carbons [17].

Of great importance for colloid chemistry and the processes of adsorption and adhesion is the formation of thick layers of a substance situated in the region of the action of surface forces at the adsorbent–liquid interface. There are not many techniques (within the scope of the NMR method) for researches into the structure of thick layers of a liquid adsorbed on the surface of solids, one of them being the technique of freezing a liquid phase [21–25]. The basis of the technique was established by the results of investigations on nonfreezing water in aqueous suspensions of adsorbents [11,26–29]. The technique consists in lowering the temperature of a suspension containing an adsorbent that was immersed into the liquid under study at a temperature below its freezing point. Then, in the absence of impurities dissolved in this liquid, its main part freezes. The unfrozen part of the liquid is the one that interacts with the adsorbent surface. The freezing point of the liquid bound to the surface is lowered due to the adsorption interactions. The stronger the adsorption interactions, the lower is the temperature at which molecules go from the adsorbed state to the solid state. Since the relaxation times for adsorbed molecules and solid bodies may differ by several orders of magnitude, at a low value of the spectrometer transmission band no NMR signal of the frozen substance is observed, and it is possible to discern only the signal of adsorbed molecules. At any temperature below the freezing point of the liquid its molecules go from the adsorbed phase to the solid phase when the free energy of adsorbed molecules and that of the frozen substance are equal. In this situation, for suspensions of adsorbents it is possible to determine such important characteristics as the decrease in the free energy caused by adsorption and the thickness of the layer that experiences the perturbing influence of the surface. Besides, it is also possible to calculate values of the free surface energy of an adsorbent, which determine the total decrease in the free energy of a substance caused by the presence of an interface.

The present review discusses the results of the ¹H NMR spectroscopy for a wide range of carbonaceous materials (heat-treated and nongraphitizable activated carbons, carbon blacks, exfoliated and oxidized graphites, porous and amorphous carbonized silicas). This technique made it possible to determine the spectral characteristics of organic molecules with diverse chemical properties, as well as of water molecules adsorbed on the surface. These characteristics are compared with the structural properties of the materials under consideration. The calculations done for the majority of the subjects of inquiry gave the values of their free surface energies in an aqueous medium as well as the characteristics of bound water layers of various types.

II. THE CONCEPT OF CHEMICAL SHIFT AND ITS DEPENDENCE ON ADSORPTION INTERACTIONS

The spectral technique of nuclear magnetic resonance is based on the phenomenon of absorption of radio-frequency energy by a system of magnetic nuclei. In an external magnetic field B_0 nuclei with a nonzero spin I are oriented discretely so that the spin projection to the field direction (m) takes on $2I + 1$ values ($m = I, I - 1, \ldots, -1$). This leads to splitting of the energy level of spin I into $2I + 1$ sublevels. The difference in energy between the sublevels is defined by the relationship

$$E_m = -\hbar\gamma B_0 \tag{1}$$

where $\hbar$ = the Planck constant and γ = the gyromagnetic ratio.

When using the NMR method, transitions between these sublevels are investigated by applying an additional high-frequency field B_1 with angular frequency ω. In this case the resonance absorption of energy takes place under the condition

$$\omega_0 = \gamma B_0 \tag{2}$$

This formula describes the general conditions for nuclear magnetic resonance. However, a local magnetic field in the position of a nucleus with spin I differs from the field B_0 generated by the NMR spectrometer magnet, which is due to the magnetic screening of the nucleus by its electron shell and by local magnetic fields created by chemical groups of a molecule under study and of nearby molecules of the medium. The difference between the magnetic field in the position of the nucleus and the NMR spectrometer field is referred to as the magnetic screening constant.

In order to characterize the quantity of magnetic screening characteristic of a concrete nucleus, use is made of the value of the chemical shift, i.e., the difference between the constant screening of a standard substance (σ_1) and the screening of the substance studied (σ). The chemical shift (δ) is measured in millionth parts of the magnetic field strength of a NMR spectrometer:

$$\delta = (\sigma_1 - \sigma) \quad \text{ppm} \tag{3}$$

The magnetic screening constant is determined by two contributions made by diamagnetic and paramagnetic screenings, $\sigma = \sigma_d + \sigma_p$ [30]. For an isolated atom in the $1s$ state the paramagnetic contribution is equal to 0, and the diamagnetic contribution is given by

$$\sigma_d = \frac{\mu_0 e^2}{4\pi 3 m_e c^2} \int r^{-1}\rho(r)d^3r \tag{4}$$

where μ_0 = the magnetic constant, e = the electron charge, m_e = the electron mass, and c = the velocity of light.

Calculations show that the change of an electric charge by $1e$ results in the change of the σ value equal to approximately 20 ppm. Thus when a theoretical estimation of a δ value is made, the main difficulty consists in calculating the paramagnetic contribution to the magnetic screening, this contribution being dependent on the overall strength of the local magnetic fields generated by all the electron systems in the position of the nucleus under investigation. The NMR spectroscopy methods have been described in detail in a number of monographs [31–34].

In the case of protons the reference substance is tetramethylsilane (TMS) whose chemical shift by convention is taken equal to 0 ppm for its sp^3 hybridization of carbon atoms. The absence of electron-donating groups in the TMS molecule means that most organic compounds have proton signals in fields that are weaker than in TMS. The chemical shift value is also dependent on participation of a molecule in intermolecular interactions that form the basis for an adsorption process. One of these interactions is the formation of hydrogen-bonded complexes exerting the most profound effect on the chemical shift value. Such complexes are formed in all cases when on the surface of an adsorbent there are proton-donating centers (e.g., hydroxyl groups) that are accessible to adsorbate molecules or, vice versa, electron-donating centers on the surface are adsorption sites for molecules having proton-donor properties. Hydrogen bond formation leads to a decrease in the electron density at the hydrogen atom taking part in bond formation [35,36] and, as a consequence, to a chemical shift decrease for this atom of hydrogen (i.e., shift of the signal towards a weaker magnetic field).

The σ value for an isolated molecule does not depend on pressure and temperature [37], while in the case of hydrogen-bonded complexes such a dependence does exist; it is due to the fact that with increasing temperature some parts of hydrogen bonds become broken and the proton signal is shifted towards stronger magnetic fields. In order to explain the temperature dependence of the chemical shift, the model of two states is typically employed [37–39], according to which the measured chemical shift can be written as

$$\delta_H(T) = (1 - P_f)\delta_{hb} + P_f\delta_f \tag{5}$$

where P_f is the portion of broken hydrogen bonds and δ_{hb} and δ_f are the chemical shifts of protons that take part and do not take part in the formation of hydrogen bonds. The recording of one averaged signal is attributed to a very short lifetime of each of these two states [40,41].

When there are adsorption interactions of other types or when the rate of molecular exchange is limited by structural properties of adsorbents, the lifetime of adsorption complexes may vary over a wide range. In conformity to the theory of exchange processes developed by Gutowsky and coworkers [42,43], for two states A and B characterized by lifetimes τ_A and τ_B and resonance frequencies ν_A and ν_B the intensity of the NMR signal as a function of frequency ν is defined by the relationship

$$g(\nu) = \frac{K\tau(\nu_A - \nu_B)^2}{\{(\nu_A - \nu_B)/2 - \nu\}^2 + 4\pi\tau^2(\nu_A - \nu)^2(\nu_B - \nu)^2} \tag{6}$$

where $\tau = \tau_A\tau_B/(\tau_A + \tau_B)$ and $K =$ a normalization factor.

If on the NMR time scale the exchange between states A and B proceeds slowly, i.e., if $\tau_A(\tau_B) \gg 1/(\nu_A - \nu_B)$, the signals corresponding to states A and B are observed separately at the frequencies characteristic of these states. In the case of a fast exchange $(\tau_A(\tau_B) \ll 1/(\nu_A - \nu_B))$, one averaged signal is observed whose resonance frequency is determined by the statistical weight of each of the states, and the frequency of this signal can be derived from Eq. (6) and represented by a formula similar to Eq. (5):

$$\nu = P_A\nu_A - P_B\nu_B \tag{7}$$

where $P_A = \tau_A/(\tau_A - \tau_B)$ and $P_B = \tau_A/(\tau_A + \tau_B)$.

The resonance curve shape for intermediate exchange rates is defined by Eq. (6). With decreasing temperature, the molecular motion speed decreases. Therefore cooling of a sample leads to changes in the shape of spectral lines for molecules participating in the exchange. Choosing experimental conditions makes it possible to transfer the exchange system from the region of a fast exchange into the region of a slow exchange and vice versa. According to Eq. (6), in the case of a slow exchange the separate recording of signals attributed to various adsorption complexes requires that differences in chemical shift values for various states be greater than the NMR signal widths for these states. The range of chemical shift for protons is relatively narrow and does not exceed 10–15 ppm. On the other hand, the width of signals for adsorbed molecules is substantially wider than in the case of liquids, which is due to a low mobility of the adsorbed molecules, and so the requirement for the separate observation of signals ascribed to different adsorption complexes often is not satisfied. This factor impedes application of ¹H NMR spectroscopy for investigation into the structure of active sites on the surface of a large number of adsorbents.

In the case of some microporous materials the above-mentioned difficulties can be resolved by using molecules of ¹²⁹Xe as the adsorbate [44–47]. The characteristic feature of this substance is a very strong dependence of its chemical shift on the intermolecular xenon–xenon interactions. Thus it has been found that when in a zeolite cell there is a single molecule of xenon the chemical shift for such molecules is equal to 80 ppm. With increasing number of neighboring molecules, the chemical shift increases by 20 ppm per each molecule sequentially adsorbed in the cell. Consequently, in terms of the relative intensity of signals corresponding to different types of adsorption complexes it becomes easy to determine the probability of their formation in a specific porous system. However, this method proved to be efficient only for microporous adsorbents of the zeolite type with a known crystal lattice structure. For the majority of carbonaceous materials there is a wide distribution of pore sizes and, besides, there are many diverse types of active sites where molecules of various chemical nature may be adsorbed. Their investigation calls for application of adsorbates of various types. It is the establishment of relationships between the structure of a carbon surface and the value of chemical shift for molecules adsorbed on the surface that is in the focus of attention of a considerable part of this review.

III. MICROTEXTURE OF CARBONACEOUS MATERIALS

Adsorption properties of carbonaceous materials are determined to a great extent by their structural characteristics. Depending on the starting raw materials, the temperature, and the method of carbonization, it is possible to produce carbonaceous materials differing both in their macroscopic parameters (grain size, density, mechanical properties) and in the characteristics of their pore structure. In conformity with the classification introduced by Dubinin [48], all porous materials can be subdivided into three groups, namely macroporous materials with pore sizes >50 nm, mesoporous materials (2–50 nm), and macroporous materials with pore sizes <2 nm. Using various starting raw materials and procedures of activation, it is possible to produce adsorbents with preassigned adsorption properties and a wide range of porosity (from microporous to macroporous adsorbents). In some cases carbonization can result in the formation of materials with an undeveloped pore structure that in a first approximation can be regarded as nonporous. The preparation of carbonaceous adsorbents of various types and the methods for researches into their characteristics are expounded in a number of monographs [49–51].

The formation of the carbon lattice is a very complex multistage process that involves thermolysis of simple organic substances or polymers and results in the development of polyaromatic structures. These structures are graphene clusters (containing 15–20 fused benzene rings) on whose boundaries there are hydrogen atoms and oxidized carbon groups. Such structures can be identified in some types of carbonaceous materials [52]. When primary carbonization products (carbonizates) are exposed to temperatures high enough for chemical reactions to proceed, these graphene clusters interact with each other, forming complex carbon aggregates. With increasing temperature and time, carbonization leads to the formation of a regular carbon lattice whose main building blocks are primary graphene clusters.

There are substantial differences in the structural characteristics of materials synthesized under various conditions of pyrolysis of gases (carbon black, carbon-containing mineral adsorbents), carbonization of liquids (cokes) and solids (chars, activated carbons). They are caused by the fact that materials of the first type are formed under conditions of free motion of graphene clusters relative to each other, which provides the possibility of comparatively easy formation of ordered carbon structures. With increasing temperature and time of carbonization, the structural order of such materials increases and, in the limit, they may be converted to graphitized forms of carbon. In the case of solid carbonized materials, their carbonization proceeds in the presence of a rigid skeleton, and the graphene clusters formed as a result of heat treatment remain chemically bound to the noncarbonized part of the material. Under such conditions a regular carbon lattice cannot be formed. The process of their structural ordering can be effected only in very severe conditions involving application of high pressures and temperatures. Such materials are referred to as ungraphitizable [53–57].

Graphitized Materials. The major successes in studying the structure of carbon black particles are related to application of transmission electron microscopy (TEM) and x-ray diffraction methods. As early as 1934, by x-ray structure investigations Warren [58] showed that carbon black particles were not amorphous (as it was believed earlier); rather they consist of ordered carbon layers whose structure resembles that of carbon layers in graphite, though these layers have regular reiterating breaks perpendicular to the graphite layers. In view of this, the term turbostratic structure was introduced [58,59], which is used by many authors to describe the constitution of graphite and some graphitized materials [49]. The turbostratic layers of carbon do not exist independently of each other. On the contrary, they are parts of tightly packed crystallites consisting of several parallel layers. The structure of such a crystallite is represented schematically in Fig. 1 [60]. Positions of fractures of graphite planes are defects of the crystal lattice which contain sp^3-hybridized carbon atoms. The characteristic dimensions of a crystallite are its true diameter (L_1) and its thickness (L_2). The angle γ characterizes the intrinsic defects of graphite planes. In the case of a turbostratic structure the interplanar spacing value in a package of graphite planes is larger than in graphite and is equal to ca. 3.44 Å [59].

On the basis of the results of investigations on a large number of commercial carbon blacks synthesized under various conditions it was established [60] that the microtexture of all the materials could be described in terms of the characteristics represented in Fig. 1. Depending on the production method of a material its parameters L_1, L_2, and γ may vary. A decrease in γ and an increase in L_1 give evidence for the ordering of the particle structure. Figure 2(a) displays a schematic view of a section of a carbon black particle, where individual crystallites are visible [60]. The surface of each of the crystallites visible in this figure has a turbostratic structure.

More recent scanning tunneling microscopy studies revealed a somewhat different structure of carbon black particles [61–63] [Fig. 2(b)]. The crystallites forming carbon black particles have a more extended structure than that shown in Fig. 2(a) and exhibit several clearly visible turbostratic bends. Here the particle surface has the shape of a curved staircase formed by overlapped crystallites. In the case of both types of carbon black particles there can be a great number of hydrogen atoms, oxidized carbon atoms, and broken carbon bonds on the boundaries of crystallites. Figure 2(c) illustrates the particle shape obtained by the atomic force microscopy technique [63]. As in the previous cases, on the surface there are visible crystallites that have the form of rectangles, but their arrange-

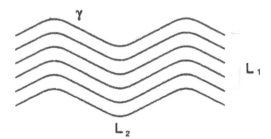

FIG. 1 Schematic representation of the constitution of a carbon crystallite where L_1 and L_2 are the sizes and thickness of a crystallite and γ is the angle that characterizes the intrinsic defects of graphite planes, in compliance with the data of Ref. 60.

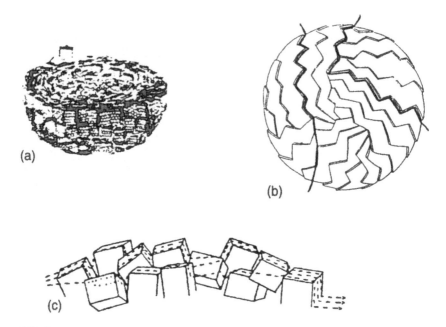

FIG. 2 Schematic representation of the constitution of carbon black particles. (a, b) On the basis of the data obtained by scanning tunneling microscopy. (From Refs. 60 and 62.) (c) On the basis of the data obtained by atomic force microscopy. (From Ref. 64.)

ment is more random than that displayed in Fig. 2(a). It is evident that the differences in the particle structure observed by various techniques reflect the diversity of synthesized carbonaceous materials and strong dependence of their structure on preparation conditions. The intraparticle pore structure of carbon blacks is practically undeveloped. Nevertheless, it is possible to expect the presence of a small number of slitlike gaps between graphene clusters.

Nongraphitized Carbonaceous Materials.   Activated carbons belong to the main type of carbonaceous adsorbents that have been used for many decades. To a great extent this is due to the high specific surface of such materials, which can amount to 1000 m^2/g and more [64–70]. Such high specific surface values provide evidence for the presence in these materials of a substantial number of slitlike pores formed by graphene planes. In contrast to the interplanar gaps in carbon black particles, the majority of these pores are accessible to adsorbed molecules. This high accessibility of slitlike gaps in activated carbons is attributed to large values of interplanar spacing in comparison with graphene clusters in carbon blacks. At present, the available TEM data give ground for construction of two models of the constitution of activated carbons, namely the model of disordered arrangement of graphene planes [71] [Fig. 3(a)] and that of "crumpled paper sheets" [72,73] [Fig. 3(b)]. In conformity with both models the pore structure is formed by gaps between graphene planes. However, in the case of the model illustrated in Fig. 3(b), the degree of surface order is considerably higher. This model is the most commonly employed one for activated carbons with a high specific surface produced by carbonization of polymeric solids and a subsequent high-temperature activation of the material. The other model is associated with a relatively low degree of lattice order and can be used when describing the structure of carbons produced by carbonization of natural materials without special procedures of activation.

The microtextural features of microporous activated carbons synthesized by pyrolysis of cellulose were studied by the TEM method [74], the method being applied to carbons differing in the temperature of their treatment in argon over the range 1000–3000 K. At low temperatures of activation the adsorbent's microtexture can be represented by a system of bound graphene planes forming a layered structure [Fig. 4(a)]. The interlayer distance is rather large owing to the presence of sp^3-hybridized carbon atoms forming chemical bonds with atoms belonging to different carbon layers. In the carbon lattice there are many defects caused by a nonplanar arrangement of some graphene clusters. In such a material there are clearly visible portions of a disordered arrangement of graphene clusters. At high temperatures of heat treatment the material order increases. Individual graphene clusters gather into dense graphite-like packages consisting of several graphene layers [Fig. 4(b)]. Graphene clusters interact with each other, forming quite extended layers. The pore structure is now formed not by individual graphene layers but by gaps between graphene clusters. The major types of pores

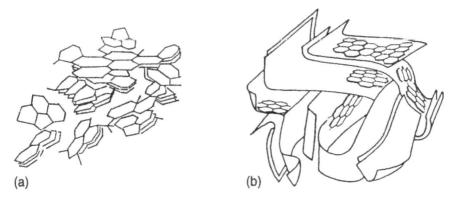

(a)   (b)

FIG. 3 Models for the structure of carbonaceous adsorbents. (a) "Disordered lattice." (From Ref. 71.) (b) "Crumpled paper sheets." (From Refs. 72 and 73.)

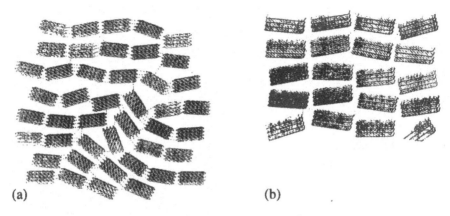

(a)   (b)

FIG. 4 Microtextural particularities of the structure of microporous activated carbons heat-treated at a low temperature (a) and partially graphitized (b) in compliance with the data of Ref. 74.

in such materials are slit- and wedge-shaped pores that may be both open and partially closed (with one pore orifice closed). The same types of pores are regarded as main types of pores in microporous carbonaceous adsorbents according to a generalized model constructed on the basis of measurements of adsorption isotherms for a wide range of substances differing in their molecular mass and heats of adsorption [75].

Oxygenous Sites on Carbon Surface.   Oxygen-containing sites on the surface of carbonaceous materials are situated predominantly on boundaries of graphene layers. This is because there is a great difference in the chemical activity of basal and prismatic faces of graphite. Thus on the basis of the data collected when cleaving crystals of graphite in high vacuum [76], it was established that the free surface energy of its faces differs by a factor of 13 and more (0.14 and 4.80 mJ/m^2 for prismatic and basal faces, respectively). The same ratio was also observed for reactivities of the crystallographic planes of graphene clusters [76]. Therefore, even when carbonization is effected in the absence of oxygen, on the boundaries of graphene clusters (formed as a result of pyrolysis of the starting material) there remains a large number of reactive groups that react readily with atmospheric oxygen after re-establishment of contact with the atmosphere.

The constitution of oxygen-containing sites in carbonaceous materials of various types was a subject of inquiry of many authors. The first systematized results were outlined in a review by Boehm [77]. In the past few years numerous researches were carried out applying the diffuse reflectance infrared Fourier transform spectroscopy (DRIFTS) technique [78–81], and their results were summarized in a paper by Fanning and Vannice [82]. Analogous results concerning the constitution of oxygen-containing sites were obtained by x-ray photoelectron spectroscopy [83–85]. The aforementioned results made the foundation for identification (Structure 1) of the types of sites on the carbon surface. These sites seem to be present to a greater or lesser extent in the overwhelming majority of carbonaceous materials. Oxygen-containing groups can play a significant role during adsorption of substances capable of forming hydrogen-bonded complexes on the surface of carbonaceous adsorbents. Of special importance may be the case of nonporous materials whose formation involved overlapping of graphene layers [Fig. 2(b)]. In such carbonaceous particles the surface portion built up by prismatic faces may be substantial.

1

Upon a thorough analysis of the data on surface morphology it can be concluded that in the case of graphitized adsorbents (carbon blacks, carbon-containing mineral adsorbents) the main types of surface adsorption sites can be graphene layers, fractures of turbostratic structures, formed mainly at the expense of sp^3-hybridized carbon atoms, oxygen-containing groups, and slitlike gaps between graphene clusters. In ungraphitized materials the main types of adsorption sites are slit- and wedge-shaped pores whose interplanar spacing may vary in a wide range, disordered carbon lattice, and oxidized carbon atoms. Depending on the method of production of carbonaceous adsorbents it is possible to increase the concentration of either type of adsorption sites.

IV. GRAPHITE AND OXIDIZED GRAPHITE

A. Exfoliated Graphite

Graphite is an interesting subject of inquiry while investigating the spectral characteristics of adsorbed molecules, since the main type of adsorption sites on its surface is the basal graphite plane. However, in this case the recording of ^{1}H NMR spectra for adsorbed molecules is complicated on account of small values of specific surface area of graphite powders and broad NMR signals due to paramagnetic relaxation of proton spins on conduction electrons. Even for adsorption of alkanes on the surface of graphitized carbon blacks [15,16] the NMR signal width can amount to several kHz, which does not allow one to make exact measurements of chemical shifts.

With the purpose of decreasing peak widths of ^{1}H NMR spectra for substances adsorbed on graphite surfaces, the authors of Refs. 86 and 87 have studied the adsorption of water, benzene, and acetonitrile on the surface of exfoliated graphite produced by heat treatment of a crystalline graphite powder intercalated with sulfuric acid. As is known, the specific surface area of such an exfoliated graphite sample is higher than that of pristine graphite, and, in addition, at the expense of reduction of the concentration of free electrons in the sample, the intensity of relaxation effects decreases. The measurements were made for suspensions of the substances under study, with the suspensions being frozen. Applying the ring current method [88,89], the authors of Ref. 90 made theoretical calculations of chemical shift values for some molecules adsorbed on graphite clusters.

In Table 1 for a number of compounds adsorbed on exfoliated graphite (EG), chemical shift values observed in experiments and those calculated by the method of ring currents are compared. The parameters for adsorption complexes were obtained by simulating their interactions with the graphite surface using the method of interatomic Scheraga's potentials [91]. For all the investigated compounds, chemical shifts for the adsorbed state are smaller than those for the condensed state. A strong screening effect for compounds susceptible to various molecular interactions enables one to conclude that it is the surface of basal planes that exerts a magnetic shielding effect on adsorbed molecules. As can be seen from Table 1, the calculated chemical shifts for acetonitrile and nitrobenzene are less than those observed in liquid phase freezing experiments. This is due to the

TABLE 1 Measured and Calculated Values of ^{1}H NMR Chemical Shifts for Water, Acetonitrile, and Nitrobenzene Adsorbed on the EG Surface

Substance	Chemical shift, ppm		
	Condensed state	Adsorbed state (freezing liquid phase)	Calculation
H_2O	5	2.5	0.5
CH_3CN	2	0	−2
$C_6H_5NO_2$	7.5	5	3.4

Source: Ref. 90.

fact that on recording ¹H NMR spectra under freezing conditions the resonance signal is an average for molecules adsorbed on the surface and for molecules that are at some distance from it but are also subjected to the perturbing action of the surface. In this case, due to a contribution of molecules that do not interact directly with areas of local magnetic anisotropy, the recorded signal is shifted into low magnetic fields as compared with a signal obtained under conditions of gas phase adsorption. For this reason, the chemical shift for water adsorbed on the graphite surface upon freezing is greater than that for water adsorbed from the gas phase. In the calculations performed the feasibility of multilayer adsorption was not considered, so that the calculated values of chemical shifts for acetonitrile and nitrobenzene are related to ¹H NMR signals in the case of gas phase adsorption.

It should be noted that positions of adsorbed molecules with respect to the magnetic axis of ring currents for a single benzene ring may vary depending on the chemical nature of an adsorbed molecule. Here, one would also expect changes in the chemical shift of the adsorbed molecules similar to those observed for protons of molecules interacting with the π-system of benzene [89]. However, for fused benzene systems or graphene clusters the integrated ring current generated by the system of conjugated rings is equal to the current flowing along the outer contour of this system. Then a molecule adsorbed at any point inside the contour will be subjected to the screening effect similar to that near the axes of a single benzene ring. It is precisely this fact that explains why the chemical shifts of protons into the region of a strong magnetic field are observed for any molecule irrespective of its coordination position on the surface.

B. Oxidized Graphite

A graphite surface (in particular, exfoliated graphite surface) contains many structural imperfections, such as oxidized groups and residual molecules of oxidizers [92]. They can be involved in the formation of H-bonded complexes with water-type molecules. Active protons in such associations should be subjected to a deshielding effect of electron-donating atoms, and, besides, in this case one could expect the appearance of ¹H NMR signals shifted to down-fields relative to the corresponding resonance lines for a condensed phase. As no such peaks were found in the spectra, and, in addition, experimentally determined chemical shifts for water are close to calculated values, it may be concluded that the concentration of these centers in EG is low in comparison with adsorption sites on basal graphite planes of a carbon surface.

To obtain ¹H NMR spectra for water molecules interacting with surface oxidized groups (carbonyls, carboxyls, hydroxyls), NMR investigations have been made on frozen aqueous suspensions of "graphite oxide" (GO) [93,94]. Graphite oxide is constituted by colloidal particles consisting of completely oxidized graphite planes [95]. In gaps among these planes there is a great deal of surface-bonded water. A ^{1}H NMR spectrum for water in such a suspension at a solid phase concentration $C_{sol} = 1.92$ wt% is shown in Fig. 5. As evident from the figure, the spectrum is composed of three signals, two of which (with $d = 7$ ppm and $d = 0$ ppm) make up a narrow component observed on the background of a very broad signal (3) with the half-width $D = 5$ kHz. The former two narrow lines may be assigned to water molecules attached to oxygen-containing and free-radical surface centers, while the third signal may be assigned to structurized water in interplanar gaps. The oxygen-containing surface centers are carbonyl, carboxyl, and hydroxyl groups. Free-radical centers are mainly located at terminal carbon atoms on graphite planes. The presence of a free radical excess in a GO suspension was corroborated by observation of an intensive ESR signal with a width of 0.2 mT and a g-factor of 2.003. Since signal 1 (Fig. 5) has a chemical shift greater than that for liquid water, it can be ascribed to water molecules incorporated into complexes of the C=O $\cdots$ HOH type, where hydrogen bonds are stronger than in pure water. Signal 2 was related [94] to water bonded to

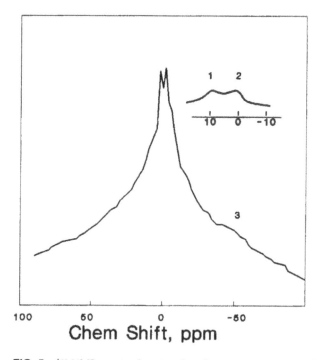

FIG. 5 ^{1}H NMR spectra for water in a frozen aqueous suspension of oxidized graphite. (From Ref. 93.)

free-radical centers. However, this assignment is not unambiguous, as GO particles may contain clusterlike structures of nonoxidized graphite. When interacting with such inclusions, water molecules are subjected to a shielding effect of ring currents in aromatic systems similar to the effect observed in EG. The value of a g-factor that is close to the g-factor for free electrons provides evidence for the validity of our argument.

The fact that in the spectra for frozen suspensions of GO there is a signal with a width of more than 5 kHz gives ground to believe that water molecules responsible for this signal have a very low molecular mobility that is substantially lower than that for water molecules bound to the surface active sites. In all likelihood these slow-moving molecules are situated in gaps among planes of disperse particles of GO in the field of surface forces. As judged by the intensity of signal 3, this part of the water in the suspensions studied is a great part of the water contained in such a suspension. The authors of Ref. 93 investigated diluted aqueous suspensions of GO whose concentrations were not higher than 1 wt%. Therefore in conformity with the data of Fig. 5 the thickness of a layer of water bound to the surface must be equal to several dozens or even hundreds of molecular diameters. In the case of such a thick water film the energy of interaction of water molecules with the surface should be low. It can be assumed that signal 3 corresponds to solid water whose constitution differs from that of ordinary ice. The mobility of water in such layers takes up an intermediate position between the mobilities of water adsorbed on the surface and water in the bulk of ice.

C. Intercalated Graphite

Nuclear magnetic resonance spectroscopy is a widely used technique for investigating intercalated graphite [96–98]. Here the major part of experimental results has been achieved for graphite intercalated with alkali metals, since nuclei of isotopes ⁷Li, ⁸⁹Ru, and ¹³³Cs are known to have a nonzero magnetic moment, which allows one to make a record of them in NMR spectra. In the case of graphite intercalated with potassium it proved possible to record well resolved ¹³C spectra for samples with various concentrations of the intercalant and to measure chemical shifts of intercalated compounds (intercalates) for various stages of intercalation. For alkali metal nuclei the chemical shift is dependent first of all on the interaction of the intercalant nucleus with conduction electrons and is predominantly determined by the value of the electric charge on the nucleus. Since in the case of the second and higher stages of graphite intercalation with alkali metals the metal atoms are in ionized state, the NMR signals are more often shifted into weak magnetic fields [97].

Graphite intercalation compounds (GIC) can form ternary GICs whose interplanar gaps contain also organic molecules. In this case they can be studied by ¹H [99–102] or ¹³C NMR spectroscopy. Such studies show that NMR signals of organic molecules situated in interplanar gaps of graphite are shifted into the region of strong magnetic field, which is due to the screening effect of π-systems of fused benzene rings. The characteristics of water molecules situated in interplanar gaps of ternary GICs ascertained by ^{1}H NMR spectroscopy were described in Ref. 86. The authors used graphite samples intercalated by the persulfate method [103]. In the course of the reaction sulfuric acid molecules were inserted into interplanar gaps of crystalline graphite, and some of them were covalently bound to structural elements of the surface. Unreacted acid molecules remained sufficiently labile. The intercalation reaction was arrested by addition of excess water to the system. Under these conditions labile acid molecules were substituted by water molecules. By prolonged washing almost all the acid can be removed from interplanar gaps.

Figure 6 displays ^{1}H NMR spectra for samples treated with washing waters at pH of 0, 2, 5, and 7 (spectra Nos. 1–4, respectively). It should be mentioned that the instrument amplification factor was adjusted to suit conditions for record-

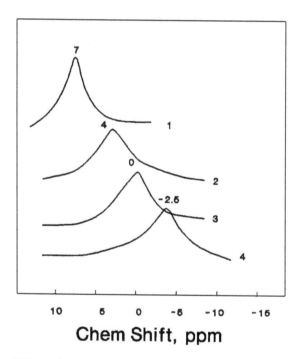

FIG. 6 ^{1}H NMR spectra for water in interplanar gaps of intercalated graphite at various values of pH for water in aqueous suspensions (1: pH = 0; 2: pH = 2; 3: pH = 5; 4: pH = 7). (From Ref. 86.)

ing the spectra; thus signal intensities in Fig. 6 do not reflect the actual concentration ratios for adsorbed molecules.

As seen in Fig. 6, chemical shifts for water molecules trapped in interplanar gaps of washed intercalated graphite decreased from 7 ppm at pH 0 to −2.5 ppm for a neutral sample. Such a down-field shift with decreasing pH value was due to proton exchange between acid and water molecules. As protons in acids give signals in very weak magnetic fields (at ca. 15–20 ppm), the presence of acids even in small amounts results in a down-field shift of the signal. After the complete removal of acid molecules from interplanar gaps in intercalated graphite the chemical shift for water protons proved to be an up-field shift. In this case the surface screening effect for water in the interplanar gaps causes the shift of the water signal by about 7.5 ppm into the region of strong magnetic field.

V. ACTIVATED CARBONS

A. Heat-Treated Activated Carbons

Although activated carbons are usually regarded as nongraphitized materials, many of them include such structural elements as more or less extended graphite planes forming a microporous structure of the material. The extent of order in such materials depends on the regularity in the arrangement of the graphite planes, the size and thickness of graphite clusters, and the spacing between planes in a microporous structure. A slit-shaped gap between two graphite planes is the most commonly encountered type of adsorption site in activated carbons subjected to a high-temperature treatment in the absence of oxygen. When a carbon material is subjected to a high-temperature treatment, graphene clusters can nucleate and grow, and the first to undergo local graphitization are microporous areas of the forming adsorbent surface. As a consequence, the local extent of graphitization in pores of different diameters may vary. The ¹H NMR spectra recorded for compounds adsorbed on such surfaces are rather complex.

Figure 7 illustrates the temperature dependence of ¹H NMR spectra for water in pores of synthetic carbon adsorbents (SCA) prepared by carbonization of porous poly(divinylbenzene) copolymers [104]. Samples of SCA 2000 treated in vacuum at 2273 K were used in the experiments. The specific surface area of this adsorbent was 360 m²/g, and the pore volume amounted to 0.56 cm³/g. The measurements were carried out under liquid phase freezing conditions. At $T <$ 273 K water in intergranular space and in the largest pores froze, and in the ¹H NMR spectrum it was possible to detect signals only for water molecules bound to the surface. The recorded spectrum was composed of three signals, signal 1 being related to bulk water, signals 2 and 3 being assigned to water attached to the adsorbent surface. On lowering the temperature, signal 1 disappeared, while

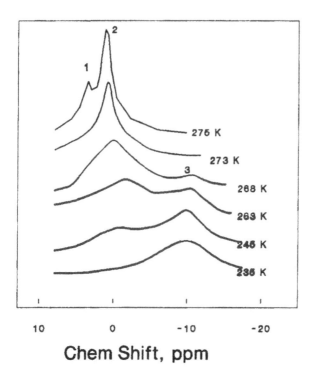

FIG. 7 Temperature-dependent variations of the shape of ^{1}H NMR spectra for water in heat-treated activated carbon SCA 2000. (From Ref. 105.)

the intensity of signal 2 decreased, and the peak was shifted to stronger magnetic fields. The intensity of signal 3, however, monotonically increased at a constant chemical shift. In Ref. 105 it was shown that the appearance of signal 3 depended on the thermal treatment of the material, i.e., on the extent of its local graphitization, rather than on the presence of molecularly adsorbed and chemisorbed oxygen in the adsorbent. This led to the conclusion that signal 3 was caused by water adsorbed on graphitized areas of the surface. A similar type of temperature dependence was also found for the following organic compounds frozen in pores of SCA 2000: acetonitrile, 1,2-dichloroethane [106], and benzene [107].

To ascertain the relationship between pore size and chemical shift of an adsorbate, use was made of the following approach [106,107]. Benzene vapor was adsorbed at a low relative pressure, $p/p_0 = 0.1$. Under these conditions the micropores of an adsorbent are mainly filled [108]. The measurements performed for a number of compounds showed that at low relative pressure the ^{1}H NMR spectroscopy detected only signal 3′ in high fields. Thus, the graphitized areas are predominantly localized in micropores of the adsorbent.

Spectral characteristics for various compounds measured by methods of liquid phase freezing and adsorption from the gas phase are summarized in Table 2. The following trends can be noted for adsorbed compounds irrespective of their capability to interact with various molecules. (1) Upon a decrease in temperature by 30 K relative to the freezing point, signal 2 was shifted to high magnetic fields by 4–7 ppm. (2) Signal 3 was temperature independent and was positioned in somewhat stronger fields as compared with signal 3′ of molecules adsorbed at low p/p_0. The latter fact implies that in liquid phase freezing experiments signal 3 resulted not only from the presence of a compound in micropores; rather it was an averaged signal caused by the presence of the compound in both micropores and pores of larger sizes. To verify the validity of this assumption for benzene, by way of example, spin–spin and spin–lattice relaxation times were determined for signals 2 and 3 after freezing of benzene in pores of SCA 2000 and for signal 3′ after adsorption at $p/p_0 = 0.1$ [107]. The resulting temperature dependence of the transverse relaxation time (T_2) is shown in Fig. 8.

As is evident from Fig. 8, the transverse relaxation time values for signals 3 and 3′ differ considerably. This is in contradiction with a suggestion made in Ref. 106 that signal 3 in liquid phase freezing experiments results exclusively from the compound filling adsorbent micropores. In such a case it could be expected that the relaxation time values for signal 3 (corresponding to completely filled pores) would be either similar to or larger than those for signal 3′. Besides, in the case of signal 3 there are benzene molecules bound to the surface as well as those situated at a distance of several molecular layers from the surface with larger T_2 values. However, the experimentally determined transverse relaxation times for signal 3 are less than the corresponding T_2 values for signal 3′ by a factor of ca. 5. Thus with allowance for the data in Table 2, signal 3 in Fig. 8 is attributed to the compound located not only in micropores but also in larger pores, presumably supermicropores.

To explain the intensity redistribution pattern for signals 2 and 3 in Fig. 7, it may be assumed that with a decrease in temperature a transfer of molecules between surfaces with different shielding properties takes place. Such reasoning would be justified for adsorption of gaseous substances when adsorbed molecules can easily diffuse through the adsorbent's pores. In our liquid phase freezing experiments, however, adsorbent pores are filled with a solid or adsorbed substance. Thus to interpret the data in Fig. 7 it is necessary to suggest that in SCA there are adsorption sites unoccupied at high temperatures, and that binding of adsorbate molecules on such sites is an exothermic process facilitated by a decrease in temperature. Still, such an approach has some considerable disadvantages, as it can be expected that instead of adsorbate molecules being transferred

TABLE 2 Proton Chemical Shifts (ppm) for Simple Organic Molecules and Water Adsorbed on SCA

		Proton chemical shift (ppm)				
		Liquid phase freezing		Adsorption at $p/p_0 = 0.1$		Extent of graphitization, α
Compound	Pure compound	Signal 2	Signal 3	Signal 3'	Calculation	
C_6H_6	7.20	4.0–0.9	−11	−6.2	−6.7	0.060
$(CH_3)_2CO$	2.07	—	—	−15.3	−11.4	0.115
CH_4	0.30	—	—	−13.6	−12.7	0.084
CH_2Cl_2	5.28	—	—	−5.1	−9.1	0.020
$C_2H_2Cl_2$	3.70	2.5–(−0.5)	−14.8	−12	−12.1	0.074
CH_3CN	1.96	−0.76–(−8.20)	−18	−13.2	−13.0	0.071
H_2O	4.50	3.0–(−1.0)	−13	—	—	—

Source: Ref. 107.

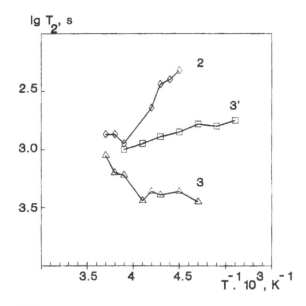

FIG. 8 Dependence of the transverse relaxation time for benzene on the reciprocal of temperature for signals 2, 3, and 3′ recorded in ¹H NMR spectra. (From Ref. 107.)

to free adsorption sites, additional molecules from the solid phase will come up to the reaction site. The presence of an unfrozen layer of the compound near the surface implies that the free energy of the molecules bound to the surface is less than that for solid benzene. Therefore the intensity of signal 2 should vary only slightly.

The variations in ¹H NMR spectra of adsorbed compounds and the nature of the temperature dependence for transverse relaxation times can be accounted for on the basis of the data collected when studying the potential energy surface (PES) for interaction of molecules with partly "graphitized" areas of a slit-shaped pore that is a model pore in SCA. The results of calculations for benzene using Scheraga's interatomic potential function [91] have shown that in terms of energetics the most preferable position is its location in a micropore with an interplanar distance of 0.53 nm. Pores of such a type correspond to curve 1 in Fig. 9. A benzene molecule lies in the plane parallel to graphite planes, the global minimum of potential energy being equidistant from both boundary planes. In such a position an adsorbed molecule is subjected to equal screening effects caused by ring currents in fused benzene rings of both graphite planes, which results in a maximum up-field shift in NMR spectra. As can be seen from Fig. 9, with increasing interplanar distance, the potential well depth diminishes (curve

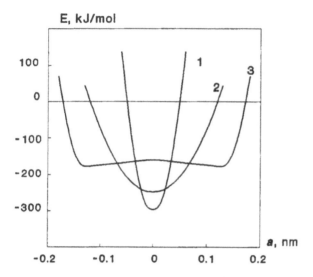

FIG. 9 Effect of the interplanar distance (a) on the curve shape for potential energy of benzene in slitlike pores formed by graphite planes (1: $a = 0.53$ nm; 2: $a = 0.62$ nm; 3: $a = 0.74$ nm). (From Ref. 107.)

2), and the interaction of an adsorbed molecule with the surface weakens. A further increase in interplanar distance leads to a change in the shape of the potential energy cross-section. A one-minimum potential curve transforms to a curve with two minima located at a distance of 0.74 nm from each plane (curve 3). A similar value of interplanar distances for one-minimum potential curves was obtained by using the Lennard–Jones potential for interaction between adsorbed molecules and a surface [109,110]. In this case an adsorbed molecule is magnetically shielded mainly by the nearest plane, which results in weakening of the observed shielding effect and in a down-field shift of the ${}^1H$ NMR signal. The calculated data show that the variations in the cross-section of PES take place in a narrow range of interplanar distances. On the other hand, it is known that graphite has a large coefficient of thermal expansion in the direction perpendicular to basal planes [111]. Then the data in Fig. 7 can be interpreted by assuming that a temperature decrease leads to an increase in the number of pores in which the interaction of adsorbed molecules with the surface can be described by a one-minimum potential curve. In ${}^1H$ NMR spectra this effect manifests itself in the enhancement of the signal 3 intensity and in the corresponding attenuation of signal 2. As a result, the effect of the second graphite plane on the molecules adsorbed in a gap becomes stronger, and their signal (i.e., signal 2) shifts to high magnetic fields.

As far as molecules adsorbed on a graphitized surface are concerned, one of the principal proton relaxation mechanisms is that involving conduction electrons delocalized over π-systems of fused benzene rings. Since the efficiency of interaction with π-systems of both planes is at its maximum when adsorbed molecules are positioned midway between the layers, for such molecules minimum T_2 values should be expected. Indeed, as follows from Fig. 8, the relaxation time for signal 3 is an order of magnitude less than for signal 2. Distinctions in kinds of interaction with conduction electrons can also explain the difference between T_2 values for signals 3 and 3′: the larger are the graphite clusters, the higher is the probability for a free electron to be delocalized there. Signal 3′ is attributed to molecules occupying gaps between relatively small graphite clusters, hence the T_2 value for them is considerably larger than for signal 3. Thus the calculated data allow one to make a consistent interpretation of temperature-dependent variations in the ¹H NMR spectra for the compounds adsorbed on SCA 2000 and to explain the particularities of relaxation parameters for the adsorbed molecules.

To ascertain the nature of anomalous chemical shifts for molecules adsorbed on SCA, the authors of Ref. 112 performed calculations of these parameters for molecules in slit-shaped pores formed by graphitized surfaces. The geometrical position of molecules in a slit-shaped pore was determined using Scheraga's function [91] for interatomic potentials. Chemical shifts were computed with allowance for a screening effect of a graphitized surface by the ring current method taking account of contributions of adsorbed molecule protons and anisotropy of a magnetic susceptibility tensor for a carbon material [18].

As the graphitelike structure is characteristic of only a certain portion of the carbon material, it is essential to know the extent of its graphitization at sites of adsorbed molecule localization. The extent of local graphitization can be taken into account by introducing a coefficient determined from a comparison between experimental and theoretical contributions of the surface and bulk solid phase to magnetic shielding. The magnetic shielding value for an adsorbed molecule proton will be determined by a term accounting for the effect of ring currents and a term representing the anisotropy of a magnetic susceptibility tensor caused by graphitization:

$$\Delta\sigma = \Delta\langle\sigma\rangle_{RC} + \alpha\Delta\langle\sigma\rangle_{AMS} \qquad (8)$$

where σ is the magnetic shielding of a proton in an adsorbed molecule, σ_{RC} is the contribution of ring currents, σ_{AMS} is the contribution of anisotropy of a magnetic susceptibility tensor for graphitized carbon atoms in the bulk of the material, and α is the extent of effective graphitization of the material.

The a value was calculated for each adsorbed compound on condition that experimental and calculated chemical shifts are equal in value. The calculated values of chemical shifts and coefficient α are listed in Table 2. In general, the theoretical values agree satisfactorily with those determined experimentally, with the exception of acetone and methylene dichloride. All the calculations were performed for a slit-shaped pore of the graphitized material with a small interplanar spacing of 0.58 nm corresponding to a one-minimum potential curve for adsorbent micropores (see Fig. 9). As can be seen, for different compounds (C_6H_6, CH_4, $C_2H_2Cl_2$, CH_3CN) the α values are close to each other. This fact corroborates the validity of the assumption that the α value is determined solely by shielding properties of the carbon material.

B. Nongraphitizable Activated Carbons

1. *Chemical Shifts of Adsorbed Molecules*

The authors of Ref. 113 carried out an investigation on the adsorption of water and a number of organic substances on the surface of activated carbons produced by carbonization of plum stones. Such carbons are widely used to remove organic substances from air and aqueous solutions. As was the case for synthetic carbonaceous adsorbents, a considerable part of the total pore volume of such adsorbents may be in slitlike micropores formed by closely spaced systems of graphite (graphene) planes. Since such carbons are usually carbonized at relatively low temperatures ($T < 1200$ K), the size of their graphene planes is typically small and does not exceed 2 nm [49]. The graphene planes can form systems consisting of 2–7 planes situated one over another, which can be regarded as graphite clusters (Figs. 3, 4). The subjects of inquiry were both the starting adsorbent (EG-0) and carbons oxidized with hydrogen peroxide (EG-$H_2O_2$1 and EG-$H_2O_2$2) and reduced with hydrogen (EG-H). The samples EG-$H_2O_2$1 and EG-$H_2O_2$2 differ in their treatment temperature.

Before adsorption of organic substances the adsorbents were placed into a measuring ampule 4 mm in diameter that was connected to a glass vacuum installation. Then the adsorbents were subjected to a thermal vacuum treatment at 500 K and 1 Pa for 3 h, following which the adsorbents were equilibrated with saturated vapours of organic substances at 293 K for 2 h (in the case of methane the adsorbate pressure in the system was equal to atmospheric pressure). When the adsorption process was completed, the ampules were sealed without contact with air and placed into the measuring element of the NMR spectrometer.

The values of chemical shift for the substances adsorbed on the surface of the starting, reduced, and oxidized carbons are summarized in Table 3. The table lists also differences in chemical shifts for the adsorbed molecules and liquid substance (as regards methane its solution in chloroform was used). As in the case considered in the previous section, the $\delta - \delta_0$ value determines the extent of the adsorbent surface effect on the chemical shifts of adsorbed substances.

As follows from the data in Table 3, the $\delta - \delta_0$ values for the various adsorbates are different. The maximum effect of the surface on the chemical shift value is recorded for adsorbed benzene (the $\delta - \delta_0$ value amounts to -9.3 ppm).

TABLE 3 Chemical Shifts for Molecules Adsorbed on Carbonaceous Adsorbents Based on Fruit Stones

Adsorbent	Water		Benzene		Acetone		Methane		DMSO	
	$\delta_{T=260}$	$\delta - \delta_0$	$\delta_{T=260}$	$\delta - \delta_0$	$\delta_{T\ 260}$	$\delta - \delta_0$	$\delta_{T\ 260}$	$\delta - \delta_0$	$\delta_{T=260}$	$\delta - \delta_0$
EG-0	-0.2	-4.3	-1.6	-8.9	-6.0	-8.2	-6.0	-5.0	3.0	0.5
EG-H	-4.4	-8.9	-2.0	-9.3	-5.3	-7.5	—	—	2.4	-0.1
EG-H$_2$O$_2$1	-2.8	-7.3	0.2	-7.0	-4.7	-6.9	-6.0	-5.0	—	—
EG-H$_2$O$_2$2	0.7	-3.8	0.0	-7.3	-5.3	-7.5	-5.5	-4.5	2.8	0.3

Chemical shift (ppm)

Source: Ref. 113.

The chemical shifts for water, acetone, and methane are somewhat lower, namely $\delta - \delta_0 = -(4.0 \div 8.2)$ ppm. In the case of dimethylsulfoxide, the $\delta - \delta_0$ value is close to zero.

The $\delta - \delta_0$ value for various adsorbates is also differently affected by the surface modification of the starting carbon. The most marked changes are observed for water. In this case the treatment of the surface with hydrogen leads to a substantial increase in the $\delta - \delta_0$ value (cf. -4.3 ppm for the starting carbon and -8.9 ppm for the reduced one). Oxidation of the carbon with a solution of hydrogen peroxide results in a decrease in the $\delta - \delta_0$ value, namely, for the EG-$H_2O_2$2 sample it becomes close to the corresponding $\delta - \delta_0$ value for the starting adsorbent. By contrast to water, in the case of benzene, $\delta - \delta_0$ values for carbons oxidized with hydrogen peroxide are lower than for the starting carbon. For other adsorbates the surface modification exerts a weak effect on the chemical shift value for adsorbed molecules.

In conformity with the principles outlined in the previous section, the results summarized in Table 3 can be interpreted in the following way. Molecules of water are adsorbed in slitlike pores that are created by graphene planes. Besides, they adsorb also through formation of hydrogen-bonded complexes containing oxidized carbon atoms. The ^{1}H NMR signal observed for adsorbed water molecules is an averaged signal for molecules of water adsorbed on these two types of sites. The treatment of the surface with molecular hydrogen results in a decrease in concentration of oxidized carbon atoms. Therefore the contribution of water molecules in slitlike pores to the averaged signal increases, which manifests itself as a shift of the signal into the region of strong magnetic field. Oxidation of the surface with hydrogen peroxide leads to an increase in the surface concentration of oxygen-containing groups, which results in an increase in the contribution of water molecules associated with these groups to the total signal.

Methane interacts with the surface by the dispersion mechanism. In doing so, it can adsorb in micropores as well as on oxidized carbon atoms and positions of junction of graphene planes. The maximum value of $\delta - \delta_0$ for methane is -6 ppm, which is by 3 ppm smaller than for water. On the assumption that graphene-plane micropores affect the chemical shifts for any adsorbed molecules in the same way, it is possible to draw a conclusion that the micropores contain about 2/3 of all the adsorbed methane molecules.

The fact that for adsorbed benzene molecules the value of the $\delta - \delta_0$ parameter is maximized seems to be caused by the specific interaction among π-systems of benzene and graphene planes. Acetone molecules having electron-donor properties are also readily adsorbed in micropores of the adsorbents studied. In this case the modification of the carbon surface practically does not affect the proportion of molecules adsorbed in micropores and on the other types of active sites ($\delta - \delta_0$ value varies by ca. 1 ppm). In contrast to other types of adsorbates, dimethyl sulfoxide molecules practically do not adsorb in micropores of these carbons even though they have strong electron-donor characteristics. This is evident from the absence of any influence of the surface on the shift for the adsorbed DMSO molecules. The major types of active sites interacting with DMSO molecules are likely to be hydroxyl and carboxyl groups of the activated carbon surface.

2. Characteristics of Adsorbed Water Layers

The ¹H NMR spectroscopy technique may be successfully applied for determining the parameters of near-surface layers of adsorbed water, and the most convenient measurement method proved to be the liquid phase freezing [22–25]. When the temperature of an aqueous suspension is lowered below 273 K, the main part of the water present in the system freezes and ceases to be recorded in high-resolution spectra due to a very small value of the proton transverse relaxation in ice. In this situation the ¹H NMR spectra exhibit only the signals attributed to that part of the water that is subjected to the perturbing action of the surface, i.e., the signals assigned to adsorbed water. Concentrations of adsorbed water in frozen suspensions (C_w) can be measured by comparing the signal intensity for unfrozen water (I) with the signal intensity for water adsorbed in a powder. Concentrations of water adsorbed in hydrated powders are determined using a calibration plot of the function $I = f(C_w)$. This plot is constructed by measuring the signal intensity for adsorbed water, while specified portions of water are being put into the ampule containing a weighed sample of an adsorbent. To prevent overcooling of suspensions, the measurements of concentration of nonfreezing water are carried out when heating the suspensions previously cooled to a temperature of 210 K. When obtaining temperature dependences of the signal intensity for carbonaceous stone adsorbents it was necessary to take into account the increase in signal intensity with decreasing temperature caused by variations of the population of nuclear energy levels [88] (the Curie law). The corrected values of signal intensity were obtained in accord with the method described in Ref. 114.

Variation of the free energy of water in an adsorption layer (G) can be determined from the temperature dependence of the free energy of ice (G_i). In doing so it was assumed that water at the interface freezes when $G = G_i$ [24,25], and the value $\Delta G = G_0 - G$ determines the decrease in the free energy of water molecules due to adsorption. The value G_0 is equal to the free energy of ice at $T = 273$ K. The thermodynamic functions of ice have been tabulated for a wide interval of temperatures [115]; therefore each value of the temperature change can be associated with a corresponding variation of the free energy of ice. In an aqueous medium, the capillary phenomena on the surface are absent, and the function $\Delta G = f(C_w)$ determines the radial dependence of the free energy of adsorption, since in the case of adsorbents with a given surface area, the adsorbed water concentration can be easily converted into layer thickness measured either in nanometers or in numbers of statistical monolayers. Values of free surface energies of adsorbents in an aqueous medium (ΔG_Σ) can be estimated by measuring the area under the curve of the function $\Delta G = f(C_w)$ extrapolated to the coordinate axes and using the formula

$$\Delta G_\Sigma = K \int_0^{C_w^{max}} \Delta G d(C_w) \tag{9}$$

where K is the scale factor and C_w^{max} is the thickness of a layer of nonfreezing water when $\Delta G \rightarrow 0$. When C_w is measured in milligrams of adsorbed water per gram of adsorbent and ΔG is measured in kJ/mol, free surface energy values will be expressed in mJ/m^2 on condition that K is equal to $55.6/S$ where S is the adsorbent surface area.

Figure 10 shows the variation in the free energy as a function of the nonfreezing water concentration in aqueous suspensions of the starting, reduced, and oxidized carbons. The curves for $\Delta G = f(C_w)$ exhibit a portion where the nonfreezing water concentration remained constant in a wide range of ΔG variations. The appearance of such a feature is attributed to the presence of micropores. In

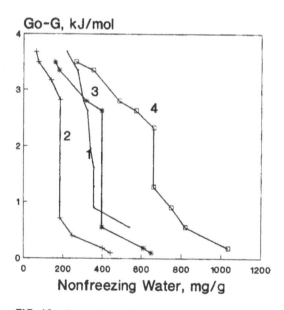

FIG. 10 Dependence of the free energy variation on concentration of nonfreezing water for activated carbons produced from fruit stones (1: starting carbon; 2: carbon reduced with hydrogen; 3: carbon oxidized with hydrogen peroxide at 523 K; 4: carbon oxidized with hydrogen peroxide at 623 K). (From Ref. 113.)

compliance with the theory of space filling of micropores [116,117], the free energy of water in micropores practically does not depend on positions of its localization within micropores. Adsorbed water freezes at a temperature when its free energy becomes equal to that of ice. If such a temperature is not reached, freezing of water does not occur, and the nonfreezing water concentration remains constant. The portion of a graph of ΔG as a function of C_w at the left of the vertical one (see Fig. 10) corresponds to freezing of water in micropores of adsorbents, and that at the right is associated with water in pores of larger diameters and weakly bound water on the outer surface of the adsorbents. Extrapolation of the graphs $\Delta G = f(C_w)$ to the ordinate enables one to estimate the maximum change of the free energy of water molecules caused by adsorption, ΔG_{max}, while extrapolation to the abscissa makes it possible to evaluate the concentration of adsorbed water. The abscissa of the vertical portion determines the concentration of water in the adsorbent micropores (C_w^{mp}). The characteristics obtained in this way, namely parameters of adsorbed water layers and values of free surface energy calculated by Eq. (9), are summarized in Table 4.

As is seen from the data in Table 4, upon treatment of the carbon surface with hydrogen, the value of its free surface energy sharply decreases. This agrees with the notion that such a treatment results in a decrease in the surface concentration of oxidized carbon atoms that form hydrogen-bonded complexes with water molecules. Besides, a sharp decrease in the concentration of water in the adsorbent micropores is also observed. This effect can be explained by lower accessibility of micropores for water molecules caused by deteriorating hydrophilic properties of the adsorbent surface. When a carbon surface is treated with hydrogen peroxide, we observe an increase in the bound water concentration at the expense of the creation of new oxidized sites that would form hydrogen-bonded complexes with water molecules.

We must note that in some case the optimum conditions for the formation of ordered layers of water on the surface call for a certain arrangement of primary

TABLE 4 Characteristics of Layers of Water Adsorbed on the Surface of Carbonaceous Stone Adsorbents in Frozen Aqueous Suspensions

Adsorbent	ΔG_{max}, kJ/mol	C_w^{max}, mg/g	C_w^{mp}, mg/g	ΔG_Σ, mJ/m²
EG-0	4.2	810	380	5.8
EG-H	4	500	190	3
EG-H₂O₂1	4	700	400	6
EG-H₂O₂2	4	1200	650	11

Source: Ref. 113.

adsorption sites. The significance of the relative positions of primary adsorption sites for the adsorption of water on the surface of microsporous and mesoporous adsorbents was shown by many authors [118–120]. Besides, in a number of cases for other types of carbons no correlation was observed between the adsorption value and the concentration of oxidized carbon atoms. The authors of Ref. 119 explained this anomalous behavior by postulating the formation of clusters of water that fill separate portions of pores with hydrophilic surfaces so that some part of the primary adsorption sites becomes inaccessible to water molecules.

VI. CARBONIZED SILICAS AND CARBON BLACKS

A. Nonporous Materials

1. Carbon Blacks

The authors of Refs. 121 and 122 used the liquid phase freezing method to investigate the spectral characteristics of adsorbed molecules in the case of carbon black particles after their immersion into water, acetonitrile, and benzene. The choice of the reagents was governed by the fact that all of them are capable of taking part in various intermolecular interactions. Hydroxyl groups of water molecules can interact with oxygen-containing surface groups by forming oxygen-containing complexes in which the water molecule acts as a proton donor. Acetonitrile molecules contain CN groups having a high dipole moment and low electron-donor capacity; therefore the main mechanism of their interaction with the surface is dipole–dipole interaction. Benzene molecules are neutral molecular species that interact with the surface by the mechanism of physical adsorption. The carbon blacks studied have a specific feature in that they belong to weakly graphitized materials whose surface contains substantial quantities of oxidized sites and adsorbed oxygen. The constitution of particles of such carbon blacks is illustrated in Fig. 2(a). Chemisorbed oxygen is present in the form of carbonyl and carboxyl groups that are predominantly situated on boundaries of graphene planes.

The ^{1}H NMR spectra for frozen suspensions of carbon black in benzene, water, and acetonitrile are shown in Fig. 11 (curves 1–3). From this figure it follows that upon freezing of a suspension the ^{1}H NMR spectrum for benzene bound to the carbon black surface consists of two signals shifted into strong magnetic fields relative to the signal for liquid benzene. With varying temperature, the chemical shifts of signals and ratios of intensities remained approximately constant. With decreasing temperature, signal width increases, which is caused by a lowering of the molecular mobility and is characteristic of the majority of substances in heterogeneous systems. As was shown in the previous sections, the shift of NMR signals into strong magnetic fields observed for substances adsorbed on carbon surfaces is attributed to the interaction of adsorbed molecules with condensed π-systems of benzene rings forming the basis of carbonaceous particles. The carbon

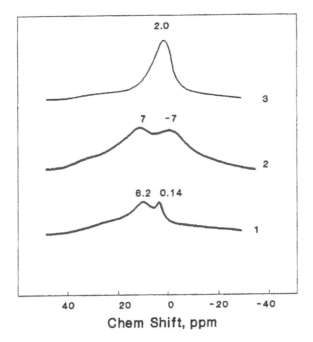

FIG. 11   ¹H NMR spectra for frozen suspensions of carbon black in benzene (1), water (2), and acetonitrile (3). (From Ref. 121.)

black surface has a considerable number of structural defects disturbing the junction among single benzene rings, and so the screening effect of the surface in the case of the physical adsorption of benzene may be relatively weak.

It is possible to give the following assignment of the ¹H NMR signals for benzene adsorbed on carbon black (Fig. 11, curve 1). The signal with a chemical shift of $\delta = 6.2$ ppm ($\delta - \delta_0 \approx 7$ ppm) is assigned to the molecules adsorbed on basal graphite planes of the surface. The weak screening effect of the surface is caused by a great number of structural defects. Besides, on the surface there is a number of slitlike pores formed by graphite planes whose interplanar spacing substantially exceeds the interplanar distance in graphite and graphitized carbons. On the basis of particle structure illustrated in Fig. 2 one can draw a conclusion that these pores are formed by gaps among single graphene packings consisting of closely spaced turbostratic graphite planes. The benzene molecules adsorbed in such pores are responsible for the ¹H NMR signal with a chemical shift of $\delta = 0.14$ ppm ($\delta - \delta_0 \approx 1$ ppm). The oxygen-containing surface groups and adsorbed oxygen do not make any appreciable contribution to the chemical shift for adsorbed benzene molecules, which corroborates the inference about the physical nature of benzene adsorption. In all probability, benzene adsorption on end faces and oxygen-containing sites can be neglected.

For the interaction between the carbon black surface and water molecules a somewhat different situation is observed (Fig. 11, curve 2). At 265 K there is only one signal, whose shift is equal to $\delta = 7$ ppm, which corresponds to a shift of 2.5 ppm into weak magnetic fields in comparison to the signal for liquid water. With decreasing temperature, this chemical shift does not change, but at 245 K on its shoulder there appears a considerably less intense signal with a chemical shift of $\delta = -3.6$ ppm. As temperature decreases, this signal is shifted into the region of strong magnetic field. When investigating the interaction of water with the oxidized graphite surface, the authors of Refs. 93 and 94 recorded a signal for water with a chemical shift of $\delta = 7$ ppm. This signal was ascribed to molecules interacting with carbonyl and carboxyl surface groups. On the other hand, for graphitized carbonaceous adsorbents and graphite it has been shown [86, 106] that the chemical shift for water adsorbed on basal graphite planes is displaced into strong magnetic fields by 5 ppm, while in the case of adsorption in slitlike pores formed by graphite planes the shift is equal to 6–18 ppm. Thus a more intense signal in Fig. 11 (curve 2) may be related to water molecules interacting with the surface through formation of hydrogen-bonded complexes with electron-donor sites of the surface. It is possible that at the interface there appears an interlayer of water structurized by a network of hydrogen bonds. In this case the descreening contribution of hydrogen-bonded complexes of the C$=$O $\cdots$ HOH and C$-$OH $\cdots$ OH$_2$ types to the chemical shift is predominant. The low-intensity signal for bound water in strong magnetic fields may be a signal assigned to water molecules situated in interplanar gaps of slitlike pores. The strong screening effect exerted by the surface on water molecules in comparison to benzene molecules seems to be caused by the accessibility of pores with a smaller interplanar gap to water molecules.

In the case of interaction between carbon black and acetonitrile the signal for molecules bound to the surface has the form of a singlet whose chemical shift does not depend on temperature and is equal to the chemical shift for liquid acetonitrile (Fig. 11, curve 3). Consequently, acetonitrile molecules interact weakly with basal graphite planes of the surface. The most probable mechanism in this case is dipole–dipole interaction with polar groups that are predominantly situated on end faces and structural defects of basal planes of graphite.

2. Carbonized Silica

Carbonized silicas (carbosils) are promising materials for application in liquid chromatography [123–125] and for the development of new types of adsorbents capable of adsorbing equally well both polar and nonpolar compounds [126]. Laboratory synthesis of carbosils is effected in special reactors containing silica.

At a temperature of 1000–1100 K, vapors of benzene [123], alcohols [127], methylene chloride [125], or other organic compounds [128] are passed through this silica, so that the pyrolysis of such vapors results in the formation of a carbonaceous coating on the starting silica. The constitution and properties of the coating depend on the carbonization conditions [129]. Conducting the pyrolysis process at various temperatures makes it possible to study its separate stages [130]. However, when synthesizing adsorbents suitable for application in practice, the degree of chemical transformations in the carbonaceous layer should be as high as possible. The microstructure of the carbon component of the carbosil surface seems to be close to that of carbon black particles, since the conditions of their synthesis are in many respects similar, although no special investigations in this field were made. However, as distinct from carbon black particles, the size of carbonaceous portions of the surface is determined to a great extent by the nature and porosity of the substrate and the sizes of its particles. Therefore the macrotexture of carbon in carbosils may differ considerably from that of carbon blacks illustrated in Fig. 2. Carbosils are complex adsorbents on whose surface there are both carbonaceous sections and sections of the starting silica, and thus their adsorption characteristics depend on the ratio of areas of each type of the surface. During adsorption various types of molecules can be adsorbed on either the silica or the carbon component of the surface.

(a) *Structure of Adsorption Sites.*   The study described in Ref. 131 was designed to investigate the behavior of two types of carbosils (CS), both synthesized through pyrolysis of methylene chloride on the surface on nonporous silica (Aerosil, specific surface of 300 m^2/g); the surface concentrations of carbon were 0.5 wt% and 40 wt% for CS1 and CS2, respectively. In order to elucidate the specificity of interactions of adsorbate molecules with silica and carbon components of the carbosil surface, the authors studied two adsorbates, water and benzene. It was assumed that water molecules (whose interaction with the surface can lead to the formation of hydrogen-bonded complexes) should in the first place interact with the silica surface, which contains a great number of hydroxyl surface groups. At the same time, benzene molecules take part in the specific interaction with the graphene layers and therefore should be adsorbed in the first instance on the carbon sections of the surface.

Figure 12 shows the NMR spectra for benzene and water adsorbed on the surface of CS1 and CS2. In the case of benzene, the spectrum for both carbosils consists of two signals displaced toward strong magnetic fields relative to the chemical shift for liquid benzene ($\delta_0 = 7.28$ ppm). The $\delta - \delta_0$ value for the more intense signal of CS1 is $\delta - \delta_0 = -2.7$ ppm, while that for CS2 is $\delta - \delta_0 = -2.3$ ppm. The second signal has approximately equal values of the chemical shift for both samples ($\delta - \delta_0 = -8$ ppm). The intensity of the signals for adsorbed benzene decreases from CS1 to CS2, which is in agreement with the

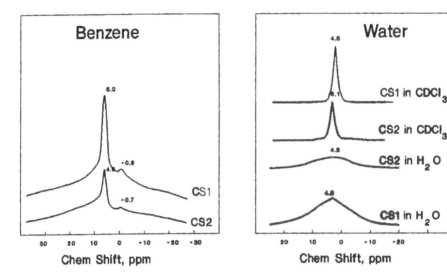

FIG. 12 ¹H NMR spectra for benzene and water adsorbed on the surface of nonporous carbosils. (In CS1 the surface carbon content is 0.5 wt%; in CS2 the surface carbon content is 40 wt%.) (From Ref. 131.)

increasing uptake of benzene molecules with increasing carbon content of the sample.

The shape of the ¹H NMR spectra for the water bound to the surface depends on the conditions of the experiment. Thus these experiments (involving freezing of a liquid phase with the adsorbent being placed into water and the temperature being lowered below 273 K) give a much wider signal than that for water at the carbosil/chloroform interfaces. As the chemical shift of liquid water is 4.5 ppm, it may be concluded that values of $\delta - \delta_0$ differing from 0 are observed only in the case of water adsorbed on CS2 in the presence of chloroform.

Figure 13 shows the temperature-dependent variation of chemical shifts for benzene and water molecules on the same adsorbents. With increasing temperature, the chemical shift of adsorbed benzene increases monotonically from 3.8 ppm at 220–230 K to 5.7 ppm at 270–279 K. The δ value for water bound to the CS1 surface practically does not depend on the temperature, whereas in the case of the CS2 surface a monotonic decrease in the δ value for adsorbed water is observed as the temperature increases.

In accordance with the data presented in Fig. 12 it is possible to draw the conclusion that, irrespective of the carbon content, the carbosil samples have at least two types of sites for benzene adsorption. On the basis of the estimated $\delta - \delta_0$ values, they may be related to the basal graphite planes (where $\delta - \delta_0$

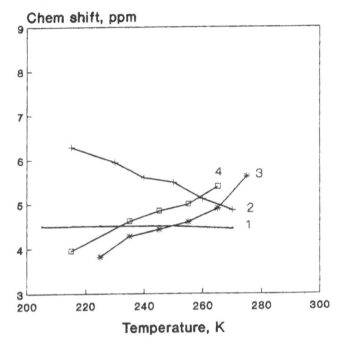

FIG. 13 Temperature dependence of variations in the chemical shift for water and benzene adsorbed on the surface of CS1 and CS2 (1: H_2O on CS1; 2: H_2O on CS2; 3: C_6H_6 on CS1; 4: C_6H_6 on CS2). (From Ref. 131.)

varies from -2 to -3 ppm) and slitlike pores between these planes ($\delta - \delta_0 = -8$ ppm). With decreasing temperature, the $\delta - \delta_0$ value for both carbosils increases because the layer of unfrozen benzene becomes thinner. As the ¹H NMR signal for adsorbed benzene is an averaged one (for molecules contacting the surface as well as for those remote from it at a distance of several molecular diameters), the $d - d_0$ value obtained in experiments with bulk freezing is slightly reduced due to the screening effect of ring currents of the graphite planes having a maximum value for the first layer of adsorbed molecules. The variations in the thickness of the adsorption layer may be regarded as responsible for changes in the $\delta - \delta_0$ values of benzene molecules bound to the surface of CS1 and CS2 samples, i.e., the higher degree of surface carbonization results in thicker layers of adsorbed benzene.

The difference in the character of the temperature dependence of chemical shifts for water adsorbed on CS1 and CS2 samples allows one to arrive at the conclusion that at different degrees of surface coverage with carbon, water molecules interact with different oxygen-containing surface groups. At a low degree of coverage, the surface has extended carbon-free fragments, where aqueous structures analogous to water adducts on the starting silica sample are formed. Thus the chemical shift for water bound to the surface depends to a small extent on the temperature and is in agreement with that for liquid water (Fig. 13, straight line 1). For the CS2 sample (high degree of coverage by carbon), the chemical shift for adsorbed water is noticeably higher than that for water on the starting silica or for liquid water. It gives evidence for the increase in the contribution of hydrogen-bonded complexes (where water molecules interact with more strong electron donor centers than those in water) to the total signal [132]. In the case of carbon surfaces, such centers are usually oxidized carbon atoms [93,94]. Then the increase in the chemical shift of adsorbed water from the CS1 to the CS2 sample may be explained in that water interacts predominantly with the oxidized carbon atoms on the surface.

(b)   Ordered Layers of Water on the Surface.   Valuable information about the processes occurring on the adsorbent–water interface can be furnished by measurements of the thickness of a layer of bound water and the value of free surface energy of an adsorbent in an aqueous medium. As it was shown in Section V.B, these characteristics can be obtained by measuring the dependence of the signal intensity for nonfreezing water on temperature; the results of such measurements make it possible to construct graphs for dependences of free energy of adsorbed molecules on the thickness of a layer of adsorbed water or on its concentration. However, while for porous carbons such dependences are limited by the adsorbent pore size, in the case of nonporous materials it is possible to investigate the whole layer of water subjected to the disturbing action of the surface. Direct experimental measurements and theoretical calculations show that at a distance of 5–6 molecular diameters the surface forces in the direction perpendicular to the flat surface show a tendency to change their sign [133–136]. This is true for both hydrophobic and hydrophilic surfaces [137]. As the surface of many industrial adsorbents is not flat, and as it is influenced by temperature changes, the averaged surface forces are repulsive forces decaying in accordance with the exponential law [138]. Their decay length may range from 0.1 to 0.3 nm. The general expression for the surface potential includes also the contribution due to the local polarization of the surface [139]. Simulations performed in Refs. 135 and 139 show that the surface affects only the liquid layer whose thickness does not exceed 3 nm. However, the authors of Refs. 140–142 recorded variations in the properties of water at considerable distances from the surface. At the same time, such parameters of liquids as density, viscosity, and solubility undergo variations near the solid–liquid interface. At the present time, however, the nature of these long-range components of the surface forces has not been completely ascertained.

Figure 14 shows plots of the thickness of the unfrozen water layer which is determined from signal intensity as a function of temperature or of changes in free energy, $273 - T = f(d)$ and $\Delta G = f(d)$, where d denotes the thickness of the unfrozen water layer in terms of the water amount per adsorbent surface unit. The thickness of the adsorbed water layer (in nanometers) or the number of statistical water monolayers can be easily derived from the data in Fig. 14 taking into account that a single monolayer corresponds to $C_w = 18$ mmol/m² and its thickness is equal to 0.3 nm. Curves 1–3 are related to water at the silica(or carbosil)–air, silica(or carbosil)–chloroform, silica(or carbosil)–water interfaces, respectively. It is evident that, under the conditions of a fixed degree of surface hydration (curves 1 and 2), freezing of water bound to the surface occurs at a temperature below 273 K. The above-mentioned phenomenon is observed predominantly for thin hydrated layers. Further, the curves in Fig. 14 indicate the effect of the medium and the distance from the surface on the free energy of water in the adsorption layer.

In Figs. 14(a,b) (curve 3) two segments can be distinguished, namely, one corresponding to a rapid decrease in thickness of the unfrozen water layer in a narrow range of ΔG changes (at temperatures close to 273 K) and the other corresponding to a relatively small decrease in d in a wide range of ΔG changes. The former is attributed to some part of the water that is adsorbed on the surface and has a specific feature that manifests itself in that during rapid changes in the thickness of a layer of unfrozen water the decrease in the free energy of water molecules due to adsorption is small. This type of bound water may be called "weakly bound water," and its existence is related to a long-range component of the radial function of the free energy variations. The second segment corresponds to strongly bound water; its existence is attributed to a short-range component of the radial function of the free energy variations. Numerical values of the layer thickness for each form of water (d^s and d^w for strongly and weakly bound water, respectively) and the maximum values of the decrease in the free energy of water due to adsorption (ΔG_s and ΔG_w) can be obtained by extrapolation of the corresponding parts of these curves to the abscissa and the ordinate.

In this case the free surface energy of an adsorbent at the silica(or carbosil)/water interface can be expressed as

$$\Delta G_\Sigma = \sum_i \Delta G_i d_i \tag{10}$$

where ΔG_i is the mean value of the free energy of water for the ith monolayer. The summation is taken over all the values of $1 < i < d_s$, where $d_s = d^s + d^w$. At $d = 1$ we obtain the ΔG_1 value corresponding to the free energy of the first monolayer of water localized in the vicinity of the surface. The maximum value of the free energy at the interface (ΔG_{max}) is defined as the $G_0 - G$ value at

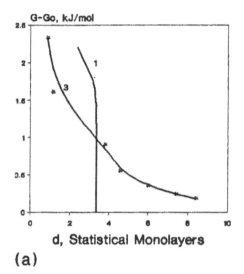

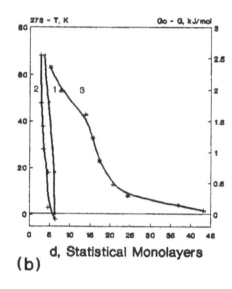

(a)

(b)

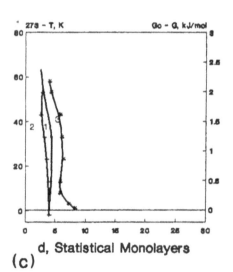

(c)

FIG. 14 Variation in the free energy of water as a function of the thickness of nonfreezing water layers for frozen aqueous suspension of Aerosil 300 (a), CS1 (b), and CS2 (c) in air, 1; in CDCl₃, 2; in water, 3. (From Ref. 131.)

$d = 0$. Being expressed in mJ/m², it characterizes a decrease in the free energy of water at the monolayer coverage of the surface. Such an approach can be applied to study correlations between surface energies, characteristics of interfacial layers of water, and the structure of the surface. The characteristics of adsorbed water layers are summarized in Table 5.

From the data summarized in Fig. 14 and Table 5 it can be concluded that for nonporous carbosil particles there is no correlation between the thickness of hydrated layers and the hydrophilic properties of the adsorbents. Thus in the case of the sample whose surface contains 0.5 wt% carbon, very large values for the thickness of adsorbed water layers and for the free surface energy of adsorbents have been recorded, which seems to be related to the formation of surface regions having some electric charge [25].

As it has been shown by Dukhin [143,144], when on the surface of disperse particles there appear oppositely charged portions, in the surrounding space there appears a long-range component of surface forces that is due to an electric field directed along the surface of the particles. This field affects the orientation of dipoles of water molecules around the disperse particles. The radius of action for the surface forces of this type is governed by the value of the electrical charges and distance between them. Since in an aqueous medium silica portions of the carbosil surface can acquire a negative charge (at the expense of a partial dissociation of surface hydroxyl groups), while carbon portions can acquire a positive charge (due to adsorption of solvated protons formed as a result of the dissociation of the electron-donating sites of the carbon portions of the surface or on the basal planes), it can be assumed that in this situation silica and carbon components of the surface carry charges of opposite signs. (The possibility of formation of a negative charge at the expense of adsorption of protons on basal planes of the carbon surface containing and not containing oxygen was considered in detail in Ref. 145.) This gives rise to polarization of adsorbent particles and the appearance of a long-range component of the surface forces. The results achieved here are

TABLE 5 Characteristics of Water Layers Adsorbed on the Surface of Carbosils and Starting Silica

Adsorbent	d^s, Number of statistical monolayers	d^w, Number of statistical monolayers	G^s, kJ/mol	G^w, kJ/mol	ΔG_Σ J/m²
CS1	25	25	3.0	0.6	820
CS2	7	3	3.0	0.8	184
Aerosil	4	6	3.5	0.6	155

Source: Ref. 131.

in good agreement with the data gathered for other nonporous materials [146–148]. In all cases only the surface polarization is a main factor that affects the thickness of adsorbed water layers. The increase in the carbon content to 40 wt% brings about a decrease in the free surface energy of the adsorbent, but in spite of this it remains significantly higher than that observed for the parent silica. It should be noted that in the case when the ΔG_Σ values are determined only by the hydrophilic properties of the surface, one can expect that the values of the free surface energy obtained for the carbosils should be lower than those obtained for the silica. In this connection, in the case of nonporous adsorbents the dominant role in the formation of the structurized interfacial water layers relates to the tendency of the adsorbent surface to undergo polarization in an aqueous medium.

The surface of carbosils whose carbon concentration ranged from 5 to 13 wt% was studied by the authors of Ref. 130 applying the electron microscopy technique. In that case on the surface there were both portions of the parent silica and portions of silica covered by a carbon layer, but any uniform coverage of the starting surface by a carbon layer was not observed. With increasing concentration of carbon, the area of the silica component of the surface noticeably decreased, and at carbon concentration of 13 wt% it became relatively small. It can be anticipated that at a carbon concentration of 40 wt% practically all the surface will be covered by a carbon layer. So, the surface of the CS2 sample may be more homogeneous than that of the CS1 sample (practically the whole surface of the silica is covered with carbon), and it can be concluded that the tendency of the adsorbent surface to undergo polarization increases as a function of the surface heterogeneity of a given material.

B. Porous Carbosils

1.   Influence of the Degree of Carbonization

The authors of Refs. [149] and [150] investigated the effect of carbon surface coverage of mesoporous silica gel Si-60 (Schuchardt Munich, Germany) on the adsorption site structure and characteristics of adsorbed water layers. A sample of about 10 g was heated at 820 K for 6 h in order to obtain heat-treated silica gel comparable with carbon-coated silica gels. Then the sample was subjected to the action of the modifying agent (CH_2Cl_2). The rate of the CH_2Cl_2 feeding was 0.6 cm^3/min for periods of 0.5, 1, 2, 3, 4, and 6 h. The silica was coated with carbon, and the products of these reactions were named CS3, CS4, CS5, CS6, CS7, and CS8, respectively. The structural characteristics of the materials studied are given in Table 6.

Figure 15 (a–c) presents the temperature dependence of the shape of ^{1}H NMR spectra for water adsorbed on the surface of CS3, CS7, and CS8 carbosils in a CDCl$_3$ medium. Figure 15(d) shows the analogous dependence for the CS8 car-

TABLE 6 Structural and Adsorption Characteristics of Adsorbents

Adsorbent	S_{BET}, m²/g	Carbon content, wt%	Carbonization time, h	V_p, cm³/g	R_p, nm
SG	343	0	0	0.75	4.4
CS3	366	0.77	0.5	0.74	4.0
CS4	339	4.37	1	0.67	4.0
CS5	299	14.89	2	0.56	3.8
CS6	258	20.32	3	0.47	3.6
CS7	223	26.7	4	0.39	3.5
CS8	163	35.0	6	0.27	3.4

Source: Ref. 150.

bosil in air. The concentration of the adsorbed water was 40–60 mg/g. In the case of CS4 and CS6 carbosils, the spectra differ insignificantly from those presented in Fig. 15(b), and that is why they are not included here. In the spectra recorded for a chloroform medium there are not only signals for adsorbed protons but also weak signals for TMS and chloroform, whose chemical shifts are 0 ppm and 7.26 ppm, respectively. With decreasing temperature, the width of the adsorbed water signal increases and its intensity decreases, which results from the decrease in the molecular mobility and partial freezing of adsorbed water.

The chemical shift of the adsorbed water signal for different adsorbents falls within a range of $4.3 < \delta < 5.5$ ppm, so it is close to the chemical shift of liquid water. The spectrum for CS8 contains the main signal of adsorbed water (signal 1) as well as a weaker signal (signal 2) characterized by a chemical shift equal to 2.1 ppm (Fig. 15(c)). With decreasing temperature, the intensity of signal 2 decreases, but to a smaller extent than that of the main signal, which provides evidence for strong adsorption interactions. Both these signals can be seen as separate only in a deuterchloroform medium. With air as the medium [Fig. 15(d)], the spectrum displays a variation in the shape of the adsorbed water signal, which points to the presence of another signal. However, recording of these signals as separate peaks is not possible due to their great widths.

The presence of two signals of water adsorbed on the surface of the CS8 sample corroborates the hypothesis that there are two kinds of active centers on the surface, and in view of the time scale of NMR the molecular exchange between these centers is a slow process (Section II). Thus the character of active centers responsible for the appearance of these signals is of interest. As the carbon concentration in the sample reaches 35 wt%, it can be assumed that the whole surface of the initial silica is covered with carbon and that both signals correspond

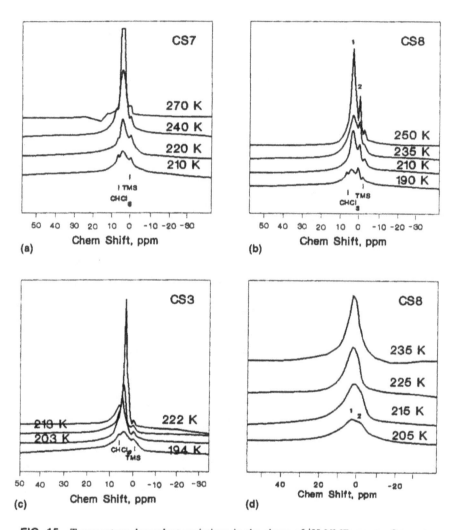

FIG. 15 Temperature-dependent variations in the shape of ^{1}H NMR spectra for water adsorbed on the surface of mesoporous carbonized silicas. (From Ref. 150.)

to water adsorbed on the carbon-containing part of the surface. The shift of the signal of adsorbed water into strong magnetic fields can be explained by two factors, namely, by the influence of local magnetic anisotropy areas of the condensed aromatic systems (Section 3), or by the changes in the structure of the hydrogen-bonded complexes forming a hydrated coating of the CS8 sample. The influence of these complexes on the chemical shift is due to the fact that the chemical shift of water is determined not only by the hydrogen bond strength [151, 152] but also by the average number (m) of hydrogen bonds per water molecule. The chemical shift depends on the m value to a much greater extent than on the hydrogen bond strength. Thus with increasing electron-donor capability of the solvent molecules the chemical shift of the dissolved water molecules increases from 1.7 ppm (chloroform or benzene) [23] to 3.2 ppm (dimethylsulfoxide) [153], while in the case of ice ($m = 4$) and liquid water ($m > 3$) the chemical shift is equal to 7 ppm [114] and 5 ppm [153], respectively. For the water adsorbed on the silica surface (Aerosil 300), the chemical shift falls within the interval from $\delta = 3$ ppm, when the concentration of water is much lower than that of surface hydroxyl groups [114, 154], to $\delta = 4.5$–4.7 ppm, for strongly hydrated surfaces [114]. In the first case, on the surface there are hydrogen-bonded complexes of water molecules with surface hydroxyl groups, and in the second case there are clusters of water adsorbed on the surface.

To ascertain which of the above-mentioned mechanisms is responsible for the appearance of signal 2, we studied the adsorption of benzene and methane on the dehydrated surfaces of carbon-coated silicas (Fig. 16). The spectra of the starting silica gel are also included. It was assumed that these adsorbates, as well

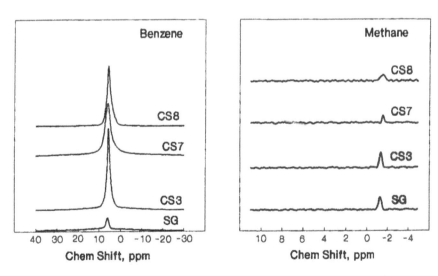

FIG. 16 ¹H NMR spectra for methane and benzene adsorbed on the surface of the starting mesoporous silica gel and porous carbosils differing in their surface carbon contents. (From Ref. 150.)

as water molecules, could adsorb on the surface fragments formed by the fused benzene rings but they could not form hydrogen-bonded complexes with electron-donor and proton-donor surface groups.

From Fig. 16 it follows that for all the carbosils studied the chemical shift of both benzene and methane is close to the chemical shift of these adsorbates in the presence of silica gel. On the carbon surface there are more adsorbed benzene molecules than on the silica surface, which is indicated by the differences in the signal intensity for benzene adsorbed on the SG and CS samples under the same conditions. However, methane adsorbs practically in the same way on both kinds of adsorbents. The signal intensity of adsorbed methane is much smaller than the signal intensity of benzene adsorbed on carbosil, and it is comparable with that of benzene on silica gel. These results can be explained by the fact that methane can interact with the surface only through the dispersion mechanism, while benzene can be bound specifically to the surface fragments formed by the condensed aromatic systems. The fact that the CS8 sample lacks signal 2 of both adsorbates gives ground to conclude that the appearance of this signal in the case of adsorption of water on CS8 is due to formation of water clusters (for which $m < 3$) in the hydration surroundings of this adsorbent. As carbon layer formation takes place in the silica gel pores, it can be assumed that signal 2 is due to the presence of water in the narrowest pores where large clusters cannot be formed.

The characteristics of water layers adsorbed on the carbosils and the parent silica gel are summarized in Table 7. It should be noted that for porous materials the thickness of an adsorbed layer cannot be much greater than the radius of the adsorbent pores. Therefore, as the pyrolysis time increases, the total pore volume decreases, which results from carbon depositing in the pores, and so there is a tendency for the adsorbed water concentration to decrease when going from SG sample to the carbosils. If the extent of carbonization is small, then C_w^s and ΔG_w

TABLE 7   The ^{1}H NMR Data for Layers of Water Adsorbed on the Surface of Mesoporous Carbosils

Adsorbent	ΔG^s, kJ/mol	C_w^s, mg/g	ΔG^*, kJ/mol	C_w^v, kJ/mol	ΔG_Σ, mJ/m^2
SG	3.5	270	1.2	730	150
CS3	3.0	320	1.5	230	110
CS4	3.2	375	2.2	155	132
CS5	4.0	175	1.7	225	109
CS6	3.0	150	1.4	125	72
CS7	3.5	220	1.2	250	139
CS8	3.7	300	1.8	100	240

Source: Ref. 150.

can be determined sufficiently accurately, but in the case of other samples they can be estimated only approximately. The appearance of inflections on the curve $\Delta G = f(C_w)$ can be caused by the adsorbent surface heterogeneity and limited pore volume. In the case of heterogeneous surfaces, different kinds of radial functions describing variations in the free energy of adsorbed water correspond to different surface fragments. When considering the total surface of a sample, the function obtained for the corresponding adsorbent is averaged.

Hydration properties of the adsorbent surface at the adsorbent–water interface can be most completely expressed in terms of the free surface energy. From the data in Tables 6 and 7 it follows that in the case when the carbon deposition time does not exceed 3 h the increases in the carbon content are accompanied by decreases in the free surface energy. The minimum value of ΔG_Σ was recorded for sample CS6 ($\Delta G_\Sigma = 72$ mJ/m²). This is in agreement with the statement that the hydration properties of the surface are determined by the concentration of primary adsorption sites that can form hydrogen-bonded complexes with water molecules.

On the amorphous carbon surface, such water adsorption sites are oxidized carbon atoms. Their concentration is usually much lower than that of hydroxyl groups on the silica surface, which should diminish the hydrophilic properties of the silica gel surface upon carbon deposition. By contrast, however, there is an increase in the hydrophilic properties for the CS7 and CS8 samples, which manifests itself in an increase of the ΔG_Σ and C_w^x values. In the case of the CS8 sample, ΔG_Σ is even higher than for the starting silica. The regularities of variations in the hydrophilic properties of the surface observed during carbonization can be explained by assuming that at the initial stage of carbonization the carbon surface is formed through development of a disordered carbon lattice. The weakening of the hydrophilic properties of the surface is brought about by a decreasing concentration of surface hydroxyl groups accessible for interaction with water molecules. With increasing carbonization time, the thickness of the carbon layer developed increases, and, in addition, its order increases too. In particular, carbonaceous surface structures can be formed whose constitution is similar to that illustrated in Fig. 2(b). Since in this case oxidized carbon atoms on the surface are predominantly situated at the boundaries of mutually overlapped graphene clusters, they are easily accessible for formation of hydrogen-bonded complexes with water molecules. At the same time, the accessibility of basal graphite planes for water molecules decreases. As a result, the hydrophilic characteristics of the surface become more pronounced.

2. Carbosils Modified with Zinc Silicate and Titanium Dioxide

Fillers designed for rubber compositions include dispersed carbon particles and metal oxides (Zn, Ti) to provide final products with the desired mechanical and wear-resistant properties as well as long service life. A promising method for producing the aforementioned fillers has been developed in recent years. It involves the synthesis of compositions containing carbon and metal oxide components. This type of material is the subject of inquiry of Refs. [155] and [156] whose authors investigated changes in the structure of boundary layers of water on the surface of porous silica under conditions of freezing a liquid phase in the processes of carbonization and impregnation of the carbonized surface with Zn silicate and Ti dioxide. The above-mentioned studies involved the determination of the free surface energy of the parent and modified silicas as well as the ascertainment of the type of radial functions of variations in the free energy of water adsorption.

Silica gel Si-6O (Merck, Germany) (SG), whose particle sizes fall in the range 0.1–0.2 mm, was taken as the parent silica for subsequent use in a process of carbonization. SG (in an amount of 5 g) and acetonitrile (0.02 moles) were used to produce carbosil free from metal oxides (CS9). Carbosils containing Ti [CS(Ti)] and Zn [CS(Zn)] were prepared using acetylacetonates of the corresponding metals. The process was carried out in glass ampules placed in an autoclave (0.3 L) and heated at a temperature of ca. 770 K for 6 hours. Our x-ray analysis revealed that samples of CS(Ti) and CS(Zn) contained titanium as TiO_2 (anatase) and Zn as Zn_2SiO_4, respectively. The amount of acetylacetonate per unit weight of SG was the same for all the carbosils. The concentration of Ti dioxide and Zn silicate formed in the process of pyrolysis was 12 wt%.

The constitution of adsorption complexes of water on the surface of these adsorbents can be determined from the ^1H NMR spectral characteristics of adsorbed water molecules. Figure 17 displays the relevant spectra. The spectrum of SG is a singlet having a chemical shift of $\delta = 4.5$ ppm; this value is close to that for liquid water and is typical for water clusters in which every molecule takes part in the formation of three or more hydrogen bonds.

Carbonization of the surface leads to significant changes in the spectra of adsorbed water. In this case water gives two signals having chemical shifts of 2.8 and 0.6 ppm. The above signals are displaced toward strong magnetic fields in contrast to the chemical shift of water on the silica surface. The presence of these two signals is strong evidence for surface inhomogeneity. As the intensity of signals in strong magnetic fields does not depend on temperature, the above signal may be attributed to water molecules that are involved in the more strongly bound surface H-complexes. In particular, the aforementioned complexes may contain silica surface hydroxyl groups as well as oxidized carbon atoms, thus allowing the formation of strong hydrogen bonds with water molecules.

Carbonization of carbosils used in this study was carried out at a sufficiently low temperature ($T = 870$ K). This condition leads to the formation of carbon layers whose graphene clusters have sizes that do not exceed 2 nm [49]. Consequently, the formation of partly graphitized areas on the carbon surface is hardly

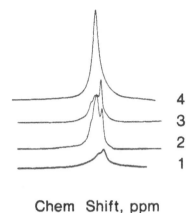

Chem Shift, ppm

FIG. 17 ¹H NMR spectra for water adsorbed on the surface of carbosils modified with titanium dioxide or zinc silicate (1: CS9, 2.2 wt% water, 277 K; 2: CS(Ti), 10 wt% water, 255 K; 3: CS(Zn), 10 wt% water, 255 K; 4: SG, 24 wt% water, 250 K). (From Ref. 155.)

probable, and the displacement of the ¹H NMR signal for the adsorbed water into strong magnetic fields may be due to water molecules that are located over the plane of the aromatic π-systems. This displacement does not exceed 5 ppm (Section 4).

It is obvious that the signals for water adsorbed on the surface may be treated as averaged over several types of adsorption complexes. The main contribution to the chemical shift is made by water molecules adsorbed on the carbon components of the surface, which are formed by the system of fused benzene rings and individual water molecules bound by hydrogen bonds to the surface hydroxyl groups. Therefore the signal of water having a chemical shift of $\delta = 0.6$ ppm should be attributed to interaction with surfaces having a higher carbon content.

The spectra of CS(Ti) and CS(Zn) samples give two signals. The first signal, being more intense ($\delta = 4.5$ ppm), is displaced toward weak magnetic fields as the temperature decreases. It is evident that this signal for both adsorbents is caused by the clusters of water adsorbed on the surface that is covered with Ti dioxide or Zn silicate formed in the course of oxidation of acetylacetonates. The x-ray analysis data, as well as the correlation between spectral parameters of adsorbed water and water adsorbed on the surface of the oxide and silicate, give strong evidence in favour of the above conclusion.

Figure 18 shows the variation in the free energy of adsorbed water as a function of the concentration of nonfreezing water for all the adsorbents studied. As C_w is proportional to the average thickness of layers of nonfreezing water, the above-mentioned dependence reflects the shape of the radial function describing

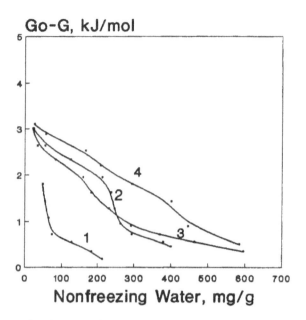

FIG. 18 Dependence of the variation of free energy on the thickness of a nonfreezing water layer for carbosil (>1), carbosil modified with titanium dioxide (2), and carbosil modified with zinc silicate (3). For comparison, the corresponding curve for the starting silica gel (4) is also presented. (From Ref. 155.)

variations in the free energy of adsorbed water. As follows from the data given in Fig. 18 (curve 4), the maximum concentration of adsorbed water for the silica gel is 700 mg/g. The volume filled with the adsorbed water is close to the total volume of pores in the adsorbent (Table 8). Therefore all the water in the pores of the starting silica is subjected to the perturbing action of the surface. The data given in Fig. 18 make it possible to conclude that the hydrophilic characteristics of the surface at first decrease as a result of carbonization and then increase in the course of impregnation with the metal oxides or silicates.

Extrapolation of the $\Delta G = f(C_w)$ curves to the water free energy axis enable us to estimate the maximum magnitude of variations in the free energy of water caused by adsorption (ΔG_{max}). The ΔG_{max} value determined in accordance with this method and the data for the free surface energy calculated using Eq. (2) are listed in Table 8. The values of maximum water layers thickness (d^{max}) perturbed by the surface, calculated on the basis of the C_w^{max} value, are also listed in Table 8. These data permit us to conclude that the free surface energy for the adsorbents studied increases in the following order:

$$(\Delta G_\Sigma)_{CS} < (\Delta G_\Sigma)_{SG} < (\Delta G_\Sigma)_{CS(Zn)} \approx (\Delta G_\Sigma)_{CS(Ti)}$$

TABLE 8 Parameters of the Water Layers Adsorbed on Parent and Modified Silicas

Material	Carbon content, wt%	S_{BET}, m²/g	Total volume of mesopores, cm³/g	ΔG_{max}, kJ/mol	ΔG_{Σ}, mJ/m²	d^{max}, nm
SG	0	394	0.744	5.9	160	1.5
CS9	4	359	0.718	6.7	26	0.6
CS(Zn)	6.3	212	0.535	5.9	204	2.6
CS(Ti)	9	192	0.424	5.9	208	2.4

ᵃ d^{max} is equivalent to C_w^{max} recalculated for a number of statistical monolayers.
Source: Ref. 155.

It should be noted that the correlation between the hydrophilic properties of the surface and the characteristics of its structure are not reflected by the revealed regularity in full measure. This is because when going from sample CS9 to CS(Zn) and CS(Ti), one can observe two trends, namely, increasing concentration of surface hydroxyl groups (owing to the formation of surface regions covered with titanium oxide or zinc silicate) and decreasing pore volume. In the case of both metal-containing samples, all the water in their pores becomes surface bound, which leads to levelling of differences in the values of the free surface energy.

VII. CONCLUSIONS

On the basis of the researches into various types of carbon materials, it is possible to make the following assignment of ¹H NMR signals for compounds adsorbed on carbon surfaces. When chemical shifts for a compound in adsorbed and condensed states are close in magnitude, adsorption takes place on a poorly ordered carbon surface with a significant number of structural imperfections in the crystal lattice. Besides, adsorbed molecules are randomly arranged relative to the axes of π-systems of fused benzene rings. Upon adsorption on graphitized surfaces, adsorbate molecules "feel" the surface screening effect. Its magnitude depends on the average position of protons relative to the axes of π-systems in basal graphite planes, and for different molecules it is equal to 4–6 ppm. In the case of multilayer adsorption, the screening effect decreases due to exchange between molecules on the surface and molecules remote from it.

When adsorbed molecules are localized in a slit-shaped gap formed by graphite planes, they experience a screening effect from both planes. The screening effect reaches its maximum for an adsorbed molecule situated halfway between the two planes. In such a case it can amount to 15 ppm and more. The magnitude of the effect depends on the dimensions of graphite clusters forming slit-shaped pores and on the local graphitization extent for a carbon material in the bulk. The larger are these graphite clusters and, consequently, the graphitization extent, the more considerable is the high-field shift of the ^{1}H NMR signal for an adsorbed molecule.

When there is an interaction with oxidized areas of the surface, the signal of protons in molecules capable of forming H-bonded complexes with the surface C=O groups can be shifted to low magnetic fields. In this case the screening effect of graphite planes does not manifest itself because most of the oxygen-containing centers are arranged along the edges of graphite planes. When oxidized carbon atoms reside on basal graphite planes, the aromaticity of benzene rings in their vicinity is disturbed, which also leads to a decrease in the surface screening effect.

In the situation of adsorption of water molecules on the carbonaceous adsorbent surface it is necessary to take into account the fact that chemical shifts of water molecules are strongly dependent on the number of hydrogen-bonded complexes involving concurrently each water molecule. Any decrease in the coordination number for water molecules as a result of the formation of adsorption complexes with the surface as well as adsorption on the carbon surface sites (e.g., systems of fused benzene rings) can lead to a shift of the signal for adsorbed water molecules into the region of strong magnetic field.

The constitution of ordered layers of water at interfaces with carbonaceous adsorbents in aqueous suspensions is governed by three major factors, namely, hydrophilic properties of the surface, porosity of the material, and the feasibility of polarization of the surface at the expense of the formation of regions carrying electric charges of opposite signs. In the general case, the thickness of an adsorbed water layer on the surface is determined by the action radius of surface forces in whose field the orientation of electric dipoles of water molecules occurs and the formation of its surface clusters takes place.

For nonporous materials it is possible to investigate the entire interfacial layer subjected to the disturbing action of the surface. In this case, the observed variations in the free energy of adsorbed water as a function of its concentration or layer thickness are derivatives of radial dependences of surface forces. With increasing surface concentration of hydrophilic sites, the thickness of a layer of bound water usually increases. Sometimes, however, the spatial arrangement of hydrophilic sites on the adsorbent surface is also important. Observations indicate that adsorbents with a particularly pronounced inhomogeneity of the surface may have a very thick adsorbed water layer (up to several dozens of molecular diameters). The free surface energy of such adsorbents may amount to 1000 mJ/m^2 and more. The appearance of such layers is proposed to be due to the presence of long-range components of surface forces as a result of the formation of surface regions having electric charges of opposite sign. In this situation the concentration of hydrophilic sites ceases to be a decisive factor in the formation of layers of bound water. If the adsorbent contains surfaces that sharply differ in the parameters of their interaction with water, the plots of radial functions describing variations in the free energy of adsorbed water may have inflections that are caused by differences in the constitution of layers of bound water over various sections of the surface. A typical representative of such systems can be carbon-coated silicas whose surface is formed by both carbonaceous regions and silica. The adsorbed water films, inhomogeneous in their properties, were observed in the case of a nonporous carbosil with a surface carbon content of 0.5 wt% and for a carbosil synthesized from a mesoporous silica gel whose surface was impregnated with titanium dioxide or zinc silicate.

The maximum layer thickness for water adsorbed in porous adsorbents cannot substantially exceed their pore radius. We have studied several such adsorbents. None of them has displayed polarization of its surface, and so no long-range component of surface forces that could be caused by such a polarization was observed. Therefore when the action radius for surface forces is smaller than the pore radius, the layer thickness for water adsorbed on the surface is governed mainly by the surface concentration of hydrophilic sites. If the surface potentials of the opposite walls of a pore are overlapped, all the water filling the pore is bound to the surface. The freezing temperature for water in such a pore is governed by the pore sizes and the force of interaction between the surface and water. The free surface energy value measured for such adsorbents determines the intensity of interactions between water molecules and the pore surface. In the case of chemically modified carbonaceous adsorbents, the comparison with values of their free surface energies in an aqueous medium makes it possible to obtain valuable information on the constitution of their surface.

ACKNOWLEDGMENTS

This work was supported by the State Committee for Scientific Research (KBN, Warsaw). Project No. 3 TO9A 03611.

The authors thank Professor L. R. Radovic for many useful remarks.

REFERENCES

1. PJE Verdegam, M Helme, J Lugtenburg, HJM de Groot. *J Am Chem Soc 119*:169–174, 1997.
2. I-S Chuang, DR Kinney, GE Maciel. *J Am Chem Soc 115*:8695–8705, 1993.
3. BS Lartiges, JY Botterro, LS Derrendinger, B Humbert, P Tekely, H Suty. *Langmuir 13*:147–152, 1997.
4. H Pfeifer, D Freude, M Hunger. *Zeolites 5*:274–277, 1985.
5. GE Maciel, DW Sindorf. *J Am Chem Soc 102*:7606–7607, 1980.

6. C Nedez, F Lefebvre, JM Basset. *Langmuir 12*:925–929, 1996.
7. Y Hiroyama, K Kume. *Solid State Commun 65*:617–619, 1988.
8. S Gradstain, J Conard, H Benoit. *J Phys Chem Solids 31*:1121–1135, 1970.
9. J Tabony, JW White, JC Delachaume, M Coulon. *Surf Sci 95*:L282–L288, 1980.
10. J Tabony, G Bomchil, N Harris, M Leslie, JW White, P Gamlen, RK Thomas, TD Trewern. *J Chem Soc Faraday Trans I, 75*:1570–1577, 1979.
11. J Tabony. *Progress in NMR Spectroscopy 14*:1–26, 1980.
12. RK Harris, TV Thompson, PR Norman, C Pottage. *J Chem Soc Faraday Trans 92*:2615–2618, 1996.
13. RK Harris, TV Thompson, P Forshaw, N Foley, KM Thomas, PR Norman, C Pottage. *Carbon 34*:1275–1279, 1996.
14. B Boddenberg, G Neue. *Molecular Physics 67*:385–398, 1989.
15. R Grosse, B Boddenberg. *Z Phys Chem (neue folge) 152*:1–12, 1987.
16. B Boddenberg, JA Moreno. *Ber Bunsenges Phys Chem 87*:83, 1983.
17. U Rolle-Kampczyk, J Karger, J Caro, M Noack, P Klobes, B Rohl-Kuhn. *J Colloid Interface Sci 159*:366–371, 1993.
18. MM Dubinin, RS Vartapetian, AM Voloshchuk, J Karger, H Pfeifer. *Carbon 26*: 515–520, 1988.
19. PC Griffiths, P Stilbs. *Langmuir 11*:898–904, 1995.
20. W Heink, J Karger, H Pfeifer, RS Vartapetian, AM Voloshchuk. *Carbon 31*:1183–1187, 1993.
21. VV Turov, VN Barvinchenko. *Colloids Surfaces B 8*:125–132, 1996.
22. VV Turov, R Leboda, VI Bogillo, J Skubiszewska-Zieba. *Adsorption Sci Techn 14*:319–330, 1996.
23. VM Gun'ko, VI Zarko, VV Turov, EF Voronin, VA Tishenko, AA Chuiko. *Langmuir 11*:2115–2120, 1995.
24. VV Turov, VM Gun'ko, VI Zarko, VM Bogatyr'ov, VV Dudnik, AA Chuiko. *Langmuir 12*:3503–3510, 1996.
25. VV Turov, R Leboda. *Adv Colloid Interface Sci 79*:173–211, 1999.
26. A Antoniou. *J Phys Chem 68*:2754–2764, 1964.
27. GG Litvan. *Can J Chem 44*:2617–2622, 1964.
28. A Bogdan, M Kulmala, B Gorbunov, A Kruppa. *J Colloid Interface Sci 177*:79–87, 1996.
29. EW Hansen, M Stocker, R Schmidt. *J Phys Chem 100*:2195–2200, 1996.
30. RO Ramseir. *J Appl Phys 38*:2553–2556, 1967.
31. Atta-ur Rahman. *One and Two Dimensional NMR Spectroscopy*. Amsterdam: Elsevier, 1989, pp 33–34.
32. JA Pople, WG Schneider, HJ Bernstein. *High-Resolution Nuclear Magnetic Resonance*. New York: McGraw-Hill, 1959, pp 165–180.
33. JW Emsley, J Feeney, LH Sutcliffe. *High Resolution Nuclear Magnetic Resonance Spectroscopy*. Oxford: Pergamon Press, 1965, pp 60–111.
34. TL James. *Nuclear Magnetic Resonance in Biochemistry*. New York: Academic Press, 1975, pp 62–92.
35. JA Pople. *Proc Roy Soc A 239*:541–550, 1957.
36. AD Buckingem. *Can J Chem 38*:300–307, 1958.
37. JC Hindman. *J Chem Phys 44*:4582–4592, 1966.

38. DE O'Reilly. *J Chem Phys 61*:1592–1593, 1974.
39. HA Resing, DW Davidson. *Can J Phys 54*:295–300, 1975.
40. GM Haggis. JB Hasted, JT Buchanan. *J Chem Phys 21*:1688–1693, 1953.
41. IP Gragerov, VK Pogorelyj, IF Franchuk. *Hydrogen Bond and Fast Proton Exchange*. Kiev: Naukova Dumka, 1978, pp 6–8.
42. HS Gutowsky, CH Holm. *J Chem Phys 25*:1228–1231, 1956.
43. HS Gutowsky, A Saika. *J Chem Phys 21*:1688–1696, 1953.
44. R Ryoo, LC de Menorval, JH Kwak. F Figueras. *J Phys Chem 97*:4124–4127, 1993.
45. A Gedeon, JL Bonardet, J Fraissard. *J Phys Chem 97*:4254–4255, 1993.
46. MG Samant, LC de Menorval, R Dalla Betta, M Boudart. *J Phys Chem 92*:3937–3943, 1988.
47. C Tsiao, DR Corbin, CR Dubowski. *J Phys Chem 94*:867–869, 1990.
48. MM Dubinin. *Chem Rev 60*:235–241, 1960.
49. VB Fenelonov. *Porous Carbon*. Novosibirsk: Boreskov Institute of Catalysis, Siberian Branch of Russian Academy of Science, 1995, pp 13–64.
50. SJ Gregg, KSW Sing, HF Stoeckli, eds. *The Characterisation of Porous Solids*. London: Soc Chem, Ind., 1979.
51. FAL Dullien. *Porous Media, Fluid Transport and Porous Structure*. New York: Academic Press, 1979.
52. JB Donnet, A Voet. *Carbon Black*. New York: Marcel Dekker, 1976.
53. PL Walker Jr. *Carbon 28*:261–279, 1990.
54. W Ruland. *Carbon 2*:365–375, 1965.
55. H Darmstadt, C Roy, S Kaliaguine. *Carbon 32*:1399–1406, 1994.
56. N El Hor, C Bourgerette, A Oberlin. *Carbon 32*:1035–1044, 1994.
57. I Rannou, V Bayot, M Lelaurain. *Carbon 32*:833–843, 1994.
58. BE Warren. *J Chem Phys 2*:551, 1934.
59. J Biscoe, BE Warren. *J Appl Phys 13*:364, 1942.
60. X Bourrat. *Carbon 31*:287–302, 1993.
61. JB Donnet, E Custodero. *Carbon 30*:813–817, 1992.
62. JB Donnet, TK Wang, E Custodero. *STM and AFM Study of Carbon Black Structure*. Proceeding of International Conference Eurofillers 95, Mulhouse (France), September 11–14, 1995, pp 187–190.
63. JB Donnet. *Carbon 32*:1305–1310, 1994.
64. I Tanahashi, A Yoshida, A Nishino. *Carbon 28*:477–482, 1990.
65. K Kuriyama, MS Dresselhaus. *J Mater Res 6*:1040–1047, 1991.
66. M Endo, H Nakamura, M Inagaki. *Twentieth Biennial Conference on Carbon*, American Carbon Society, University of California, Santa Barbara, CA, 1991, p. 654.
67. VM Linkov, RD Sanderson, GK Ivakhnyuk. *Carbon 32*:361–362, 1994.
68. K Gergova, N Petrov, S Eser. *Carbon 32*:693–702, 1994.
69. MG Lussier, JC Shull, DJ Miller. *Carbon 32*:1493–1498, 1994.
70. JJ Freeman, JB Tomlinson, KSW Sing, CR Theocharis. *Carbon 31*:865–869 1993.
71. RH Bradley, B Rand. *J Colloid Interface Sci 169*:168–176, 1995.
72. HF Stoeckli. *Carbon 28*:1–6, 1990.
73. M Huttepain, A Oberlin. *Carbon 28*:103–111, 1990.

74. K Kaneko, C Ishii, M Ruike, H Kuwabara. *Carbon 30*:1075–1088, 1992.
75. DA Wickens *Carbon 28*:97–101, 1990.
76. J Abrahamson. *Carbon 11*:337–362, 1973.
77. HP Boehm. *Adv Catal 16*:179, 1966.
78. JJ Venter, MA Vannice. *Carbon 26*:889–902, 1988.
79. BJ Meldrum, CH Rochester. *J Chem Soc Faraday Trans 86*:861–865, 1990.
80. JJ Venter, MA Vannice. *Appl Spectrosc 42*:1096, 1988.
81. MA Vannice, PL Walker Jr, C Jung, C Moreno-Castilla, OP Mahajan. *Proc. 7th International Cong. Catalysis* (edited by T. Seiyama and K. Tanabe) New York: Elsevier, 1981, p 460.
82. PE Fanning, MA Vannice. *Carbon 31*:721–730, 1993.
83. JM Thomas, EM Evans, M Barber, P Swift. *Trans Faraday Soc. 67*:1875, 1971.
84. A Ishitani. *Carbon 19*:269, 1981.
85. Y Nakayama, F Soeda, A Ishitani. *Carbon 28*:21–26, 1990.
86. VV Turov, KV Pogorelyi, LA Mironova, AA Chuiko. *Dokl AN USSR N 3*:128–131, 1991.
87. VV Turov, KV Pogorelyi, LA Mironova, AA Chuiko. *Zh Fiz Khim 65*:170–174, 1991.
88. A Abragam. *The Principles of Nuclear Magnetism*. Oxford: Clarendon Press, 1961, pp 462–466.
89. JW Emsley, J Feeney, LH Sutcliffe. *High Resolution Nuclear Magnetic Resonance Spectroscopy*. Oxford: Pergamon Press, v. I, pp. 456–474.
90. KV Pogorelyi, VV Turov, LA Mironova, AA Chuiko. *Dokl AN USSR N4*:128–131, 1991.
91. FA Momany, LM Carruthers, RF McGuire, HA Scheraga. *J Phys Chem 78*:1595–1620, 1974.
92. T Morimoto, K Miura. *Langmuir 1*:658–662, 1985.
93. GA Karpenko, VV Turov, NI Kovtyukhova, EI Bakay, AA Chuiko. *Teor Eksp Khim 26*:102–106, 1990.
94. GA Karpenko, VV Turov, AA Chuiko, in *Extended Abstracts of International Carbon 90 Conference, Paris 16–20 July, 1990*, pp 676–677.
95. G Hennig. *Progr Anorg Chem 1*:125–205, 1959.
96. D Davidov, H Selig, in *Intercalation in Layered Materials*. Proc. 10th Course Erise Summer Ach/July 5–15, 1986, London, pp. 433–456.
97. H Estrade-Szwarckopf. *Helvetica Physica Acta 58*:139–161, 1985.
98. D Kang, K Nomura, K Kume. *Synthetic Metals 12*:395, 1985.
99. MF Quinton, L Facchini, F Beguin, AP Legrand. *Rev Chim Min 19*:407–419, 1982.
100. AL Blumenfeld, EI Fedin, YuV Isaev, YuN Novikov. *Synthetic Metals 6*:15, 1983.
101. MF Quinton, AP Legrand, L. Facchini, F Beguin. *Synthetic Metals 23*:271–276 1988.
102. MF Quinton, F Beguin, AP Legrand, in *Intercalation in Layered Materials*. Proc. 10th Course Erise Summer Ach/July 5–15, 1986, London, pp 457–459.
103. C Bockel, A Thomy. *Carbon 19*:142, 1981.
104. TN Burushkina, VG Aleinikov, NA Kislitsin. *Adsorbtsiya i Adsorbenty 7*:15–23, 1979.

105. VV Turov, VI Kolychev, EA Bakay, TN Burushkina, AA Chuiko, *Teor Eksp Khim* 26:111–115, 1990.
106. VV Turov, VI Kolychev, TN Burushkina. *Teor Eksp Khim* 28:80–84, 1992.
107. VV Turov, KV Pogoreliy, VI Kolychev, TN Burushkina. *Ukr Khim Zh* 58:470–475, 1992.
108. MM Dubinin, in *Adsorbtsiya i Poristost'*, 1972, p 127.
109. A Matsumoto, J-X Zhao, K Tsutsumi. *Langmuir* 13:496–501, 1997.
110. K Kaneko. *Colloids Surf A* 109:319–333, 1996.
111. DR Lide, *Handbook of Chemistry and Physics*, 75th ed. Boca Raton: CRC Press, 1995, pp 12–159.
112. VV Turov, KV Pogorely, TN Buruschkina. *React Kinet Catal Lett* 50:279–284, 1993.
113. R Leboda, VV Turov, W Tomaszewski, J Skubiszewska-Zieba. *Carbon* (in press).
114. DR Kinney, IS Chuang, GE Maciel. *J Amer Chem Soc* 115:6786–6794, 1993.
115. VM Gluscko. *Handbook in Thermodynamic Properties of Individual Substances.* Moscow: Nauka, 1:309, 1978.
116. BP Bering, MM Dubinin, VV Serpinsky. *J Colloid Interface Sci* 21:378–393, 1966.
117. MM Dubinin, HF Stoeckli. *J Colloid Interface Sci* 75:34–50, 1980.
118. PL Walker Jr, I Janov. *J Colloid Interface Sci* 28:449, 1968.
119. RSh Vartapetian, AM Voloshchuk, MM Dubinin. *Izv. AN SSSR* N1, 1981, 44–48.
120. T Iiyama, K Nishikawa, T Suzuki, K Kaneko. *23rd Biennal Conference on Carbon*, University Park, PA, USA, American Carbon Society, 18–23, July 1997 v 1, pp 96–97.
121. VV Turov, VI Bogillo, EV Utlenko. *Zh Prikl Spektrosk* 61:106–113, 1994.
122. VV Turov, R Leboda, VI Bogillo, J Skubiszewska-Zieba. *Langmuir* 11:931–935, 1995.
123. NK Bebris, AV Kiselev, YuS Nikitin, II Frolov, LV Tarasov, YaI Yaschin. *Chromatographia* 11:206–211, 1978.
124. H Colin, G Guiochon. *J Chromatogr* 126:43 1976.
125. R Leboda. *Chromatographia* 14:524–528, 1981.
126. R Leboda, A Dabrowski, in *Adsorption on New and Modified Inorganic Sorbents* (A Dabrowski and V A Tertykh, eds.) Ser. Studies in Surface Science and Catalysis, Vol. 99. Amsterdam: Elsevier Science, 1996, pp. 115–146.
127. R Leboda, A Waksmundzki, Z Suprinowicz, M Waksmundzka-Hajnos. *Chem Anal* 21:165–175, 1976.
128. R Leboda. *Polish J Chem* 54:2305–2312, 1980.
129. E Tracz, R Leboda. *J Chromatogr* 34:346–358, 1985.
130. R Leboda. *J Thermal Anal* 32:1435–1448, 1987.
131. VV Turov, R Leboda, VI Bogillo, J Skubiszewska-Zieba. *J Chem Soc Faraday Trans* 93:4047–4053, 1997.
132. RS Drago, NA O'Brayn, GC Vogel. *J Am Chem Soc* 92:3924–3929, 1970.
133. P Attard, JL Parker. *J Phys Chem* 96:5086–5093, 1992.
134. A Delville. *J Phys Chem* 97:9703–9712, 1993.
135. SB Zhu, GW Robinson. *J Chem Phys* 94:1403, 1991.
136. JN Israelachvili, RM Pashley. *Nature* 306:249–250, 1983.
137. JN Israelachvili, PM McGuiggan. *Science* 241:795–800, 1988.

138. JN Israelachvili, H Wennerstrom. *Langmuir* 6:873–876 1990.
139. SJ Marrink, M Berkowitz, HJC Berendsen. *Langmuir* 9:3122–3131, 1993.
140. BV Derjaguin, NV Churaev, VM Muller. *Surface Forces*. New York: Consultants Bureau, 1987, pp 59–96.
141. ML Gee, TW Healy, LR White. *J Colloid Interface Sci 131*:18–23, 1989.
142. RM Pashley, PM Guiggan, BW Ninham, DF Evans. *Science 229*:1088–1089, 1985.
143. SS Dukhin. *Adv Colloid Interface Sci 44*:1–134, 1993.
144. SS Dukhin, J Lyklema. *Langmuir 3*:94–98, 1987.
145. CA Leon Leon, JM Solar, V Calemma, LR Radovic. *Carbon 30*:797–811, 1992.
146. VV Turov, IF Mironiuk. *Colloids Surfaces A 134*:257–263, 1998.
147. VM Gun'ko, VV Turov, VI Zarko, EF Voronin, VA Tishchenko, VV Dudnik, EM Pakhlov, AA Chuiko. *Langmuir 13*:1529–1544, 1997.
148. VI Bogillo, VV Turov, A Voekel. *J Adhesion Sci Technol 12*:1531–1547, 1999.
149. B Charmas. *Studies on the Structure and Energetic Properties of the Carbosil Surface*. Ph.D. thesis, Maria Curie-Sklodowska University, Lublin, 1999.
150. R Leboda, VV Turov, B Charmas, VM Gun'ko. *J Colloid Interface Sci* (in press).
151. JC Davis, KK Deb. Adv. *Magn Resonance 4*:1, 1969.
152. GC Pimentel, AL McClellan. *Ann Rev Phys Chem 22*:3247, 1971.
153. AJ Gordon, RA Ford. *The Chemist's Companion*. New York: John Wiley, 1972, pp 275–283.
154. Yu I Gorlov, VV Brey, AV Samoson, AA Chuiko. *Teor Eksp Khim 24*:235–139, 1988.
155. VV Turov, R Leboda, J Skubiszewska-Zieba. *J Colloid Interface Sci 206*:58–65, 1998.
156. R Leboda, J Skubiszewska-Zieba, VI Bogillo, VV Turov. *Composites Interface 6*: 35–47, 1999.

3

Electrochemical Studies of Phenomena at Active Carbon–Electrolyte Solution Interfaces

Stanisław Biniak

Nicolaus Copernicus University, Toruń, Poland

Andrzej Świątkowski

Military Technical Academy, Warsaw, Poland

Maciej Pakuła

Naval Academy, Gdynia, Poland

I. INTRODUCTION

Active carbon (AC) is the collective name for a group of porous carbons produced
by the carbonization and activation of carbonaceous (organic) materials. These
carbons are prepared in order to exhibit a highly ramified porous structure and
an extensive surface area (typically about 1000 m^2/g). Active carbon is most
commonly applied in the form of powder and granules (both extruded and pel-
letized), though other new types such as fibers, cloth, and felts are also used.
From the point of view of structure, active carbon consists of aromatic sheets
and strips, with gaps of variable molecular dimensions between them; these are
the micropores. During activation, the spaces between the graphitelike crystallites
become cleared of various carbonaceous compounds and disorganized carbon,
and meso- and macroporosity is developed. The highly ramified porous structure
results in a very large adsorptive power. In recent years these topics have been
reviewed several times [1–6].

A. Topic Presentation

Owing to their widespread use in electroanalysis, electrosynthesis, electroadsorption, and electrocatalysis, electrodes prepared from various forms of carbon materials have been extensively investigated, particular attention being paid to the
phenomena occurring on the carbon surface [7,8]. Significant efforts have been
directed towards an understanding of the relationship between the surface structure and electron transfer (ET) reactivity. Several different types of carbon have
been employed in these studies, including pyrographite, glassy carbon, carbon
fibers, and various forms of active carbons. The most important factors affecting

ET reactivity mentioned in the literature are (1) the microstructure (and porosity) of carbon materials, (2) the presence of surface functional groups (mainly oxides) and (3) surface purity (dependent on the electrode's history). If we are to understand the phenomena controlling the reactivity of a carbon surface on a molecular level, knowledge of the chemical structure of the surface is essential. Of all the methods available for studying electrode processes, cyclic voltammetry (CV) is probably the most widely used. It involves the application of a continuously time-varying potential to the working electrode. This causes the electroactive species to be oxidized or reduced on the electrode surface and in solution (faradaic reactions); owing to double-layer charging species may be adsorbed in accordance with the potential and capacitive current. Cyclic voltammetry studies of the electrochemical behavior of carbon materials in electrolyte solution have drawn attention to the close correlation between the shape of the cyclic curves and the chemical structure of the carbon electrode surface.

Cyclic voltammetric methods have been used for investigating surface oxygen compounds present on the surface of unmodified carbon materials, and in some cases, previously oxidized materials (electrochemically, oxygen rf-plasma, air, and steam), such as carbon blacks [9,10], glasslike carbon [11–15], graphite [16,17], carbon fibers [18–21], pyrolytic carbon [22,23] and active carbon [24–28].

As can be seen from these examples, the carbon materials most frequently studied possess a relatively well defined type of surface geometry. Despite their extensive applications in electrochemistry, active carbons, with their highly ramified porous structure, are rarely investigated.

In this chapter, attention is given to the cyclovoltammetric studies of phenomena at the interface between chemically and electrochemically modified active carbon and an aqueous or nonaqueous electrolyte solution. Interactions between selected heavy metal ions and an active carbon surface are also discussed. Before the various structural and CV measurement results are considered, a review, together with some pertinent details, will be given in every section.

B. Specific Properties of Active Carbon in Relation to Other Carbon Materials

A powdered active carbon electrode consists of a continuous matrix of electrically conducting solid that is interspersed with interconnecting voids or pores whose characteristic dimensions are small compared to the size of the electrode. The electrochemical reactions in such electrodes occur predominantly in the pores, which represent the major fraction of the total surface area. The external surface area is relatively small with respect to the area of the pore walls. It is the high interfacial surface area available for electrochemical reaction that provides the major advantage of porous electrodes over smooth electrodes (e.g., glasslike car-

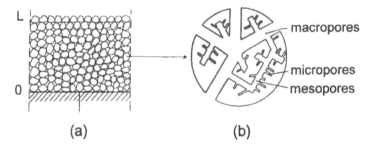

FIG. 1 Schematic representation of the powder active carbon electrode (a) and the porous structure of a carbon granule (b); L, sedimentation bed thickness. (Inspired by Refs. 4 and 29.)

bon, pyrolytic graphite). Figure 1 shows a schematic illustration of the powdered carbon electrode and the porous structure of carbon granules.

When the electrode is completely immersed in the electrolyte solution, only a two-phase interface (i.e., liquid–solid) is present in the electrode structure. In form it may be either a consolidated powdered active carbon or a confined but unconsolidated bed of carbon particles. These are used for flow-through porous electrodes in many electrochemical systems. The other mode of operation is the gas-diffusion electrode, in which the electrode pores contain both the electrolyte solution and a gaseous phase. Numerous publications [29–31] have reported on a theoretical analysis of flow-through porous electrodes and gas-diffusion electrodes, which takes into account the physicochemical characteristics of carbon electrode materials. There does not seem to be a uniform explanation for the effects of structural and chemical heterogeneity in carbons.

II. PROPERTIES OF ACTIVE CARBONS IMPORTANT FOR THEIR ELECTROCHEMICAL BEHAVIOR

Any discussion of the electrochemical behavior of active carbon electrode material must consider relevant information about the carbon/electrolyte interface, where the carbon surface porous structure and chemistry play an important role. Depending on the type of precursor and its preparation, the structure of the carbon skeleton (graphitelike crystallites and a nonorganized phase composed of complex aromatic-aliphatic forms) is significant as regards the electronic properties of active carbon [32–35]. The crystallites are composed of a few (about three) parallel plane layers of graphite, the diameter of which is estimated to be about 2 nm, or about nine times the width of one carbon hexagon [1]. The regular array of carbon bonds on the surface of the crystallites is disrupted during the activation process, yielding very reactive free valences. The crystallites are randomly oriented and extensively interconnected. The structure of an active carbon may be visualized as a stack of poorly developed aromatic sheets (crystallites), distributed and cross-linked in a random manner, separated by disorganized carbonaceous matter and inorganic matter (ash) derived from the raw material. The anisotropic crystallite alignment is associated with the presence of voids.

A. Pore Structure

The extent of activation will determine the porous structure of the final carbon material. During activation the spaces between the crystallites become cleared of less organized carbonaceous matter and, at the same time, part of the carbon is also removed from the crystallites. The resulting channels through the graphitic regions and the interstices between the crystallites of active carbon, together with fissures within and parallel to the graphitic planes, constitute the porous structure, which has a large surface area (usually from 500 to 1500 m^2/g). The pores belong to several size range groups (Fig. 1b). A number of standards for grouping pore size ranges have been used in the past. According to Dubinin [36,37], adsorbent pores can be classified into micropores with a linear size up to $x < 0.6$–0.7 nm, supermicropores where 0.6–$0.7 < x < 1.5$–1.6 nm, mesopores where 1.5–$1.6 < x < 100$–200 nm and macropores where $x > 100$–200 nm. The linear size of a pore is the half-width in the slitlike pore model, and the radius in cylindrical or spherical pores. The classification proposed by Dubinin is based on the difference in the mechanisms of adsorption and capillary condensation phenomena occurring in adsorbent pores. The finest pores, i.e., the micropores, are commensurate with the molecules adsorbed. When the dispersion force fields of the opposite micropore walls are superposed, the adsorption energies in the micropores are greatly increased. They are the most significant for adsorption owing to their very large specific surface area and their large specific volume. At least 90–95% of the total surface area of an active carbon can be made up of micropores. While the curvature of the mesopore surface hardly affects adsorption, mono- and polymolecular adsorption does occur there, which acquires a clear-cut physical significance in that the mesopore volume becomes filled by the capillary condensation mechanism. As a consequence, a hysteresis loop appears on the desorption branch of the isotherm, and its interpretation gives an idea of the distribution of the mesopores in the adsorbent. The supermicropores form the transitional porosity region, above which the characteristic features of the micropores gradually degenerate and the mesopore properties begin to manifest themselves. Finally, macropores remain practically unfilled by capillary condensation because of their relative large width, and hence they act as broad transport arteries in the adsorption process. The limit between mesopores and macropores corresponds to the practical limit of the method for pore-size determination based on the analysis of the hysteresis loop. According to the IUPAC recommendation reported by Sing et al. [38], the pores in active carbon may be classified into three groups:

1. Micropores, the width of which (distance between the walls of a slit-shaped pore) does not exceed 2 nm. A more precise classification would distinguish two types of micropores: narrow (up to 0.7 nm) and wide (from 0.7 to 2.0 nm) [39].
2. Mesopores, the widths of which lie between 2.0 and 50 nm.
3. Macropores, whose width exceeds 50 nm.

This classification, which is not entirely arbitrary, is now widely accepted and implemented.

The porous structure of active carbons can be characterized by various techniques: adsorption of gases (N_2, Ar, Kr, CO_2) [5,39] or vapors (benzene, water) [5,39] by static (volumetric or gravimetric) or dynamic methods [39]; adsorption from liquid solutions of solutes with a limited solubility and of solutes that are completely miscible with the solvent in all proportions [39]; gas chromatography [40]; immersion calorimetry [3,41]; flow microcalorimetry [42]; temperature-programmed desorption [43]; mercury porosimetry [36,41]; transmission electron microscopy (TEM) [44] and scanning electron microscopy (SEM) [44]; small-angle x-ray scattering (SAXS) [44]; x-ray diffraction (XRD) [44].

The principal purpose in characterizing the nature of the porosity is to estimate surface area, pore volumes, pore size distribution (or potential energy distribution), average pore size, pore shape, and the diffusion paths controlling rates of adsorption. The utility of an activated carbon depends in great measure on the specific extent and size distribution of its micropore volume. The most commonly employed methods to characterize these structural aspects of the porosity are based on the interpretation of adsorption isotherms (e.g., N_2 at 77 K). The popularity of these methods has been aided by the availability of good automatic instruments for obtaining the raw data and by developments in adsorption theory leading to better interpretation. The theoretical advances include the work of Horvath and Kawazoe [45]. In recent years improved methods of adsorption isotherm analysis based on the molecular approach [principally, density functional theory (DFT) and molecular simulation] have been suggested [46–51]. DFT pore size distribution analysis offers several advantages over classical methods. It is a quantitatively more accurate method for predicting adsorption behavior by pores of well-defined geometry. Moreover, it is a valid and accurate description for small pores and a description of the full adsorption isotherm (not just the capillary condensation pressure), as well as other properties such as heats of adsorption. It can be used for supercritical conditions and accounts for the effects due to pore shape. Finally, it offers the possibility of systematic improvement, through the use of more sophisticated potential models and more flexible models for pore structure. Generally, authors have improved the reliability of pore size distribution (PSD) determined from nitrogen isotherms at 77 K, especially for pores in the nanometer range. Recently the analysis of PSD was extended by taking into consideration other gases such as argon and carbon dioxide and higher temperatures for the typical active carbons [51].

One method of characterizing surface roughness is the use of surface fractal dimensions, D. For a smooth surface $D \sim 2$ (in the case of carbon black), and with increasing roughness D approaches 3 (for activated charcoals) [52–54]. At intermediate values, $2 < D < 3$, the surface interpolates in a natural way between a plane and a volume. The first experimental study of a fractal surface by multilayer adsorption was published in 1989 [53] and has been followed by numerous investigations since then. An important rationale for investigating fractal surfaces by multilayer adsorption is the wide availability of gas adsorption instruments and the ease with which adsorption isotherms can be measured up to quite high relative pressures. Pfeifer and Liu [54] have shown that the fractal dimension of a pore wall surface, D, is related to pore size distribution and pore radius. The use of fractal dimensions, as applied to active carbons, is still being developed [55–56].

B.   Chemistry of Carbon Surfaces

As was mentioned above, the electrochemical properties of carbon materials are strongly influenced by the presence of heteroatoms on their surfaces. The importance of surface chemistry to the electrochemical behavior of nonporous carbon electrodes is well established and has been reviewed extensively [7,57–62]. The chemical structure of active carbons is peculiarly complicated because the reactivity of carbon atoms with unsatisfied valences at edge sites is greater than that of atoms in the basal planes. Consequently, the chemical properties of the carbon material vary with the relative fraction of edge sites and basal-plane sites on its surface. Because active carbons have a large porosity and numerous disordered spaces, heteroatoms are readily combined on the surface during the manufacturing processes (carbonization, activation, and demineralization). The concentration and type of surface functional groups present on a carbon surface can vary tremendously, and numerous reviews are available that discuss these subjects [1–7,63–66]. Various functional groups containing oxygen, nitrogen, and other heteroatoms have been identified on active carbon surfaces. The most important of these is oxygen. Whereas nitrogen and sulfur originate from a natural or artificial precursor, oxygen can also be taken up during carbon formation or storage. Much more oxygen is chemisorbed on heating carbon materials in an atmosphere containing oxygen (or an oxidizing agent such as NO_x, O_3) or by treatment with oxidizing media such as solutions of HNO_3, $Na_2S_2O_8$, $NaOCl$, or H_2O_2.

Carbons with a low oxygen content show basic properties in aqueous suspensions; they have a positive surface charge and anion exchange (or proton consumption) behavior. The basic properties are ascribed to the presence of basic surface oxides, but it has been shown that the π-electron systems of the basal planes of carbon are sufficiently basic to bind protons from aqueous solutions of acids [67–71]. Even though this basic character has been associated with the presence of some oxygen-containing surface species at the edges of carbon crystallites (e.g., pyrone- and chromene-type structures) [72–76], the main contribution to surface basicity is often from oxygen-free Lewis base sites on the basal planes, i.e., within the graphene layers making up the carbon crystallites. Most of the treatments suggested for obtaining active carbon with basic surface properties consist in heating material in different (inert or reducing) atmospheres or under vacuum in order to remove the oxygen-containing surface groups [77–82]. Thus, for example, carbon subjected to high-temperature heat treatment [above 730°C ($\sim$1000 K)] in an inert atmosphere and subsequently exposed to air below 230°C ($\sim$500 K) is known as H-carbon [83]. When such carbon is exposed to dry oxygen after cooling to room temperature, some oxygen is chemisorbed. After immersing this carbon in aqueous acid, a further quantity of oxygen is consumed, and approximately one equivalent of acid per chemisorbed oxygen atom is bound at the same time [84]. Some hydrogen peroxide is formed during this chemisorption process, but the carbon surface catalyzes its decomposition and thus it breaks down rapidly [85]. Using pH-metric titration, two types of proton-binding centers were found on a carbon film [86]; one corresponded to a base with mean basicity constant $pK_b = 6.6$, while the second site was a very weak base ($pK_b > 11$). Water is a sufficiently strong acid, so elevation of the alkaline pH of the carbon dispersion in pure water and neutral electrolyte aqueous solutions was observed [87]. Treatment with ammonia, performed typically at 430–930°C (700–1200 K), removes the surface oxides but may also introduce basic nitrogen-containing groups (e.g., amine) onto the carbon surface [88–90]. The ether-type oxygen (e.g., from g-pyrone-like structures) can easily be replaced by nitrogen in the reaction with ammonia giving pyridine- or acridine-type nitrogen [89].

The acidic surface properties of active carbons are due to the presence of acidic surface functional groups. Such materials exhibit cation exchange abilities and are known as L-carbons. The acidic surface oxides have been the subject of many studies summarized in several reviews [63–66,91–94]. There are numerous methods of qualitatively and quantitatively determining surface functional groups [95–107], and attempts have been made to study the surface groups by spectroscopic methods, especially by infrared (IR) [91,100–103] and x-ray photoelectron spectroscopy (XPS, ESCA) [102]. Figure 2 presents several IR-active functional groups that may be found at the edges of and within graphene layers after the oxidative treatment of active carbon [101]. Apart from C≡C aromatic (a) moieties, a number of oxygen-containing functional groups can be identified (b–n). Chemisorbed CO_2 may exist as carboxyl-carbonate structures (b) and (c). Carboxyl groups can be isolated (d) or may yield 5- or 6-membered ring carboxylic

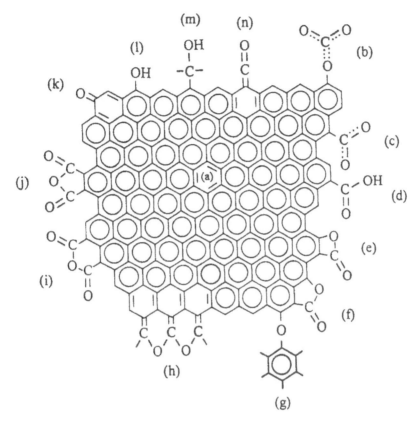

FIG. 2 IR-active functionalities on carbon surfaces: (a) aromatic C=C stretching; (b) and (c) carboxyl-carbonates; (d) carboxylic acid; (e) lactone (4-membered ring); (f) lactone (5-membered ring); (g) ether bridge; (h) cyclic ethers; (i) cyclic anhydride (5-membered ring); (j) cyclic anhydride (6-membered ring); (k) quinone; (l) phenol; (m) alcohol; and (n) ketene. (From Ref. 101.)

anhydrides (i and j, respectively) if they are close together. In close proximity to hydroxyl groups, carboxyl groups may condense to 4- or 5-membered ring lactones (e and f respectively). Single hydroxyl groups would be of a phenolic (l) or alcoholic (m) nature, depending on the character (aromatic or aliphatic) of the carbon atom substituted. The existence of carbonyl groups is more plausible; they could appear either isolated or arranged in quinone-like systems (k). Additionally, the presence of other IR-active species such as ketenes (n) was suggested [101]. Usually the active carbons contain more oxygen (detected by elemental analysis) than can be explained by the quantitative determination of surface oxides [104]. This oxygen is usually ascribed to xanthene- or ether-type species (g and h), which are very difficult to detect.

The active carbon surface groups containing carboxylic or phenolic moieties may react more or less weakly acidically. Evidence for their existence has been found by standard acid–base titration methods, supplemented by classical chemical detection methods such as esterification and differentiation of the methyl esters and ethers by their variable resistance towards hydrolysis. The content of active hydrogen (by methylation in nonaqueous solvents) was usually found to be low compared to that of carboxylic and phenolic groups deduced from neutralization adsorption behavior [103]. Obviously, the amounts of detected groups depend strongly on the degree of surface hydration (hydrolysis of anhydride–lactone surface systems) of the active carbon sample [103]. The individual functionalities, such as carboxyl or phenolic groups, will exhibit a spread of their dissociation constants, depending on the structural environment (e.g., neighboring groups, size of graphene layers). Fortunately, the acidity constants of these groups differ over several orders of magnitude, and it has been established that the various types of groups can be distinguished by their neutralization behavior in aqueous media. L-type active carbons immersed in pure water or in neutral electrolyte aqueous solution are acidic, which indicates that carboxyl groups have dissociated to carboxylates, with the counter ions dispersed in the diffuse double layer. The problem with direct potentiometric titration (as ion-exchangers) is that ion exchange equilibria are established only very slowly, particularly at higher pH values [105]. Hence, the titration curve does not show discernible end points that could be used to discriminate between various types of functional groups. Some authors [105–109] have suggested that, rather than showing the inapplicability of the method, the smoothness of titration curves illustrates a continuous distribution of acidic strength.

The classical method proposed by Boehm et al. [63] assumed that different categories of acidic groups, classified as free (isolated) carboxylic acids, lactones, or phenols, are neutralized (titrated) by appropriate inorganic bases. The most convenient way of determining free carboxyl groups is to perform a neutralization adsorption experiment with 0.05 M $NaHCO_3$ while carboxyl groups in lactone-like bonds are neutralized with 0.025 M Na_2CO_3 solution. The phenolic hydroxyl groups on the carbon surface react with strong alkali (e.g., NaOH) in the same way as free phenols. A recent experiment employing fully automatic, very slow potentiometric titration has yielded titration curves overlapping quantitatively with Boehm's titration data [106].

If a base even stronger than NaOH, sodium ethoxide C_2H_5ONa in ethanol, is used, additional groups are detected. It has been shown [63] that equivalent quantities of Na^+ and $C_2H_5O^-$ are bound by carbon in this reaction. This was explained as follows. Not only is the acidic group neutralized (equivalent to NaOH consumption), sodium salts of hemiacetals from surface carbonyl groups are also formed. Pairs of carbonyl groups arranged at the periphery of the graphene layers in such a way that a system of conjugated double bonds can formally be drawn in will behave in the same way as a quinone group.

The oxides on an active carbon surface provide hydrophilic sites, and water adsorption is greatly enhanced after oxidative modification of carbon materials. It has been reported that active hydrogen-bearing groups are primary adsorption sites for water molecules in a 1:1 ratio [110]. Well-oxidized carbons become hydrophilic and are easily dispersed in water, whereas nonmodified carbons exhibit hydrophobic properties [111]. Thus the oxygen surface complex on active carbon can also be characterized by a combination of different techniques such as water adsorption analyzed by the Dubinin–Astakhov method [112] or immersion calorimetry [113,114]. Heats of immersion of active carbons into water were reported by Puri et al. [98] to vary linearly with respect to combined oxygen evolved as carbon dioxide; this linear relationship was not so obvious with respect to total oxygen. The measurements of the heat of immersion into water and aqueous solutions of NaOH and HCl give the respective enthalpies of hydration and neutralization of the functional groups. A good correlation has been found for the calorimetric technique with neutralization methods [113].

The various functional groups differ in thermal stability. They decompose to give oxides of carbon [64,92]. The evolution of CO_2 is associated with carboxylic groups and their derivatives, such as lactones and anhydrides, while CO evolution is attributed to the decomposition of carbonyl and hydroxyl groups. Carboxyl-carbonate and isolated carboxyl groups are least stable and begin to decompose above 300°C (570 K), yielding mostly CO_2 (the first CO_2 peak in TPD—temperature programmed desorption). At the higher temperature, the CO_2 peak usually shows a shoulder at about 550°C (820 K) and a maximum at about 660°C (930 K). The TPD of the composite peak shows two maxima at around 630°C (900 K) and 800°C (1070 K). There have been many attempts to ascribe CO_2 and CO formation during vacuum pyrolysis to definite functional groups [80,81,115–117]. Unfortunately, the probability of secondary reactions of the evolved gases is very substantially enhanced with porous carbon. Especially in micropores, CO_2 could react with the carbon surface to give CO, and at lower temperatures CO could well react with surface oxygen complexes to yield CO_2 [116]. Recently [117], the intermittent TPD (ITPD) method was applied to the decomposition of oxygen groups (into CO_2) naturally present at the surface of the microporous active carbon. Decomposition of these groups occurs in at least 7 distinct stages. More recently [118], the amounts of different functional groups on modified active carbons were estimated by deconvolution of TPD spectra—the results obtained agree well with the qualitative observation by infrared spectroscopy.

In spite of the well-known experimental difficulties (energy absorption, scattering effect) that occur in IR studies of carbon materials [91] and the ambiguous interpretation of the resulting spectral features, the technique has played a significant role over the last three decades in elucidating the heterogeneous surface composition of different types of carbons [91,100–103,117–121]. Some IR transmittance spectra (halide pellet or suspension techniques) with very finely dispersed carbons, as well as spectra obtained with carbon films [86,98,91, 99,100,103], have confirmed the results of the chemical methods [89,118,119]. The IRS (internal reflectance spectroscopy) technique enables spectra to be obtained directly from the pure surface of a sample [102], but these spectra are relatively poor and their applications are infrequent. Fourier transform infrared spectroscopy (FTIR) offers a considerable advantage over conventional methods. The use of an interferometer allows much more energy to reach the detector and improves the quality of the spectrum recorded for highly absorbing materials such as carbons. The digital form and high accuracy of spectral data make FTIR suitable for computerized manipulation and routine measurements [89,90,120–122].

High quality IR spectra of different carbon surfaces were obtained by photothermal beam deflection spectroscopy (IR-PBDS) [123,124]. This technique was developed with the intention of providing an IR technique that could be used to study the surface properties of materials that are difficult or impossible to examine by conventional means. Recently, diffuse reflectance infrared Fourier transform spectroscopy (DRIFTS) has been successfully applied to study the effect of different pretreatments on the surface functional groups of carbon materials [101,125–128]. Several studies aiming to improve the characterization of the carbon electrode surface and the electrode–electrolyte interface have been carried out using various in situ IR techniques [14,128–132]. The development of in situ spectroelectrochemical methods has made it possible to detect changes in the surface oxides in electrolyte solutions during electrochemical actions.

X-ray photoelectron spectroscopy (XPS) is a suitable method for supplying detailed information on the surface chemistry of carbons in general [17,20,24,102,133,134] and active carbon [89,117,135,136]. By using XPS, it is possible to measure directly the concentration of carbon, oxygen, and other elements on or near the surface layer. In addition, it is possible to detect the different classes of surface functionalities in the outermost atomic layers of powdered active carbon. XPS data can supplement the results obtained from chemical titration techniques or other analytical methods [63,89]. Typical XPS survey scans for a nonmodified active carbon sample contain the peaks for carbon (C 1s) and oxygen (O 1s). The abundance of surface oxygen is, however, quite different in the samples with a large BET surface area. The absolute binding energies (BE) of the C 1s lines are at 284.8 eV and do not provide any information about the sample. The presence of chemisorbed electronegative elements gives rise to small component peaks on the high BE side. Peak positions shifted by 1.6, 3.0, and 4.2 eV from the main peak are ascribed to C—O (ether and hydroxyl), C=O (carbonyl), and COO (carboxyl) moieties, respectively. The position of the O 1s signal is influenced by the type of bonding of the oxygen. Two main signals, usually near 533 and 531 eV, are associated with C—O and C=O moieties, respectively. Unfortunately, the oxygen species are primarily present at the gas–solid interface; their quantitative analysis is hindered by the porous structure of the active carbon samples. Additionally, the differences in chemical structure within active carbon granules were noted for unmodified and oxidized carbon substrates [137].

C. Electrical Properties

The electrical properties of active carbon (e.g., conductivity, thermoelectric power, work function) are directly related to the material structure. Precursor materials usually containing only σ-bonds between carbon atoms in the sp^3 state are generally insulators (conductivity less than 10^{-13} $\Omega^{-1}m^{-1}$). When π-bonds associated with groups of carbon atoms in the sp^2 state are present, electrons are delocalized and are available as charge carriers. The increasing proportion of conjugated carbon in the sp^2 state during precursor carbonization progressively changes the material from an insulator to a good conductor. The Mrozowski model [138] presents an interpretation of the changes occurring in chars (cokes) from the temperature of formation up to graphitization. The removal of hydrogen and low-molecular-weight hydrocarbons between 500 and 1200°C (770–1470 K) gives unpaired σ-electrons on the peripheries of the condensed ring systems. The π-electrons can jump from the π-band into the σ-state, forming a spin pair [138]. This removes an electron from the π-band and creates a hole in the filled band, leading to p-type conductivity. The large number of holes thereby created accounts for the great increase in electrical conductivity.

Active carbon is a composite of amorphous and microcrystalline substructures and exhibits semiconducting properties. An interpretation of the mechanism of electronic conduction in active carbon is more complicated than that for crystalline (graphitized) carbon. The conduction processes are very different because the current carriers in active carbon are assumed to be localized by disorder, which introduces randomness into the potential-energy bands for electrons.

The electrical resistance of a packed bed of carbon particles is a function of the contact resistance between the particles (major fraction) and the intraparticle resistance. A reasonable comparison of the electrical resistivities of carbon powders measured by different investigators is often difficult because the resistivity varies with the pressure that is applied to compact the powder particles. The pressure increase leads to a resistivity decrease as a result of the lower contact resistance between carbon particles [139].

The surface oxygen content also affects the resistivity of the carbon particles. Thus active carbon oxidation, as in other powdered carbon materials, enhances, whereas heat treatment in an oxygen-free atmosphere diminishes electrical resistivity [140,141]. Intuitively, this may be explained by surface oxygen groups at the edges of the microcrystallites, which are realized as flat graphitelike elements. The smaller the oxygen content, the lower is the barrier for electrons from such a microcrystalline element to the next [140]. The resistance enhancement of oxidized activated carbon cloth samples is evidently the result of formation of surface oxygen groups during oxidative treatment [133]. Recently, correspondence with electronic energy band models of the electronic properties of active carbon was obtained by constructing electronic levels from the energy data [142,143]. The dependence of the conductivity and of the thermoelectric power on the temperature shows that the mechanism of the electronic conduction resembles neither that of metals nor that of semiconductors. It is characterized as "hopping" between electronically coherent domains and resembles the process in conducting polymers [143].

According to the conduction band model [138], electrons in a carbon material occupy a continuous range of free energies. The electron that is easiest to remove is at the top of the conduction band. This energy is termed the Fermi level energy and is approximately equal (and opposite in sign) to the work function, the amount of energy required to remove an electron from the bulk material to a vacuum. The quantity that is measured is the contact potential difference (CPD) between the carbon sample surface and a gold reference surface in a vacuum or strictly described atmosphere (Volta potential). This relative measurement can be achieved by various methods [67,111,144–147]. Generally speaking, the work function is strongly dependent on the surface composition—a monotonic change in the Volta potential with a change in oxygen content in a series of carbon blacks [67] and active carbons [111] was reported. The work function is correlated with the aqueous solution properties of active carbon as slurry pH, acid–base uptake, redox ability and double layer capacity [67,111,145–147].

D. Surface Phenomena in Electrolyte Solutions

When a powdered active carbon electrode is immersed for the first time into an electrolyte solution, it exhibits large electrochemical capacities, about 100 F/g and more, due to the huge pore surface and electrolyte ion adsorption [7,8,67,148–151]. The ratio of the number of ions adsorbed to the total number of ions present in the solutions, in the case of high surface electrodes, is several orders of magnitude higher than that for ordinary electrodes. Since a great many of the pores are very narrow, this double layer capacity cannot be predicted easily. However, the relative capacities and the BET surfaces give results of about 0.1 to 0.3 F/m^2, which is a reasonable range for the ordinary double layer capacity [148]. The tendency to lower values in the case of active carbon compared with a graphite electrode may be attributed to the fact that a considerable number of the pores in active carbon have radii of the order of or less than the Debye length for the ionic distribution into the solution, $1/\kappa$ (where κ is the Debye–Hückel parameter), which is a measure of the thickness of the Gouy–Chapman layer [148]. The small pore diameters will not only influence the formation of the double layer but also hamper the mobility of ions in accessible pores, as has already been discussed by Soffer and coworkers [151,152]. It is generally assumed that the total area of the internal structure of the porous carbon electrode is completely wetted by electrolyte. This may not be the case with high-surface-area carbons containing micropores that are not accessible to electrolyte. Besides the usual electrolytic double layer capacity, chemical surface groups with redox behavior at the inner surface of the pores would be capable of storing charges; this has been discussed in the literature [70,153–156]. Cyclic voltammetry studies indicate that the double layer capacitance of oxidized carbon changes relative to carbon samples either not modified or heat-treated in an inert atmosphere. This point needs to be explored further, because the results obtained in various works for different carbon materials are divergent and difficult to explain and reconcile [9,152,157,158].

A porous carbon electrode immersed in an electrolyte solution under conditions such that no charge transfer is allowed across its interface comes to the state where its electrical double layer has zero electronic charge [8,78,157]. The immersion potential (IP) measured versus a reference electrode in open-circuit mode should therefore correspond to the potential of zero charge (PZC) [159]. The pH point of zero charge varies in response to the net total (external and internal) surface charge of carbon particles. The external surface charges of carbon particles immersed in electrolyte solution is characterized by the zeta potential (ζ), representing approximately the electric potential at the beginning of the diffuse part of the double layer. In the presence of a large quantity of inert electrolyte, the value of ζ drops to zero—the electrode potential is called the isoelectric point (IEP). This is, in general, not equal to the point of zero charge, as the values of the latter are affected by the presence of specifically adsorbed species. Hence the difference PZC minus IEP can be interpreted as a measure of surface charge distribution of porous carbons [8,79]. The point of zero charge can be determined by mass titration, the isoelectric point by a measure of the electrophoretic mobility of carbon particles versus the pH of the solution. Therefore the combination of two relatively routine techniques, electrophoresis and mass titration, is an attractive tool for characterizing the surface chemistry of active carbons [79].

Considerable attention has also been paid to the potentiometric response of powdered active carbon electrodes, which in considerable part depends on the type and concentration of functional groups on the surface [7,70,160,161]. The response of a carbon electrode to ionic species in aqueous solution arises from the adsorption behavior of surface functional groups. In addition, physically and/or chemically adsorbed gases (mainly CO_2 and oxygen) affect this process significantly.

Therefore the mechanism of interaction of the active carbon surface and the electrolyte solution ions has not been unequivocally explained. When the carbon electrode potential (E) in an aqueous electrolyte solution is measured, there is a constant growth of potential with time [161]. According to the early suggestion of Frumkin et al. [162–164], this is due to electroreduction of the adsorbed oxygen. They explain the bonding of ions by the active carbon surface by their electrochemical adsorption. Garten and Weiss suggest a mechanism of electrolyte adsorption based on the assumption that surface chromenelike groups occur on the carbon surface [165]. According to this model, chromene-type groups or free radical sites were assumed to generate resonance-stabilized carbonium ions by oxidation–reduction reactions in electrolyte solution with oxygen and proton consumption. This also explained the carbon basicity. Later studies indicated that basic carbons containing some oxygen were capable of simultaneously adsorbing equivalent amounts of oxygen and acid from solution [84]. The process was described in terms of pyrone-type structures, even though small amounts of hydrogen peroxide were detected. More recent studies point out that oxygen-free carbon sites (π-electron-rich regions) can adsorb protons from solution with enough strength to render the surface positively charged [69,71].

The oxygen-containing surface groups are responsible for the potential response of carbon electrodes to hydrogen ion concentration in solution. Potential measurements of carbon materials in buffer solutions show that the relationship $E = f(\text{pH})$ is linear in the pH range 2.0–7.0 and can be described by a linear equation with a slope between 20 and 60 mV/pH, depending on the nature of the electrode material [160,161].

III. PHYSICOCHEMICAL PROPERTIES OF ACTIVE CARBONS USED FOR ELECTRODE PREPARATION

A. Materials Presentation

The four kinds of active carbons used in our laboratory were obtained from different commercial sources and in accordance with supplier information were produced from various precursors. Ash was removed from the raw carbon using concentrated hydrofluoric and hydrochloric acids by Korver's method [166]. Carbon samples were subjected to surface modification by oxidation in air or with concentrated HNO_3, annealing in a vacuum or an ammonia atmosphere. Afterwards, all carbon samples were desorbed under vacuum (10^{-2} Pa) at 150°C (423 K) for 3 h. The procedure used for carbon purification ensured the removal of nitric acid and nitrogen oxides (after nitric acid oxidative modification) or physically sorbed NH_3 (after heat treatment in ammonia). The samples thus prepared were handled in ambient air, but in order to minimize weathering effects, all further measurements were carried out on freshly dried samples. The names and sources of commercial carbons as well as the conditions of surface modifications are listed in Table 1.

B. Porosity

The main porous structure characteristics (Table 2) were determined on the basis of benzene vapor adsorption isotherms using McBain–Baker sorption balances at 20°C (293 K), i.e., the specific BET surface area (S_{BET}) [39], the surface area of mesopores (S_{me}), and the parameters of the Dubinin–Radushkevich equation (the volumes of the micropores and supermicropores, W_{01} and W_{02}, and the characteristic energies of adsorption, E_{01} and E_{02}) [36,37]. In addition, the total micropore volume ($\sum W_0$) and geometric micropore surface area (S_g) [168] were calcu-

TABLE 1 Active Carbons and Methods of Modifying Them

Name of commercial active carbon and supplier	Symbol of carbon sample	Methods of modification
CWZ	CWZ—NM	Deashing with conc. HF and HCl
HPSDD, Hajnówka, Poland	CWZ—Ox	NM sample oxidation with conc. HNO_3 at 80°C (353 K), 4 h, desorbed at 150°C (423 K)
CWN2	CWN2—NM	Deashing with conc. HF and HCl
ZEW, Racibórz, Poland	CWN2—Ox	oxidation of NM sample in air stream at 300°C (673 K), 6 h (in fluid-bed process)
RKD3	RKD3—NM	Deashing with conc. HF and HCl
Norit, The Netherlands	RKD3—Ox	NM sample oxidation with conc. HNO_3 at 80°C (353 K), 5 h, heated in H_2 at 175°C (448 K)
D43/1	D—H	Previously deashed sample heat-treated under vacuum (10^{-2} Pa) at 830°C (1000 K) for 3 h
Carbo-Tech, Essen, Germany	D—Ox	Deashed sample oxidized with conc. HNO_3 at 80°C (353 K), 4 h, desorbed at 150°C (423 K)
	D—N	Deashed sample heat-treated in a stream of NH_3 for 2 h at 900°C (1170 K)

Source: Refs. 27, 28, 89, and 167.

TABLE 2 Surface Porosity Characteristics of the Active Carbons Investigated

Carbon sample	S_{BET}, m²/g	S_{me}, m²/g	W_{01}, cm³/g	E_{01}, kJ/mol	W_{02}, cm³/g	E_{02}, kJ/mol	ΣW_0, cm³/g	S_g, m²/g
CWZ—NM	630	122	0.232	26.3	—	—	—	510
CWZ—Ox	455	112	0.142	27.9	—	—	—	330
CWN2—NM	845	135	0.315	23.4	—	—	—	615
CWN2—Ox	1100	175	0.284	24.5	0.149	15.5	0.433	770
RKD3—NM	970	67	0.265	22.1	0.177	13.6	0.442	695
RKD3—Ox	805	51	0.252	24.7	0.147	14.1	0.399	685
D—H	985	29	0.232	24.8	0.214	15.2	0.446	750
D—Ox	970	30	0.269	26.1	0.116	15.1	0.385	730
D—N	985	30	0.226	28.4	0.208	12.8	0.434	755

Source: Partly Refs. 27, 28, and 167.

lated; these magnitudes are also given in Table 2. The authors realize that the BET surface area does not strictly reflect the real surface of an active carbon with a high micropore content. However, this inaccuracy notwithstanding, it is generally used in the literature as a useful reference point [39]. The estimated pore volumes of the carbons indicate that they possess a well-developed pore structure. The surface oxidative modification procedures used in our work caused changes in the porosity dependent on the kind of active carbon tested. It is known that nitric acid treatment often causes the specific surface area and micropore volume to decrease [87,96,119,169]. We observed this effect in all carbon samples after HNO$_3$ oxidation. After oxidation, the total micropore volume decreases, which indicates that oxygen-containing groups could render large parts of micropores inaccessible to adsorbate molecules. By contrast, air (oxygen) treatment at 400°C (673 K) leads to an increase in S_{BET} as a result of the additional burn-off of carbon (CWN2). The total pore volumes and pore areas are much the same for D—H and D—N carbons, but are different for D—Ox carbon.

C. Surface Chemistry

The surface chemical properties of the carbon materials were characterized as follows: measurement of pH of carbon slurries (in 0.1 M NaCl solution) [89]; neutralization with bases of different strength and dilute HCl according to Boehm's method [63,66]; determination of total oxygen/nitrogen content by elemental analysis (with an accuracy of 0.2%) [170]; mass loss of carbon samples after heat treatment in a vacuum. Additionally, the number of primary adsorption centers (a_0) was determined from water vapor adsorption isotherms according to the Dubinin–Serpinsky method [171], as was the heat of immersion in water for selected samples [111,172]. The results of these operations are presented in Table 3. For all samples transmission Fourier Transform Infrared (FTIR) spectra and X-ray photoelectron spectra (XPS) were recorded.

1. Functional Groups

A number of the oxygen-containing functional groups formed during the oxidative modification were determined by neutralization with bases of increasing strength and calculated according to Boehm's method [63,66]. Oxidation with nitric acid or dioxygen caused large amounts of oxygen to become fixed on the carbon surface (Table 3); part of the surface oxygen existed in the form of well-defined functional groups. At the same time, the number of basic groups (dilute inorganic acid consumption) and the pH of the carbon slurry in NaCl solution decreased markedly. According to Boehm, only the strongly acidic carboxylic groups can be neutralized by NaHCO$_3$, whereas those neutralized by Na$_2$CO$_3$ are believed to be lactones, though more likely they are lactols [66]. The NaOH additionally neutralizes weakly acidic hydroxylic (phenolic) groups. These stable

TABLE 3 Summary of the Chemical Properties of the Active Carbons Investigated

Carbon sample	pH[a]	Base/acid consumption, meqv/g					Total O/N, %	Mass loss,[b] wt%	a_0, mmol/g	H_{um}, J/g
		HCO_3^-	CO_3^{2-}	OH^-	$C_2H_5O^-$	H_3O^+				
CWZ—NM	—	0.09	0.10	0.33	0.51	0.24	2.9	4.2	1.39	—
CWZ—Ox	—	0.84	1.29	2.00	2.10	0.02	9.7	12.9	3.85	—
CWN2—NM	—	0.04	0.07	0.18	0.21	0.20	2.6	4.3	0.98	—
CWN2—Ox	—	0.90	1.19	1.97	2.00	0.09	6.2	11.0	3.61	—
RKD3—NM	4.65	0.26	0.48	0.59	0.78	0.25	2.5	5.1	1.12	38.2
RKD3—Ox	3.30	0.73	1.06	1.63	1.84	0.12	10.3	16.0	3.73	86.1
D—H	10.01	0.00	0.01	0.13	0.22	0.42	0.6	3	0.38	—
D—Ox	3.08	0.72	1.10	1.66	2.05	0.13	10.8	18	3.94	93.4
D—N	10.36	0.00	0.03	0.10	0.32	0.63	0.4/1.9	4	0.94	66.2

[a] 1 g AC in 100 cm^3 0.1 M NaCl (pH = 6.68).
[b] Heat-treated in vacuum at increased temperature until 1400°C (from DTG experiment); mass loss mainly as CO and CO_2.
Source: Partly from Refs. 27 and 89.

protogenic groups are responsible for the ion-exchange capacity of the oxidized carbon. Using a still stronger base, sodium ethoxide $NaOC_2H_5$ in ethanol, carbonyl groups (neutral in aqueous solutions) are additionally detected. The amounts of hypothetical surface functionalities in relation to unit surface area and to mass unit are shown in Table 4. These ways of identifying functional groups on active carbon surfaces are still recorded in the literature [173,174] and provide a good characterization of carbon samples prepared under different conditions. Changes in the concentrations of surface functional groups after oxidative modification depend on the kind of active carbon involved. The highest increases were observed for carboxyls (from 2.5 to 20 times depending on its concentration in the nonmodified material), while for lactones and hydroxyls the increase was much smaller. For carbonyl groups, insignificant changes in surface concentration were noticed in all cases; however, the acid consumption (basicity) decreased by two to ten times after carbon sample oxidation (Table 4).

Acid or base strength of an organic compound, such as carbon, is strongly influenced by its chemical environment and above all by the effects that local chemical structure can induce in a functional group [107]. These effects influence polarization of the hydrogen–oxygen, and carbon–oxygen bonds in the surface groups, altering dissociation energies. Hydroxyl groups bonded to aliphatic structural carbon (alcoholic) behave as strong bases in an acidic environment because of the electron-donor inductive effect of the adjacent groups. On the contrary, hydroxyl groups directly bonded to the edges of aromatic layers (phenolic) are much less basic because of the electron-withdrawing resonance effect exerted by the aromatic rings. The ability of these groups to dissociate hydrogen ions via delocalization of the negative charge generated on the carbon surface can lead to keto/enol equilibrium (for single group): $>C(H)-C(H)=O \Leftrightarrow \geq C=C(H)-OH$, or to conjugation of neighboring $C=O/C-OH$ structures [91].

Pairs of carbonyl groups arranged at the periphery of the graphene layers in such a way that a system of conjugated double bonds can be formally drawn in will behave in a similar way to quinone groups. According to some authors [69,75,82], the existence of pyronelike structures incorporated in the carbon matrix is partly responsible for carbon's basicity. As pyrones are very slightly basic ($pK_b \approx 13$), the recorded pH values of carbon suspensions (near 10) indicate the presence of relatively strong basic sites with a pK_b of about 4. These sites may be the result of the adsorption of molecular oxygen and the form of superoxide ions O_2^-, which can act as a strong Brønsted base [70,86,175]. Acid buffering capacity of basic active carbon were proposed to be due to the combination of the redox reactions and proton transfer to/from the surface and the supporting electrolyte [70]. The more pronounced basic properties of the ammonia-treated carbon (D—N) are due to the presence of additional basic sites—probably nitrogen structures incorporated into the carbon matrix [176–179].

TABLE 4 Concentration of Surface Functional Groups on the Carbons Investigated

Carbon sample	Concentration of surface functional groups, $\mu mol/m^2$ (mmol/g)				
	Carboxyl	Lactone	Hydroxyl	Carbonyl	Basic
CWZ—NM	0.14 (0.09)	0.01 (0.01)	0.36 (0.23)	0.29 (0.18)	0.39 (0.24)
CWZ—Ox	1.84 (0.84)	1.01 (0.45)	1.56 (0.71)	0.23 (0.10)	0.04 (0.02)
CWN2—NM	0.05 (0.04)	0.04 (0.03)	0.13 (0.11)	0.04 (0.03)	0.24 (0.20)
CWN2—Ox	0.82 (0.90)	0.26 (0.29)	0.71 (0.78)	0.27 (0.03)	0.08 (0.09)
RKD3—NM	0.27 (0.26)	0.23 (0.22)	0.12 (0.11)	0.20 (0.19)	0.26 (0.25)
RKD3—Ox	0.91 (0.73)	0.42 (0.33)	0.71 (0.57)	0.26 (0.21)	0.15 (0.12)
D—H	0.00 (0.00)	0.01 (0.01)	0.12 (0.12)	0.09 (0.09)	0.42 (0.42)
D—Ox	0.73 (0.72)	0.39 (0.38)	0.58 (0.56)	0.40 (0.39)	0.13 (0.13)
D—N	0.00 (0.00)	0.03 (0.03)	0.07 (0.07)	0.22 (0.22)	0.63 (0.63)

Source: Partly from Ref. 27.

2. Hydrophilic–Hydrophobic Properties

The hydrophilic and hydrophobic properties can be characterized from an analysis of water vapor adsorption isotherms over a range of low and medium relative adsorbate pressures as well as on the basis of heats of immersion of carbon samples in water. The water adsorption isotherms were determined using the volumetric method (microburettes) at 25°C (298 K). Hence, with the aid of the Dubinin–Serpinsky equation [171], the concentration of primary centers for water adsorption (a_0) could be calculated. The values of a_0 for the carbon samples examined (Table 3) were closely dependent on the chemical structure of their surface. An almost threefold increase in a_0 was observed following carbon oxidation. Similar changes in the value of a_0 due to the oxidation of carbon materials have been reported in the literature [111,180,181]. The correlation between the wettability of carbon materials and the activity of porous electrodes has also been investigated [182]. The electrode prepared by using hydrophilic active carbon as a carrier exhibited some degree of activity due to the hydrophilic property of active carbon. When a carbon is immersed in water its surface oxides interact, showing basic, acidic or amphoteric behavior. This process can be complicated by the slow attainment of hydration equilibrium [107] and presence of oxygen entrapped in carbon's micropores [70].

3. FTIR Spectroscopy

FTIR spectra of the carbon samples were obtained using a Perkin-Elmer FTIR Spectrum 2000 spectrometer. The active carbon–KBr mixtures in a ratio of $1:300$ were ground in an agate mortar, desorbed under vacuum (10^{-2} Pa), and finally pressed in a hydraulic press. Before the spectrum of a sample was recorded, the background line was obtained arbitrarily and subtracted. The spectra were recorded from 4000 to 450 cm^{-1} at a scan rate of 0.2 cm/s, and the number of interferograms at a nominal resolution of 4 cm^{-1} was fixed at 25. Figures 3 and 4 present the FTIR spectra obtained for the relevant carbon samples. The measurements applied here (KBr pellet technique) make it impossible to compare quantitatively the FTIR spectra obtained for different carbons, but they do indicate which individual chemical structures may or may not be present in the carbon [90,91,123].

In the spectra of the carbon materials the band of stretching OH vibrations (3600–3100 cm^{-1}) was due to surface hydroxylic groups and chemisorbed water. The asymmetry of this band at lower wave numbers indicates the presence of strong hydrogen bonds. The same shape of O—H stretching vibrations of ammonia-treated carbon (D=N, Fig. 4) indicates that N—H (amine) structures are not formed during ammonia treatment. The presence of absorption bands characteristic of —CH$_3$ and —CH$_2$— structures (2960, 2920 and 2850 cm^{-1}) in all the spectra suggests the existence of some aliphatic species on the carbons.

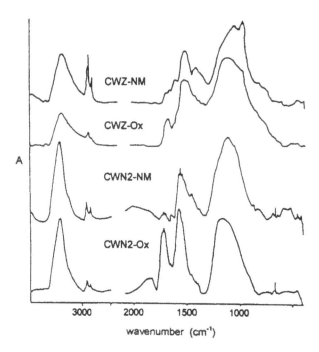

FIG. 3 FTIR spectra of CWZ and CWN2 (NM and Ox) carbons.

Below 2000 cm^{-1}, the spectra for nonmodified carbons (CWZ—NM, CWN2—NM, RKD3—NM) show absorption typical of surface and structural oxygen species (functional groups and inactive heteroatoms incorporated into the solid matrix). The spectra are similar to the reported spectra of other carbons derived from a wide variety of sources [89–91,116–126,183–186]. The presence of bands at 1740 cm^{-1} (i), 1630 cm^{-1} (ii), and 1560 cm^{-1} (iii) can be respectively attributed to stretching vibrations of C=O moieties in (i) carboxylic, ester, lactonic or anhydride groups; (ii) quinone and/or ion-radical structures; (iii) conjugated systems like diketone, keto-esters and keto-enol structures [89–91,116–126,183–187].

Since H$_2$O is sorbed on the surface of active carbons with the participation of both specific (hydrogen bonds, chemisorption due to surface oxide hydration) and nonspecific interactions (physical adsorption), the bands in the 1500–1600 cm^{-1} region can also be described by OH binding vibrations. The complicated nature of the adsorption bands in the 1650–1500 cm^{-1} region suggests that aromatic ring bands and double bond (C=C) vibrations overlap the aforesaid C=O stretching vibration bands and OH binding vibration bands. Another broad band in the 1470–1380 cm^{-1} range consists of a series of overlapping absorption bands

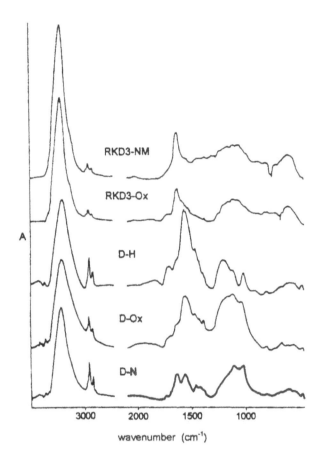

FIG. 4   FTIR spectra of RKD3 (NM and Ox) and D43/1 (D—H, D—Ox, and D—N) carbons.

ascribable to the deformation vibration of surface hydroxyl groups and in-plane vibrations of C—H in various C=C—H structures. The partially resolved peaks forming the absorption band in the 1260–1000 cm^{-1} region can be assigned to ether-like (symmetrical stretching vibrations), epoxide, and phenolic (vibrations at 1180 cm^{-1}) structures existing in different structural environments.

The spectra of the oxidized carbons (CWZ—Ox, CWN2—Ox, RKD3—Ox, D—Ox) are quite similar to those obtained for various carbon materials oxidized with nitric acid or dioxygen [89–91,116–126,173,174]. The relative decrease in intensity of C—H moiety bands (near 2900 cm^{-1}) after oxidation may indicate that oxygen surface species are also formed at the expense of the aliphatic residues in the carbon material. After oxidation the band characteristic of carbonyl moieties in a carboxylic acid (1750–1700 cm^{-1}) increases, and there is simultaneously a considerable increase in band intensity of these carbonyl functional groups in different surroundings. Additionally, a new band characteristic of carbonyl moieties in carboxylic anhydrides appears in the 1850–1800 cm^{-1} region. Oxidation also changes the shape of the overlapping bands in the "fingerprint" region (1400–1000 cm^{-1}), mainly enhancing absorption in the 1250–1000 cm^{-1} range (maximum near 1180 cm^{-1}). This suggests an increase in the number of ether and hydroxylic structures on the carbons investigated.

The spectra of the annealed carbons (D—H, D—N) (Fig. 4) confirm the observation that heat treatment under vacuum or in ammonia diminishes the content of oxygen surface complexes, especially that of strongly acidic surface groups. There the band typical of carboxylic structures disappears and the bands of the remaining surface oxygen complexes are much reduced.

4. XPS Measurements

XP spectra were obtained with an EscaLab 210 (V. G. Scientific Ltd.) photoelectron spectrometer using nonmonochromatized Al K$_\alpha$ radiation (1486.6 eV), the source being operated at 15 kV and 34 mA. Prior to XPS measurement, the powdered carbon samples were dried for 2 h at 100°C (373 K). The vacuum in the analysis chamber was always better than $5 \cdot 10^{-10}$ Pa. The high-resolution scans were performed over the 280–294 and 527–540 eV ranges (C 1s and O 1s spectra, respectively). In order to obtain an acceptable signal-to-noise ratio the spectral region was scanned 20 times. After subtraction of the base line (Shirley-type), curve fitting was performed using the nonlinear least-squares algorithm and assuming a mixed Gaussian/Lorenzian peak shape of variable proportion (mainly 0.3). This peak-fitting was repeated until an acceptable fit was obtained (error 5%). The positions of the deconvoluted peaks (binding energy, BE) were determined from both literature data [24,117,133–136,188] and empirically derived values.

Figure 5 and 6 show the high-resolution XPS spectra in the C 1s and O 1s regions for all the carbon samples in question. There were marked differences between the experimental (dots) and synthesized (continuous) lines in all the spectra. The main peak position range (BE) and the relative peak area (r.p.a.) for separate peaks were estimated and collected in Table 5. The C 1s spectra of a nonmodified carbon sample (Fig. 5: CWZ—NM, CWN2—NM and Fig. 6: RKD3—NM) shows a shoulder on the high-energy side and comprises the main peaks for binding energy (BE) at 284.5 (I), 286.0 (II), 288.0 (III) and 290.0 (IV) eV corresponding to C≡C (graphite), C—O (C—OH, C—O—C), C=O (carbonyl and/or quinone-like), COO (carboxylic acid, lactone, anhydride) surface moieties, respectively [89,135,187–192]. Furthermore, an additional peak

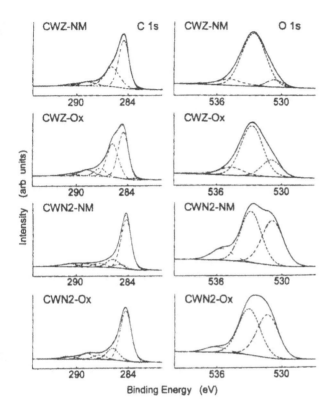

FIG. 5 High-resolution XPS spectra of C 1s and O 1s regions of CWZ and CWN2 (NM and Ox) carbons.

(IIa) at 287.0 eV can be separated for some carbons, which has been assigned to carbon in keto-enolic and/or ion-radical equilibria (C—O in the free-radical semiquinone group). The presence of such stable structures has been observed on the surface of carbon materials [91]. These assignments agree very well with the extensive XPS studies made on commercially available carbon used as catalyst supports [24,135,189]. The O 1s spectrum for the carbon samples displays one main peak (II) corresponding to the C—O (532.0–533.0 eV) moiety in different surface oxygen-containing functional groups. The two other peaks at 530.5 eV and 535.0 can be assigned to C=O species and chemisorbed oxygen and/or water, respectively [89]. Differences in relative peak areas of surface species are partially correlated with the analytically detected surface oxide concentration (Table 4). The number of groups detected from XPS data is only approximate;

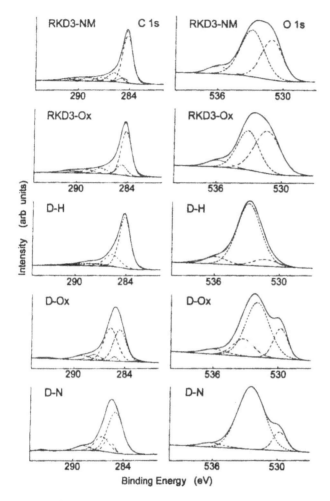

FIG. 6 High-resolution XPS spectra of C 1s and O 1s regions of RKD3 (NM and Ox) and D43/1 (D—H, D—Ox, and D—N) carbons.

the main information of this method is derived from the external part (and also the macropore surface area) of the active carbon material to a depth of ca. 10 monolayers [189,192].

Oxidative modification enhanced the quantity of oxygen (and therefore of surface oxygen-containing groups) more than threefold (Table 3). The magnitude of the total O 1s peak (in arbitrary units) for oxidized carbon samples (CWZ—Ox, CWN2—Ox, RKD3—Ox, D—Ox) changed slightly after oxidation: the shoulder

TABLE 5 Relative Peak Area (r.p.a.) of the Principal C 1s and O 1s Peaks Deduced from High-Resolution XPS Spectra of Active Carbons Investigated

Peak	No	Binding energy range, eV	Relative peak area, %								
			CWZ–NM	CWZ–Ox	CWN2–NM	CWN2–Ox	RKD3–NM	RKD3–Ox	D–H	D–Ox	D–N
C 1s	I	284.2–284.8	59.3	43.7	68.1	62.5	68.3	60.4	74.1	31.5	62.2
	II	285.0–286.5	30.9	34.0	10.40[a]	16.3[a]	14.3[a]	20.3[a]	18.0	47.2[a]	10.1[a]
	IIa	286.7–287.2	—	7.8	4.5	6.9	4.6	8.7	—	—	16.1
	III	288.3–289.0	7.3	8.7	4.9	8.8	4.5	5.2	3.6	9.2	7.0
	IV	290.0–291.0	1.8	4.1	2.5	4.0	2.1	4.5	—	8.6	—
O 1s	I	530.0–531.0	7.8	14.6	42.4	45.5	40.1	51.6	10.9	20.1	14.1
	II	531.8–533.0	81.4	71.7	46.7	42.7	52.7	41.4	77.8	61.0	82.2
	III	533.6–536.0	10.8	13.7	11.0	5.9	7.2	7.0	11.4	16.9	3.5

[a] The superposition of two peaks observed in this BE range.

on the peaks of oxidized samples clearly suggests the presence of larger quantities of oxygen-containing species. The fitted spectrum exhibits a graphitic carbon peak relatively diminished to an extent dependent on the type of carbon material. Likewise, the relative peak intensities of the C—O, C=O and COO moieties increased (Table 5). The O 1s peak of oxidized carbon samples consists of three peaks (as for nonmodified carbon samples): the main one (II) at 532.5 eV was relatively smaller, the next two larger. The considerable intensity of the peak attributed to adsorbed water in the O 1s region for both the nonmodified and the oxidized carbon indicates that their surfaces are highly hydrated. Similar results were obtained from photoemission characterization of various carbon surfaces activated by oxidative treatment [188–192].

Heating under vacuum (D—H carbon sample) leads to a drop in the concentration of surface oxygen and a change in its distribution. There is a relative increase in the number of hydroxyl and ether groups at the expense of the carbonyl groups. The surface oxides appear as a result of the chemisorption of oxygen or water (at room temperature) at the reactive centers formed during heating. They are responsible for the basic character of carbon [86,91]. Annealing in ammonia (D—N carbon sample) also reduces the surface oxide concentration and changes the shape of the XPS C 1s and O 1s signals, which indicates the formation of surface nitrogen moieties [89]. The formation of pyridinic and pyrrolic structures as well as pyridonelike groups is discussed in the literature [89,193]. These functionalities are responsible for the more pronounced basic character of the ammonia-treated carbon. For D—H and D—N carbon samples a smaller O 1s peak ascribed to adsorbed water (at 534.0 eV) is observed, which confirms the hydrophobic nature of the surface of these carbons.

IV. VOLTAMMETRY OF POWDERED ACTIVE CARBON ELECTRODES (PACE)

A. System Description

1. Apparatus

Cyclic voltammetric measurements were performed using the typical three-electrode system and electrochemical cell presented in Figure 7a [194]. A powdered active carbon electrode (PACE), Pt wire, and saturated calomel electrode (SCE) were used as working, counter, and reference electrodes, respectively. All potentials are reported against this reference electrode. Deaerated (Ar or H_2 bubbling) aqueous solutions of Na_2SO_4, $NaNO_3$, H_2SO_4, HNO_3, NaOH, and Britton–Robinson buffer, in concentrations of 0.05 or 0.1 M, were used as electrolytes. Anhydrous acetonitrile or methanol with dissolved $LiClO_4$ (0.1 M) was used for nonaqueous solutions. The chemicals were analytical grade and the water doubly distilled. A salt bridge was used for the nonaqueous solutions. Cyclic voltamme-

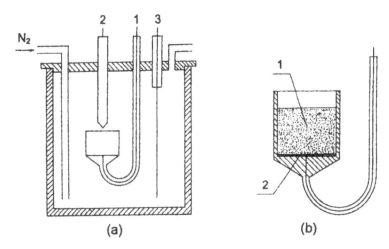

FIG. 7 (a) Overall view of the electrochemical cell: 1, PACE; 2, SCE; 3, Pt reference. (b) Cross-section of the powdered active carbon electrode: 1, carbon powder; 2, Pt contact. (From Refs. 27 and 194.)

tric curves (CVs) were recorded for different sweep amplitudes, after establishment of electrochemical equilibrium (no change in CV curve shape). The sweep parameters used (rates, amplitudes) are marked each time in the figure captions.

2. Electrode Preparation

The working PACE design is illustrated in Figure 7b [194]. A Pt plate (surface area 0.78 cm²) provided electrical contact with the powdered carbon material. After prior vacuum desorption (10^{-2} Pa), the powdered carbon (grain size 0.075–0.060 mm, mass 20–100 mg) was placed in an electrode container and drenched with a deaerated solution to obtain a 3–5 mm sedimentation layer. Although the electrode design ensures that the shape of the CV curves recorded is reproducible, estimation of the specific capacitance is not now possible because the potential and current distributions in the porous and powdered electrode bed are unknown [29–31]. The main difficulty here is the estimation of the electrochemically active part of the external surface area of the powdered electrode material [195–198]. Electrochemical processes occur at the carbon particle surface of the open spaces more often than at the outer planar surface, which results in three-dimensional electrochemical activity rather than the planar responses characteristic of solid electrodes. Obviously, less than the total BET surface area is available for charging in electrolyte solution. However, the extensive surface area within the micropores constitutes a large fraction of the electrode/electrolyte interface and therefore plays a major role in the overall potential distribution within the electrode

[31]. Thus the potential in the macropores is directly affected by transport and charging phenomena in the micropores accessible to electrolyte. These properties of active carbon render it a difficult material to use as an electrode. The large electrochemically active surface area leads to considerable double-layer charging currents, which tend to obscure faradic current features. The network of micropores in the electrode material might be expected to result in a significant ohmic effect, which would further impair the potential resolution (IR drop on electrode material) obtainable by PACE voltammetry. CV curves recorded with different masses (and sediment layer thicknesses) of powdered samples of selected carbons in various electrolyte solutions are presented in Fig. 8 as an example [194]. Where amounts of material were greater than 20 mg, the CVs recorded were of the same shape.

It is expected that practical problems in using PACEs could arise as a result of electrical contacts between the sediment layer and the current collector, and with the mechanical durability of the electrode bed. The comparison of CVs for a Pt electrode and selected PACE (with the same current scale) is shown in Fig. 9 to illustrate the influence of Pt contact on the voltammograms recorded. Similar results were obtained using graphite, glassy carbon, or Au as contact materials

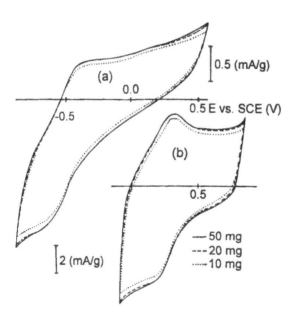

FIG. 8 CV curves with different masses of powdered carbon as the working electrode for samples:(a) CWZ—Ox in 0.05 M Na$_2$SO$_4$(pH = 7.08); (b) CWN2—Ox in 0.05 M H$_2$SO$_4$(pH = 1.03). Sweep rate v = 3.33·10^{-2} V/s. (From Ref. 194.)

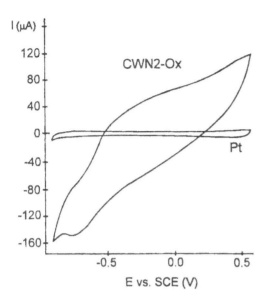

FIG. 9 Comparison of CV curves for clean Pt and PACE as working electrode for CWN2—Ox (20 mg) in the buffer solution (pH = 6.0). Sweep rate $v = 3.33 \cdot 10^{-2}$V/s. (From Ref. 194.)

[194]. In spite of the much smaller background currents for contact materials, these currents have been omitted (without subtraction) in further discussion of the results. The curves usually remained constant after the electrode had been immersed for several hours in the electrolyte solution and after the first few sweeps, indicating that the electrochemically active electrode area remained constant during the potential scans.

B. The Influence of Carbon Surface Modification on the Shape of Cyclic Voltammograms in Aqueous Solution

1. CV Windows (Potentials of O_2 and H_2 Evolution)

Carbon electrodes have an effect on the potentials at which oxygen or hydrogen evolution is observed in aqueous electrolyte solutions. Figure 10 presents a number of cyclic voltammograms recorded in solutions with different pH for selected carbons [194]. The estimated potentials for oxygen and hydrogen evolution at the carbon electrodes in question are set out in Table 6. The evolved gaseous products can destroy the sediment layer of the electrode material: this was observed in the case of hydrogen evolution on carbon particles collected on a hang-

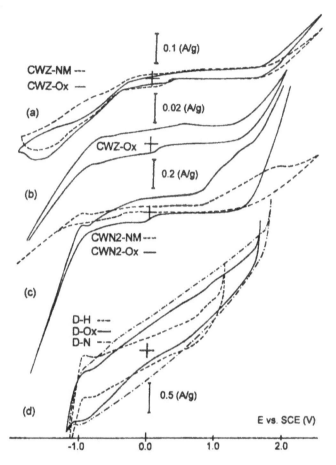

FIG. 10 CV windows for selected systems: (a) CWZ (NM and Ox) in 0.05 M H$_2$SO$_4$;
(b) CWZ—Ox in 0.05 M Na$_2$SO$_4$;(c) CWN2 (NM and Ox) in 0.1 M NaOH; (d) D43/1
(H, Ox, and N) in 0.05 M Na$_2$SO$_4$. Sweep rate $v = 3.33 \cdot 10^{-2}$V/s.

ing mercury drop electrode [199]. Further measurements (besides the study of
oxygen generation) were performed for sweep potential amplitudes and are included in estimated CV windows.

2.   Potential Sweep Rate

At least one anodic or cathodic peak (wave) can be observed in the CV curves
recorded in aqueous electrolyte solution. These peaks (waves) may be due to
oxidation or reduction of surface oxygen functional groups similar to what was
observed for various carbon electrodes [7,9,14,16,24,26,132]. On the basis of

TABLE 6 Potentials of Hydrogen and Oxygen Evolution for Active Carbon Electrodes in Different Electrolytes (V vs. SCE)

Electrolyte active carbon	0.05 M H$_2$SO$_4$		0.05 M Na$_2$SO$_4$		0.1 M NaOH	
	H$_2$	O$_2$	H$_2$	O$_2$	H$_2$	O$_2$
CWZ—NM	−0.40	+1.62	−0.85	+1.25	−1.20	+0.70
CWZ—Ox	−0.41	+1.48	−0.84	+1.22	−1.28	+0.68
CWN2—NM	−0.35	+1.48	−0.76	+1.45	−1.16	+0.80
CWN2—Ox	−0.42	+1.53	−0.94	+1.36	−1.08	+0.76
RKD3—NM	−1.30	+2.10	−1.10	+1.35	−0.98	+0.64
RKD3—Ox	−0.46	+1.81	−1.08	+1.18	−1.26	+0.88
D—H	−0.40	+1.80	−0.85	+1.00	−1.03	+0.70
D—Ox	−0.40	+1.66	−1.00	+1.60	−1.22	+0.42
D—N	−0.41	+1.70	−1.10	+1.76	−1.80	+0.74

Source: Partly from Ref. 27.

these results the assumption that the quinone and hydroquinone groups are responsible for the redox characteristic of carbon electrodes seems reasonable. The experimental evidence indicates that the observed oxidation or reduction peaks (waves) were due to electrochemical reactions of surface functionalities such as quinone/hydroquinone couple (the signal current is independent of the type of electrolyte and its concentration). While the nature of these functional groups cannot be determined unequivocally, an indication of the surface reaction can be obtained from the voltammograms. By way of example, Fig. 11 shows the CVs for selected PACE (CWN2—Ox) in an acidic solution recorded at different sweep rates [194]. Beside the increase in capacitive current, the presence of a well-shaped cathodic peak and a rather poorly shaped and broad anodic peak are observed at all sweep rates from 5 to 100 mV/s. Additionally, a shift in the cathodic peak potential towards smaller values (increasing the difference between cathodic and anodic potentials) is noted with sweep rate increase. Similar results of current–potential studies for a graphite electrode in acid solution were obtained earlier (Fig. 12) [16].

Variations in cathodic peak current with the square of the sweep rate [$i_{pc} = f(v^{1/2})$] for selected PACE/electrolyte systems are shown in Figs. 13a and 13b [194]. In all cases the relation is linear and the correlation coefficient tends to unity, especially in the lower ranges of the potential sweep rates. In agreement with numerous researchers [7,9,14,16,24,26,132], we also suggest that the cathodic peak is due to the reduction of quinonelike groups, whereas the anodic wave is due to the oxidation of hydroquinonelike groups in different environments on the active carbon surface of the electrode. These reactions are faradic

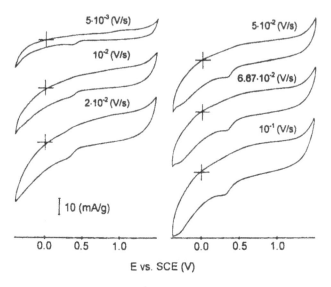

FIG. 11 CV curves of CWN2—Ox in 0.05 M H₂SO₄ (pH = 1.03) recorded with various sweep rates.

in nature and are controlled by the rate of the charge transfer process. Additionally, at lower potential sweep rates, the influence of the background solution concentration diminishes, and capacitive currents decrease. The majority of further measurements were carried out at a sweep rate of $3.33 \cdot 10^{-3}$ V/s, which yields reproducible results and a reasonable experimental time.

3. Thermal Decomposition of Surface Functionalities of PACE Materials

Active carbon oxidized with conc. nitric acid (CWZ—Ox) was annealed under vacuum (10^{-2} Pa) at several temperatures (300, 400, 500, and 950°C) during 4 hours. The concentration of surface oxide groups in the samples was obtained by Boehm's method, and the total oxygen content was determined by elemental analysis [170]. Obtained results are set out in Table 7 [194]. Thermal desorption at gradually increasing temperatures leads to a reduction in the surface acidic group concentration and total oxygen content (from 9.7% in the oxidized sample to 0.6% in the sample desorbed at 950°C). Likewise, the basicity of carbon samples rises—the acid consumption increasing from 0.019 mM/g (for the oxidized sample) to 0.532 mM/g (for the sample annealed at 950°C). For all samples the cyclovoltammetric studies were carried out in various electrolyte solutions [194].

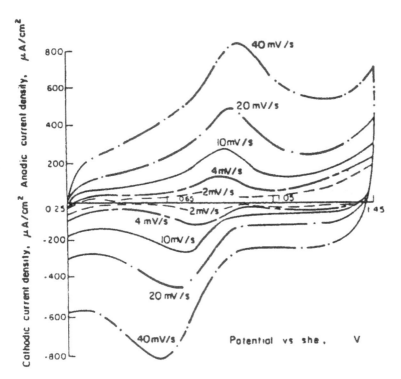

FIG. 12 Current–potential curves of graphite electrode in 0.01 H_2SO_4 solution at 30°C. Sweep rates 2 mV/s, 4 mV/s, 10 mV/s, 20 mV/s, 40 mV/s. (From Ref. 16.)

Figure 14 shows the cyclic voltammograms for PACE prepared from the carbon samples thus obtained and recorded in acidic (0.05 M H_2SO_4, pH = 1.07) aqueous solution [194]. With increasing heat treatment temperature (HTT), capacitive currents drop distinctly with HTT until 400°C but rise rapidly above 500°C. This means that the removal of acidic functional groups (with active hydrogen) leads (in a first stage) to decreasing capacitive (double layer charging) currents. Similarly, the increase in differential double layer capacities (measured by galvanostatic charging and AC impedance technique) of carbon fibers with rising air-oxidation temperature (increase in the extent of surface oxidation) was reported recently [200]. Again, electrochemical oxidative pretreatment of glassy carbon [201] and carbon fiber [202] electrodes caused the double layer capacitance to rise (in particular for 0.3 V to 0.6 V potential range) [202]. Faradically active surface functional groups can contribute to the measured charge and are consequently included in the calculated capacitance values (i.e., a pseudocapacitance) [202]. For HTT above 500°C, the residual oxygen is removed, but struc-

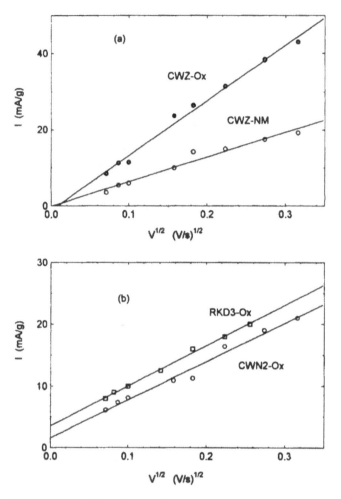

FIG. 13 Cathodic peak current dependence on the square root of the sweep rate for the systems: (a) CWZ (Ox and NM) in 0.05 M H_2SO_4; (b) CWN2—Ox in 0.05 M Na_2SO_4 and RKD3—Ox in 0.05 M H_2SO_4. (From Ref. 194.)

tural changes in the active carbon surface (free radical creation) [68] again raises the double layer capacity. Simultaneously, the reduction in both cathodic and anodic peaks (waves) on CVs occurs for HTT above 400°C. The faradic (charge transfer) process disappears altogether when the surface quinone/hydroquinone systems undergo thermal destruction. These are rather new findings and need to be explored further for a greater number of carbon samples.

TABLE 7 Physicochemical Properties of the Thermally Treated Active Carbon CWZ—Ox

HTT °C	S_{BET} m²/g	Surface functional groups, mmol/g					Total oxygen[a] %
		—COOH	—COO—	>C—OH	>C=O	Basic	
200	455	0.84	0.46	0.71	0.10	0.02	9.7
300	486	0.70	0.21	0.72	0.16	0.09	8.6
400	515	0.49	0.10	0.64	0.21	0.10	6.8
500	545	0.16	0.05	0.41	0.49	0.21	5.2
950	595	0	0.01	0.02	0.02	0.53	0.6

[a] From elemental analysis.
Source: Ref. 194.

C. Cyclovoltammetric Studies in Nonaqueous and Mixed Solutions

To explain how surface oxidation and the nature of the solvent influence the electrochemical processes occurring at the solid–liquid interphase, the CV curves for all the carbons studied were recorded in neutral aqueous, nonaqueous (acetonitrile), and mixed (1:1) solutions of 0.1 M $LiClO_4$. The cyclic curves for CWN2 carbons are presented in Fig. 15 by way of example [27]. For unoxidized (NM) carbons in aqueous solutions (Fig. 15a) no faradic processes are observed in the potential range investigated (from −0.9 to 1.0 V vs SCE). The chemical or electrochemical oxidation of the carbon surface causes anodic and cathodic peaks to appear [7,11–26,202–204]. We also confirmed this effect in PACEs prepared from oxidized carbons (Fig. 15b). The voltammogram obtained for a carbon electrode in an aqueous organic solvent (Fig. 15c) shows the partial disappearance, and in a nonaqueous solvent, the complete disappearance (Fig. 15d), of the peaks on the cathodic and anodic waves. The charge transferred in the anodic and cathodic processes decreases when the polarity of the solvent and the oxonium/hydroxyl ion concentration in the background aqueous solution do so.

Comparison of the cyclovoltammetric curves recorded for carbon fibers (CF) in both aqueous and nonaqueous solutions provided evidence of a considerable decrease in double electric layer capacity with increase in CF carbonization temperature (particularly between 1100 and 1400°C) [21,204]. This observation applies to both unmodified and oxidized carbon fibers. Oxidative modification leads to the appearance of a broad anodic peak with a potential between 0.1 and 0.3 V in neutral aqueous solutions (see Fig. 16a). The absence of any peak on CV curves recorded in acetonitrile solutions (Fig. 16b) suggests that functional

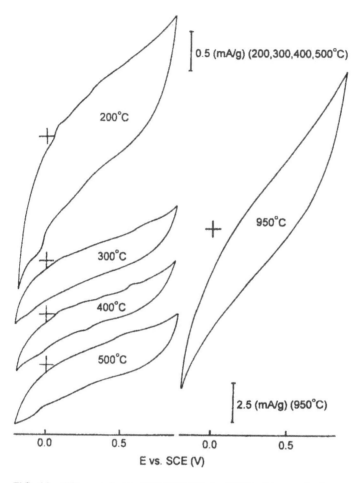

FIG. 14 CVs recorded in 0.05 M H_2SO_4 for CWZ—Ox samples heat-treated under vacuum at temperatures 200°C, 300°C, 400°C, 500°C, 950°C. Sweep rate $v = 3.33 \cdot 10^{-2}$ V/s. (From Ref. 194.)

groups on the surface of the electrode material exhibit no electroactivity in non-protic systems. On the CV curves recorded in methanol solutions (Fig. 16c), marked anodic peaks ($E \approx 0.5$ V) and poorly shaped cathodic peaks ($E \approx 0.0$ V) are observed for oxidized carbon fibers. In a low protic environment, the oxidation of functional groups is believed to occur at a higher potential than in aqueous solutions. The electron transfer character of the process responsible for this peak can explain the lack of such a shift in cathodic peak potential. Therefore,

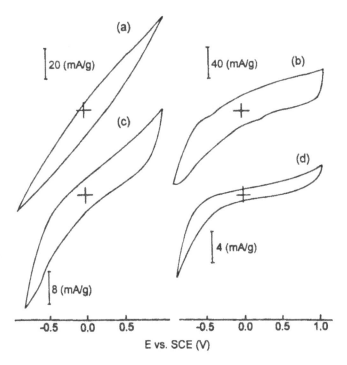

FIG. 15 CVs of PACEs: (a) CWN2—NM in 0.1 M LiClO$_4$/H$_2$O; (b) CWN2—Ox in 0.1 M LiClO$_4$/H$_2$O; (c) CWN2—Ox in 0.1 M LiClO$_4$/(H$_2$O + AN (1:1)); (d) CWN2—Ox in 0.1 M LiClO$_4$/AN. Sweep rate $v = 3.33 \cdot 10^{-2}$V/s. (Adapted from Ref. 27.)

the presence of protons in the electrolyte solution influences electrode processes in which chemisorbed oxygen species participate.

D.   Influence of pH

To define the possible mechanism of the surface electrochemical reactions, the cyclic voltammograms for oxidized carbons were recorded in blank aqueous electrolytes over a wide pH range. For CWN2—Ox and CWZ—Ox, aq.H$_2$SO$_4$-aq. NaOH solutions of constant ionic strength were used [27]. The RKD—Ox carbon was studied in Britton–Robinson buffer. The CVs for different pH values recorded for CWZ—Ox carbon are shown in Fig. 17. In all the curves there are more or less well-defined cathodic and anodic peaks (or waves), the potentials of which are pH dependent. Some sections of the cathodic waves recorded in solutions of different pH are presented in Fig. 18. It can also be seen that the

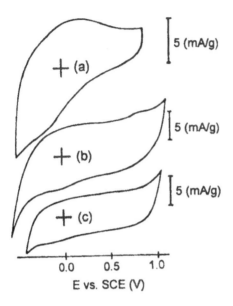

FIG. 16 CVs of CF1400—Ox electrode in (a) 0.05 M Na_2SO_4 aqueous solution; (b) 0.1 M $LiClO_4$ methanol solution; (c) 0.1 M $LiClO_4$ acetonitrile solution. Sweep rate $v = 3.33 \cdot 10^{-2}$ V/s. (Adapted from Ref. 21.)

height of the recorded reduction waves depends on the pH too. From these waves, the half-wave potentials can thus be estimated to within 5 mV.

To characterize the influence of the oxonium ion concentration on the redox reactions responsible for the observed waves, the relations between half-wave potentials and pH values are presented (Fig. 19). These relationships are linear for all the PACEs studied (from previously oxidized carbons), and the slopes (calculated by the least-squares method) take the respective values -57 ± 3, -51 ± 2, and 63 ± 2 mV/pH for CWZ—Ox, CWN2—Ox, and RKD3—Ox. A similar dependence of the estimated anodic peak potential on the pH of aqueous electrolytes for the oxidized carbon fiber (CF 2400ox) is illustrated in Fig. 20 [21]. A fairly good straight line is obtained within the entire range of pH values with an average slope of -69.4 mV/pH. One contributing factor to the discrepancy in slopes (obtained and theoretical) could be the incomplete resolution of the overlapping peaks exhibiting differing electrochemical reversibility. One would expect to find hydroxylic groups (quasi-hydroquinones) in different local environments on the carbon surface. Carbon fibers are thought to possess localized graphitic planes, at the edge of which one could expect various electroactive functionalities to form during oxidation. Slopes of the relationship $mV = f(pH)$, close to those given above, were detected for the cathodic peaks obtained on cyclic

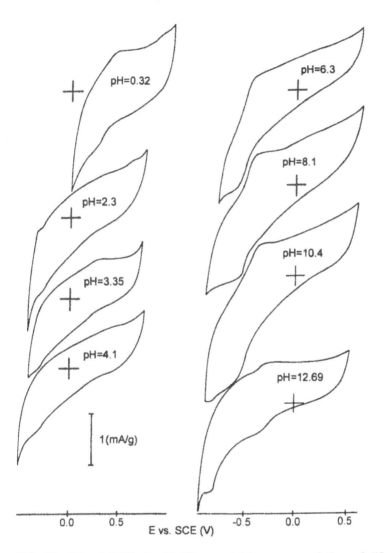

FIG. 17 CVs of CWZ—Ox PACEs recorded in aqueous solutions of different pH. Sweep rate $v = 3.33 \cdot 10^{-2} V/s$. (From Ref. 194.)

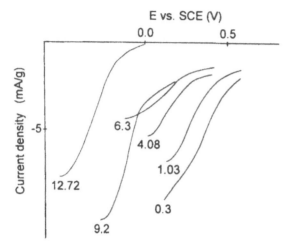

FIG. 18 Cathodic scans (section) of CVs of CWN2—Ox recorded in aqueous electrolytes of different pH. Sweep rate $v = 3.33 \cdot 10^{-2}$V/s. (From Ref. 27.)

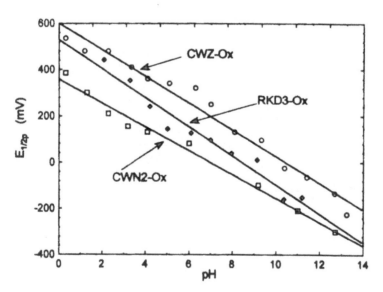

FIG. 19 Nernst plots of cathodic peak potentials vs. pH for oxidized carbon electrodes: (a) CWZ—Ox; (b) CWN2—Ox; (c) RKD3—Ox. (Adapted from Refs. 27 and 194.)

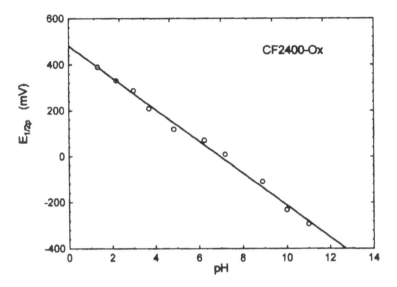

FIG. 20 Anodic peak potential of CF2400—Ox vs. pH of aqueous solutions. (From Ref. 21.)

voltammograms of a glassy carbon electrode after oxidative modification in air at 800°C (Fig. 21) [15] and after oxygen rf-plasma treatment (Fig. 22) [11]. The respective pH relationships of these peak potentials were at −63 mV/pH [15] and −62 mV/pH [11]. Strict relationships between pH and the anodic (−62.7 mV/pH) and cathodic (−65.3 mV/pH) peaks for a preanodized glassy carbon electrode were also found [205].

According to the general electrode reaction,

$$Ox + mH^+ + ne = Red \tag{1}$$

when [Ox] = [Red], the Nernst equation predicts that the electrode potential for this process (E) is related to the formal electrode potential (E^0) by

$$E = E^0 - \frac{m}{n} (0.059) \text{ pH} \tag{2}$$

When $m = n$, a theoretical slope of −59 mV/pH for the $E = f(\text{pH})$ relationship is expected. The Nernst slopes obtained in our work [21,27] were in good agreement with theoretical values when equal numbers of electrons and protons are transferred. In particular, good agreement with theoretical values was obtained for carbons oxidized with conc. nitric acid (CWZ—Ox, RKD3—Ox). We suggest, therefore, that this oxidizer yields surface functional groups with a chemical

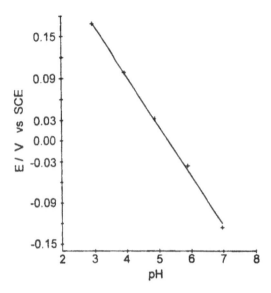

FIG. 21 Nernst plots for a glassy carbon electrode activated in air at 800°C. Slope = −0.072 V/pH, intercept = 0.383 V. (From Ref. 15.)

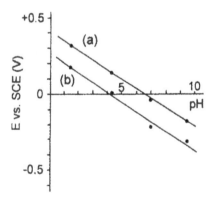

FIG. 22 pH dependence of the peak potential of the surface quinone peaks: (a) quinone peak I; (b) quinone peak II. (From Ref. 11.)

structure similar to those in the model system [156,206–208]. We would expect to find quinones of different types (e.g., 1,2- and 1,4-), as well as different local environments on the carbon surface. These would very likely exhibit significant variation in their reduction potentials.

1,2-quinone   1,4-quinone

The presence of other cathodic and anodic peaks points to electrochemical activity on other oxygen species existing on the carbon surface (see Table 4). Additionally, they may be overlapped by a significant capacitive current [153]. However, it should be remembered that the real chemical structure of an oxidized carbon surface [101] depends on the hydrolysis of lactone-, ester- or ether-like anhydrous systems and the ionization of some functionalities at extreme pH values (acidic or basic environments) [91]. These phenomena influence the surface density of species that can take part in charge-transfer processes, which explains the observed differences in height of reduction peak in different environments (see Fig. 18). These relationships can account for the reactions, e.g. [7,14,148],

$$>C{=}O + H^+ + e^- \rightarrow \geq C{-}OH \tag{3}$$

$$(>C{=}O)_2 + 2H^+ + 2e^- \rightarrow (\geq C{-}OH)_2 \tag{4}$$

Comparison of the surface density ($\mu mol/m^2$) of the functional groups on active carbons oxidized by nitric acid in relation to carbon oxidized in air (see Table 4) shows a significantly higher (about 10 times) concentration of quinoid (carbonyl) structures. The gentler Nernst slope for CWN2—Ox is indicative of reduction with partial ionization of the hydroxyl group, according to the reaction

$$>C{=}O + e^- \rightarrow \geq C{-}O^- \tag{5}$$

This is probably due to the higher relative concentration of hydroxyl groups on this carbon (ca. 25 times) and the different distribution of their acidic strength. Another factor contributing to the discrepancy between the obtained and theoretical slopes could be the incomplete resolution of the overlapping peaks,

TABLE 8 Cyclovoltammetric Study Results of Selected Organic Compounds in Aqueous Solution (pH ≈ 7, All Potentials Reported vs. SCE)

Studied compound	Electrode materials	E_f, mV	ΔE_p, mV	$\Delta E/pH$, mV/pH	$E_{p,a}$, mV	$E_{p,c}$, mV	v, V/s	Reference
	Pt	+48	30	−57.5			1	[206]
	PG	+80	44	−60			0.02	[12]
	PG, AC	+50		−60			0.25	[207,208]
1,4-benzoquinone	GC	+125		−63			0.1	[201]
	CF	+115		−64			300	[209]
	CCEs	+35	30				0.1	[210]
	GC	+100					0.1	[211]
1,2-benzoquinone	PG, AC	+124		−60			0.25	[208,209]
2-hydroxy-1,4-benzoquinone	Pt	−151	30	−88.3			1	[206]
	GC		34	−59			0.5	[11]
1,4-naphthoquinone	CCEs	−190	30			−220	0.1	[210]
	Pt	−151	30	−57.8				[206]
1,2-naphthoquinone	GC		34				0.5	[11]
	Pt	−400	30	−82.2		−100	1	[206]
2-hydroxy-1,4-naphthoquinone	GC					−530	0.5	[11]
9,10-antraquinone	GC	−220				−220	0.5	[11]
9,10-phenanthrequinone	CCEs						0.1	[210]
	CPE		130		+132		0.01	[212]

Compound	Electrode						Conc.	Ref.
1,4-dihydroxybenzene (hydroquinone)	CPE	+92	68				0.05	[213]
	PG	+60	100				0.05	[214]
	GC				+127		0.05	[215]
1,2-dihydroxybenzene (m-pyrocatechol)	GC	+200	45				0.1	[216]
	GC	+217	50				0.1	[211]
	CF	+175			+465	−130	300	[209]
4-methyl-1,2-dihydroxybenzene	CPE	+124					0.1	[217]
	GC	+120					0.1	[211]
3,4-dihydroxybenzene carboxylic acid	CPE	+150					0.4	[217]
	CPE	+143			+248	+39	0.1	[218]
pyrocatechol carboxylic acid	PG	+175	25	−58			0.001	[219]
	PG	+211	25	−58			0.001	[219]
dihydroxybenzaldehyde: 3,5-DHB					+245		0.1	[220]
2,5-DHB					+20			
3,4-DHB					+120			
2,3-DHB					+150			
2,4-dimethylphenol	GC			−56	+470		0.05	[221]

Electrode materials: Pt, platinum; PG, pyrographite; AC, active carbon; GC, glasslike carbon; CF, carbon fiber, CCEs, ceramic carbon electrodes; CPE, carbon (graphite) paste electrode.

Source: Refs. 12, 201, and 206–221.

which exhibit differing electrochemical reversibilities. The systems studied are extremely complex, both chemically and electrochemically, so the mechanism of electrochemical activity and the role of surface species require a comparison with the model systems well known from organic electrochemistry.

E. Models

Cyclic voltammetry is a method frequently used for studying the electrochemical behavior of organic compounds containing functional groups with redox properties, such as carbonyl (quinone) [12,199,206–210], hydroxide (phenol, hydroquinone) [211–221], carboxyl [217–219], or aldehyde [220]. Table 8 gives some results of electrochemical studies for a number of organic compounds with functional groups characteristic of an active carbon surface. This gives values of formal potentials (E_f), differences in peak potentials (ΔE) in neutral (pH = 7) aqueous solution as well as the potential–pH response (ΔE/pH) determined on various electrode materials. The results in Table 8 give an idea of the influence of the position of carbonyl/phenol groups and the number of aromatic rings on the electrochemical parameters estimated for the systems under investigation. The approach of carbonyl groups from the para to the ortho position leads to a positive shift in the formal potential, regardless of the number of aromatic rings. An increase in the number of aromatic rings causes the formal potential to decrease, as is illustrated schematically in Table 9.

Substitution of phenolic groups in 1,4-benzoquinone also leads to a decrease in formal potentials, as can be seen in the 2-hydroxo-1,4-benzoquinone/1,2,4-trihydroxybenzene system (Table 8). However, the addition of a carboxyl group has no such effect. Therefore, the peak potential in neutral aqueous solution (0.1 M Na_2SO_4) determined for oxidized carbon samples, set out in Table 10, confirms the presence of surface quinone-type structures conjugated with hydroxyl functionalities and π-electrons of the aromatic system.

TABLE 9 The Influence of Ring Number and Substitution Position of Quinone Groups on the Formal Peak Potential (mV) of Quinone/Hydroquinone Couples

Quinone	1,4-Substituted	1,2-Substituted
	$\approx +80$	$\approx +125$
	≈ -170	≈ -100
	≈ -530	≈ -220

Source: Inspired by data from Table 8.

TABLE 10   Peak Characteristics of the Oxidized Carbons Investigated

Carbon sample	$E_{p,c}$, V	$E_{p,a}$, V	ΔE, V	E_f, V	$\Delta E/\Delta pH$, mV/pH
CWZ—Ox	−0.58	−0.34	0.24	−0.46	−57.7
CWN2—Ox	−0.52	−0.28	0.24	−0.40	−51.6
RKD3—Ox	+0.08	+0.28	0.20	+0.18	−62.8
D—Ox	−0.55	−0.30	0.25	−0.42	—
CF2400—Ox	−0.10	0.00	0.10	−0.05	−69.4

Source: Partly from Refs. 21, 27, and 194.

The general shape of the potential–pH diagrams for organic quinones (Fig. 23 [206]) is similar to the potential–pH relationship for the quinone/hydroquinone system on an active carbon surface (Fig. 19). The values of $\Delta E/pH$ for oxidized carbons given in Table 10 show that like many other carbon electrodes, the tested powdered active carbons exhibit a potential–pH response with Nernstian behavior (i.e., 59 mV/pH unit).

F.   Oxygen Generation

The preliminary oxidation of active carbon ought to protect the surface from permanent and uncontrollable oxidation during electrochemical processes. Hence both nonmodified and chemically oxidized active carbons were used in our stud-

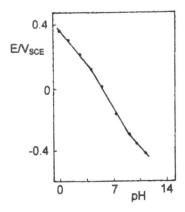

FIG. 23 $E_{1/2}$–pH diagram for 2-hydroxy-1,4-benzoquinone. Line is drawn for least-squares fit. (From Ref. 206.)

ies. The electrochemical behavior of the functional groups created by chemical and electrochemical oxidation can thereby be compared.

The CV curves of such carbons in neutral, acidic, and basic electrolytes were recorded for various anodic amplitudes. Examples of curves for chemically oxidized carbon (CWN2—Ox) plotted in a neutral environment are shown in Fig. 24. The respective CVs recorded in acidic and basic environments for CWZ—NM are presented in Figs. 25a and 25b. All the CV curves exhibit a more or less well-defined peak or pair of peaks, which has been ascribed to a surface system of quinone-hydroquinone species (a, a' on Fig. 24). As can be seen, the anodic part of the curves records the presence of a rather poorly shaped broad peak, or a system of overlapping peaks. The anodic peak (a') marked in Fig. 24 could have been due to the overlapping of electrochemical oxidation processes and/or of double layer charging processes. These peaks are not subject to any distinct changes during anodic potential scanning; only small and slow changes are visible on curves recorded for higher anodic turnaround potentials. Anodic polarization creates surface species that undergo reduction to yield a large ca-

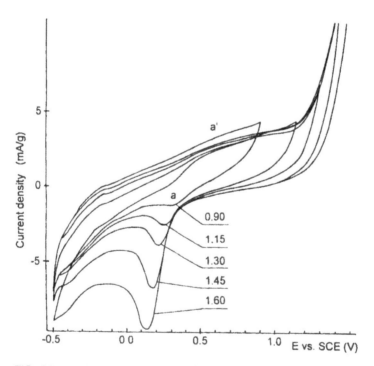

FIG. 24 Cyclic voltammograms of CWN2—Ox active carbon in 0.05 M Na$_2$SO$_4$ recorded for different sweep amplitudes. (From Ref. 28.)

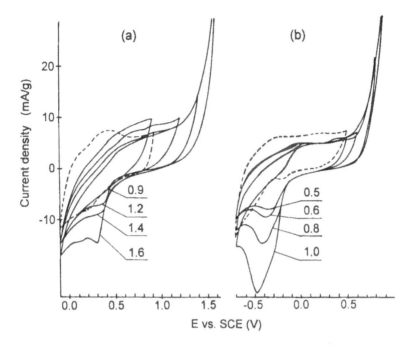

FIG. 25 Cyclic voltammograms of CWZ—NM active carbon in 0.05 M H_2SO_4 (a) and 0.1M NaOH (b) for different sweep amplitudes. The dashed lines show the CVs recorded for chemically oxidized CWZ carbon. (From Ref. 28.)

thodic peak. Only a small part of these reduced forms yields an anodic response. Electroactive surface species (*e.g.*, quinone-like) are probably formed, but their participation in the overall anodic processes is rather small.

The above CVs (Figs. 24 and 25) display well-formed reduction peaks independent of the blank solution and the type of active carbon materials. The combined shape of the cathodic peaks indicates that surface species participate in electrochemical processes in different local environments, or with various structures but convergent peak potentials. The effect of anodic polarization is more readily observed in a basic environment than in an acid solution. Similarly, a positive shift of cathodic peak potential with a decrease in anodic sweep potential limit takes place. Similar results were obtained for studies of electrochemical oxidation of graphite [17] and glass-like carbon [222] electrodes. There was considerable enlargement of both anodic and cathodic peaks after anodic polarization in 20% sulfuric acid (Fig. 26) [17].

Generally, for all the PACEs prepared from nonmodified or oxidized carbons the cathodic peak (a) is relatively well formed and its size depends strongly on

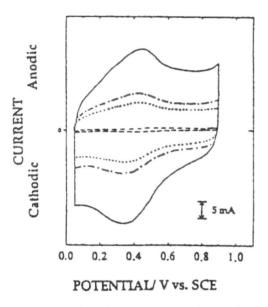

FIG. 26 Cyclic voltammograms of an ATJ graphite electrode at a scan rate of 10 mVs⁻¹ before (---) and after electrochemical oxidation in 20% H_2SO_4 at a current density of 12.5 mA/cm for a charge of 5 (···), 10 (— · —) and 15 (—) C/cm. (From Ref. 17.)

the anodic polarization amplitude. Above we presented the changes in cathodic peak potentials with the pH of the aqueous electrolyte. The equivalence in numbers of protons and electrons in cathodic reactions was confirmed (the Nernst slope was nearly −60 mV/pH). The behavior of this peak in relation to the anodic sweep potential limits E_a for oxidized carbons in acidic solutions and for nonmodified carbons in basic solution is shown in Figs. 27 and 28. In addition, Fig. 28 (dotted line) shows a section of the reduction curves of chemically oxidized carbons for comparison. To the same end, the voltammograms of oxidized carbons are presented in Fig. 27 (dotted lines). The cathodic peak potentials of chemically oxidized carbons lie within the potential range of electrochemically generated species, and this may be evidence for the structural similarity of the electroactive species formed in two oxidative procedures. The complex nature of the reduction of species generated in a basic solution (a double peak at least) suggests that more complicated surface structures are formed.

The decrease in the anodic potential limit leads to changes both in the height of the cathodic peaks (I_h) and in their areas. The cathodic peak areas (surface charge density, Q_s) were estimated in the manner described in the literature [222]. The enlarged sections of the voltammograms were used for designating base lines

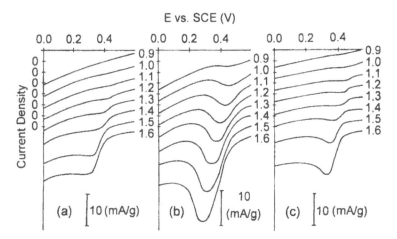

FIG. 27 Reduction peaks of the surface oxide species on CWN2—Ox (a), CWZ—Ox (b), and RKD3—Ox (c) recorded for different sweep amplitudes (anodic potential limits) in 0.05 M H_2SO_4. On the ordinate axis only 0 is marked, indicating vertical shift of the curves. (From Ref. 28.)

(capacity currents) and calculating peak areas. The intensity of reduction processes (cathodic peak magnitude) depends on the kind of active carbon the PACE was prepared from. In an acidic environment reduction occurs most rapidly using CWZ—Ox carbon; it is characterized by the highest mesopore content in the total surface area (BET) (see Table 2) and the highest surface concentration of functional groups (Table 4). Moreover, in a basic environment the cathodic peak separates into two overlapping peaks. This is particularly apparent at higher values of the anodic turnaround potential (E_a). The estimated dependence of cathodic peak currents (I_h) and their areas (charge transferred during electrochemical processes, Q_s) in relation to anodic turnaround potentials for the whole system studied are shown in Figs. 29a and 29b, respectively. The absence of well-shaped cathodic reduction peaks of dioxygen (oxygen maximum) on curves recorded before the evolution of anodic oxygen may indicate that the oxygen species generated are irreversibly attached to the carbon surface. A considerable increase in cathodic peak area is observed when the anodic potential limit exceeds the oxygen evaluation potential on these carbons (Table 6). The species formed undergo reduction at potentials dependent on the pH of the electrolytes. These potentials coincide with the reduction potential of the surface quinoidlike groups generated by chemical oxidation. After alternation of polarization, the reduced surface structures are gradually oxidized over a wide potential range. The anodic peaks formed are extended and superposed, which makes it impossible to recognize

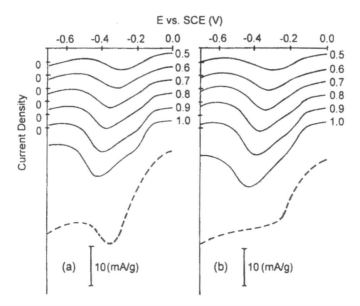

FIG. 28 Reduction peaks of the surface oxide species on CWN2—NM (a) and RKD3—NM (b) recorded for different sweep amplitudes (anodic potential limits) in 0.1 M NaOH. On the ordinate axis only 0 is marked, indicating vertical shift of the curves. The dotted lines show the sections of CVs recorded for chemically oxidized carbons. (From Ref. 28.)

their potentials or areas. The data presented in Figs. 29 and 30 show that changes in both the heights and the sizes of cathodic peaks depend strongly on the kind of electrolyte solution, the type of carbon modification, and the porous structure of the electrode surface. I_h and Q_s increase significantly in both acidic and basic environments when the oxygen evolution potential is exceeded. However, depending on the modification procedure, differences in the slopes of the plots of I_h and Q_s against E_a are noticeable. In acidic supporting electrolytes these relations indicate that reduction of electrochemically generated oxygen species works best on chemically oxidized carbon CWZ. The residual carbons, both oxidized and nonmodified, display a considerably lower activity in this process (for RKD3—NM no activity is observed). In a basic environment, the behavior of these carbons in the reduction of electrochemically generated oxygen species is distinctly different: here nonmodified carbons are better, and the best is CWN2—NM. The oxidized carbons in this supporting electrolyte display a similarly low activity.

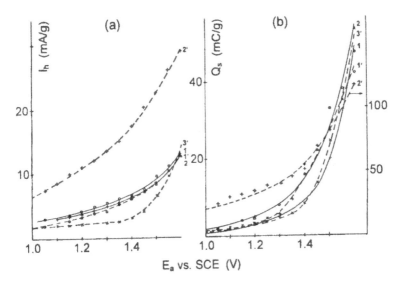

FIG. 29 Dependence of the cathodic peak current (I_h) and surface charge density (Q_s) on the anodic sweep potential limit (E_a) for reduction peaks recorded for PACEs in 0.05M H_2SO_4: 1, 1', CWN2; 2, 2', CWZ; 3, 3', RKD3 (1–3, nonmodified; 1'–3', oxidized). (From Ref. 28.)

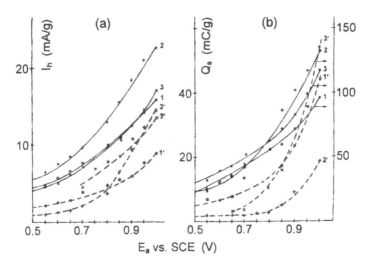

FIG. 30 Dependence of I_h and Q_s on E_a for PACEs in 0.1 M NaOH. The electrode materials are as in Fig. 29. (From Ref. 28.)

The results indicate that the activity of carbons used to reduce electrochemically generated oxygen species depends to a large extent on surface development, in particular on the contribution of the mesopore surface to the total specific surface area. This is reasonable for a basic supporting electrolyte. In an acidic environment, however, this process depends additionally on the previous chemical oxidation of the electrode material surface. These results may constitute a basis for the selection of active carbons for use as electrode materials.

V. THE APPLICATION OF CYCLIC VOLTAMMETRY FOR THE STUDY OF INTERACTIONS BETWEEN HEAVY METAL IONS AND AN ACTIVE CARBON SURFACE

Because of their large surface area and their high degree of surface reactivity, active carbons are regarded as very good adsorbents for the removal of heavy metal ions from aqueous phases [223–232] and can be used in a number of possible technological and analytical applications. An understanding of the interaction of metal ions with the carbon surface is essential for the control of the sorption/desorption processes applicable to active carbon sorbents as well as for the recognition of catalytic properties of carbon materials promoted by adsorbed metal species [233]. The mechanism of cationic species adsorption has been widely discussed but is not yet adequately explained [223–239]. The heavy metal adsorption equilibrium has been modeled by using the empirical adsorption isotherms (linear, Langmuir, Freundlich) [238] or by applying mechanistic models such as the surface complex formation [229] and surface ionization models [237]. Removal of heavy metal cations from water is influenced by various factors, such as solution concentration and pH, contact time, carbon dosage, and the sorbent surface modification procedure. Most investigators have reported an increase in adsorption with increasing pH of the medium [223–232]. Generally, the process depends on the cationic species to be adsorbed, the adsorbent, and the adsorption conditions. A variety of interactions between metal ions and the carbon surface could be occurring, e.g., the formation of surface complexes [223]:

$$2(\underline{C}OH) + M^{2+} \Leftrightarrow (\underline{C}O-)_2M^{2+} + 2H^+ \tag{6}$$

ion-exchange processes with the participation of a strongly acidic surface group [234]:

$$n[H^+]_{solid} + M^{n+} \Leftrightarrow [M^{n+}]_{solid} + nH^+ \tag{7}$$

and redox reactions with a change of metal valence [235].

In this section we present some results of investigations into the influence of surface oxygen and/or nitrogen groups on the adsorption of ions from aqueous solution on modified active carbons. Furthermore, we have attempted to ascertain the real nature of the adsorbed ionic species–modified carbon surface interactions using spectral (XPS and FTIR) and electrochemical (cyclic voltammetry) measurements. A significant effort has been directed towards achieving an understanding of the relationship between surface structure and electron transfer (ET) reactivity for carbon electrodes [23,60]. The most important phenomena affecting ET reactivity mentioned in the literature are (1) the microstructure (and porosity) of carbon materials, (2) the presence of surface functional groups (mainly oxides), and (3) surface purity (e.g., the presence of sorbed metal species). Unfortunately, many of the common electrode preparation procedures affect more than one of these variables. If we are to understand the phenomena controlling the reactivity of a carbon surface on a molecular level, knowledge of the chemical structure of the surface is essential.

A. Fe^{3+}/Fe^{2+} Couple Behavior

Using cyclic voltammetry, the electrochemical behavior of powdered active carbon electrodes (PACEs) were studied for the Fe^{3+}/Fe^{2+} couple in acidic (0.01–0.1 M nitric acid) aqueous solutions [227]. Surface chemical differences in the active carbon affect its capacity to sorb iron ions. Oxidative modification of the CWZ—NM carbon surface increases the equilibrium quantity of Fe^{3+} adsorbed from aqueous solution [0.05 M $Fe(NO_3)_3$, pH = 2] by a factor of more than three (from 0.48 mM/g to 1.60 mM/g). The higher number of adsorbed iron ions recorded following oxidative modification correlated best with the larger number of surface carboxyl and hydroxyl functional groups. The surface modification procedures applied for D-43/1 carbon alter the apparent surface area of the samples obtained (S_{BET}) only slightly (see Table 2). Fe^{3+} adsorption on these samples was determined from nitrate solutions of pH 1–3 (Fig. 31). The quantity of adsorbed ions depends strictly on the pH of the solution and the modification procedure (Fig. 32). The higher the pH, the greater the likelihood of iron being precipitated as $Fe(OH)_3$. Cation uptake increases linearly (1.2 mM/pH) over the pH range examined for all carbon samples. The recorded similarities in adsorption capability of carbons with different surface chemistries suggest that the ionic metal species present in the aqueous solution (aqua and hydroxy complexes, hydroxide ions, and electronegative complexes) can interact with the carbon surface in various ways [234,235]. To discover the state of the adsorbed species, some independent measurements of the surface layer of adsorbent were carried out. All carbon samples were studied by FTIR and XPS in powdered form following iron uptake.

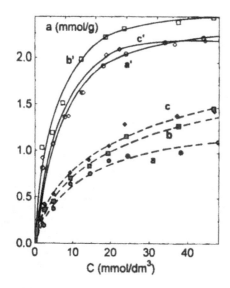

FIG. 31   The adsorption isotherms of Fe^{3+} ions on modified D43/1 carbons (a,a′, D—H; b,b′, D—Ox; c,c′, D—N) from nitrate solution with various pH: a–c, 1.0; a′–c′, 2.7.

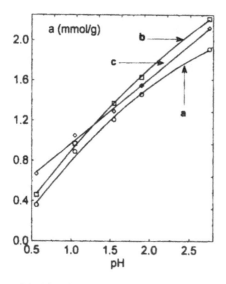

FIG. 32   The sorption isotherms of Fe^{3+} ions on modified D43/1 carbons (a, D—H; b, D—Ox; c, D—N) from nitrate solution as a function of pH.

In the FTIR spectra of the selected active carbon materials (Fig. 33), the band of OH stretching vibrations (around 3400 cm^{-1}) was due to surface hydroxylic groups, chemisorbed water, and adsorbed iron-containing species. Additionally, a decrease in the carboxylic band (1740 cm^{-1}) following Fe(III) ion adsorption was commonly recorded for carbon materials with oxygen-containing functionalities. For some carbon a new band at wavelength 1380 cm^{-1} was ascribed to NO_3^- ions adsorbed at basic centers of carbon and/or as an ion compensating for adsorbed iron-containing species. The relative increase in overlapping bands in the 1250–1000 cm^{-1} range (maximum at 1120 cm^{-1}) may be confirmation of additional hydration of the adsorbent surface. These observations suggest that

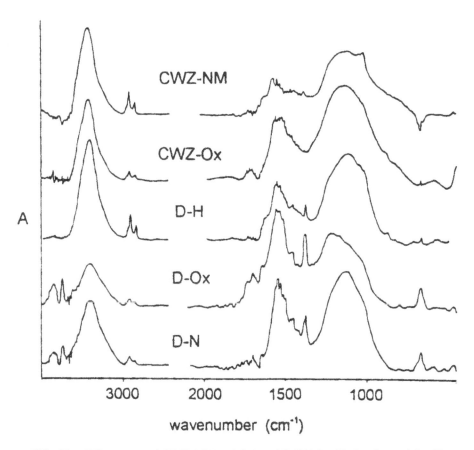

FIG. 33 FTIR spectra of CWZ (NM and Ox) and D43/1(D—H, D—Ox, and D—N) carbons with adsorbed Fe^{3+} ions from nitrate solution.

the adsorption of iron (III) hydroxide species ($FeOH^{2+}$, $[Fe(OH)_2]^+$) and iron (III) aqua-complexes may take place with the participation of surface functionalities. This adsorption may in part be due to ion exchange, as expected. We believe that the formation of surface iron complexes with oxygen functional groups is dominant in sorption processes on oxidized carbons (CWZ—Ox, D—Ox). In carbon samples where the surface functionalities (D—H, D—N) are of a basic character, other types of interactions between metal species and carbon surface presumably occur (π-d interactions, hydroxide, and oxyhydroxide creation). Examples of possible surface structures formed during the adsorption of ionic iron species on these carbons can be given schematically. For D—H carbon (Scheme 8 a–d),

$$>C—H + Fe(OH)^{2+} \rightarrow >C—H \tag{8a}$$
$$\cdot Fe(OH)^{2+} \quad \text{(dipole–dipole, } \pi\text{-}d \text{ interactions)}$$

$$>C\bullet + Fe(OH)^{2+} \rightarrow \oplus>C \cdot Fe(OH)^+ \tag{8b}$$

$$>C—O\bullet + Fe^{3+} + H_2O \rightarrow >C—O—FeOH^+ + H^+ \tag{8c}$$

$$>C=O + Fe^{3+} + 2\,H_2O \rightarrow >C—O—Fe(OH)_2 + 2\,H^+ \tag{8d}$$

For CWZ—Ox and D—Ox carbons (Scheme 9 a–c),

$$>C—COOH + Fe(OH)^{2+} \rightarrow >C—COOFe(OH)^{2+} + H^+ \tag{9a}$$

$$(>C—COOH)_2 + Fe(OH)_2^+ \rightarrow (>C—COO)_2Fe^+ + 2H_2O \tag{9b}$$

$$>C—OH + Fe(OH)^{2+} \rightarrow >C—O\,Fe(OH)^+ + H^+ \tag{9c}$$

For D—N carbon (Scheme 10 a–c),

$$>C—NH_2 + Fe(OH)^{2+} \rightarrow >C—NH_2\cdot Fe(OH)^{2+} \tag{10a}$$

$$>N: + Fe(OH)^{2+} + H_2O \rightarrow >N—Fe(OH)_2^+ + H^+ \tag{10b}$$

$$>N: + Fe(OH)^{2+} + H_2O \rightarrow \oplus>NOH\cdot Fe(OH)_2 + H^+ \tag{10c}$$

where the symbols $\bullet$ and $\oplus$ denote the ionoradical and positive holes in the carbon structure, respectively. In the interests of clarity, the coadsorbed counterions (mainly NO_3^-) are not marked.

Figure 34 shows the C 1s (a–d), O 1s (e–h), and Fe $2p_{3/2}$ (i–j) XPS peaks of CWZ (NM and Ox) samples with adsorbed iron ions. There were marked differences between the experimental (dotted) and synthesized (continuous) lines in all the spectra. The position of deconvoluted peaks (dashed lines) were determined according to both literature data [188,191,240–247] and empirically derived values. The relative areas (%) of the fitted peaks were also calculated. Several peaks attributable to carbon, oxygen, nitrogen, and iron were present. The XP survey spectra of the initial modified carbons (before adsorption) were discussed

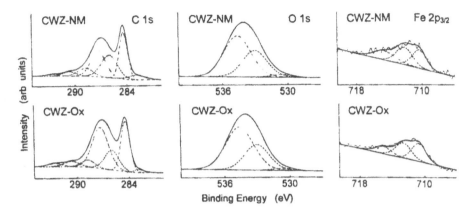

FIG. 34 High-resolution XP spectra of C 1s, O 1s, and Fe $2p_{3/2}$ regions of CWZ carbons with adsorbed Fe^{3+} ions from nitrate solution. (From Ref. 227.)

in an earlier section. Various surface groups can participate in iron ion sorption (Scheme 8–10).

The iron ions adsorbed on the carbon surface studied by XPS give Fe 2p lines at binding energies of near 725 eV ($2p_{1/2}$) and 711 eV ($2p_{3/2}$) and influence the intensity and binding energies of the C 1s and O 1s spectra [241–244]. The C 1s spectrum in the CWZ—NM carbon sample following Fe^{3+} ion adsorption consists of a number of peaks, some of which have shifted towards the high-energy side. The new small peak (282.8 eV, 5.1%) can be attributed to the C—Fe bond. The C=C peak (284.5 eV, 27.6%) has shrunk slightly. Moreover, the presence of two peaks at 285.9 eV (26.1%) and 287.3 eV (27.1%) in the chemical shift range of C—O (C—O—C) and C—O—(Fe) moieties, respectively, confirms the suggestion that adsorbed, highly hydrated (aqua-complex) Fe^{3+} and $FeOH^{2+}$ ions interact with oxygen-containing surface functional groups as is shown in Scheme 9. Changes in peak shape in the chemical shift range corresponding to C=O and COO groups (288.8 eV, 8.2% and 290.7 eV, 3.9%) can be ascribed to iron carboxylate formation, resulting in isolated Fe^{3+} cations. These results are confirmed by changes in the O 1s spectrum: only the peaks at 534.7 eV (61.4%) and 532.8 eV (36.0%) in the chemical shift range corresponding to surface C—O moieties in different surface groups and adsorbed water are seen. The Fe $2p_{3/2}$ spectrum comprises the peaks at 710.3 eV (36.5%), 712.1 eV (36.9%) and 714.5 eV (26.6%). The spectrum has no shoulder at lower binding energies, which suggests that a reduced chemical form of iron like Fe^{2+} (708.7–707.1 eV), or even metallic Fe (707.2–706.7 eV) or iron carbide (706.9 eV) are absent [242,244], as expected. The BE values of the first two peaks (710.3 and 712.1 eV) are near to the literature data for Fe^{3+} (710.8–711.8 eV); the third one (714.5 eV) may be ascribed to iron ions complexed with electronegative surface ligands [245].

The results obtained for CWZ—Ox carbon subsequent to the adsorption of Fe^{3+} ions are similar (Fig. 34). The C 1s spectrum consists of several peaks, some of which have shifted towards the high-energy side. The C=C peak (284.5 eV) has become the smallest (27.3%) in relation to that of other carbon samples; furthermore, there is an increase in peak size, ascribed to C—O—(Fe) moieties (287.3 eV, 40.2%) as well as peaks in the chemical shift range of carboxylate species (288.8–290.7 eV, 10.7%). C—O moieties (285.9 eV, 17.0%) ascribed to structural C—O—C groups are also present. The O 1s spectrum consists of three peaks: the two main peaks at 532.8 eV (33.6%) and 534.7 eV (53.7%) ascribed as above, and an additional small one at 530.5 eV (1.9%) ascribed to isolated C=O moieties. The Fe $2p_{3/2}$ spectrum comprises the peaks at 710.3 eV (43.6%), 712.1 eV (36.9%), and 714.5 eV (19.4%). A relative increase in the peaks typical of Fe^{3+} is correlated with enhanced adsorption and ion exchange following oxidation. This suggests that iron ions could have been adsorbed at different active centers (surface oxides) and at a variety of sites surrounding the adsorbed iron. Recently, similar results of XPS studies were reported for some other metal-loaded active carbon materials [243].

The CWZ (—NM and —Ox) and D43/1 (—H, —Ox and —N) active carbons described and characterized above were used as powdered electrode materials in cyclic voltammetry experiments. An aqueous solution of 0.1 M or 0.01 M HNO_3 as background electrolyte and iron (III) nitrate as depolarizer were employed. Additionally, for comparison of electrochemical behavior, some measurements with the use of commercial carbon electrodes (made from pyrographite and glassy carbon) were carried out. By way of example, we show CV curves for the active carbon electrode materials studied, recorded in the absence (Fig. 35a–d) and in the presence (Fig. 35a′–d′) of Fe^{3+} ions (0.05 M) in the background electrolyte (0.01 M HNO_3). These background currents depend on the kind of carbon samples used as electrode materials as well as on the surface area of electrodes in contact with the solution. The estimated capacitive surface (geometric) current densities in the absence of electroactive species obtained from a–d CVs (Fig. 35) at potential $E = 0.75$ V are 256.4, 134.6, 66.1, and 13.9 $\mu A/cm^2$ for CWZ—NM, CWZ—Ox, pyrographite (RG), and a glassy carbon (GC) electrode, respectively. The capacitance of powdered carbon electrodes calculated from these data ($v = 5$ mV/s) is near 20 times higher that of GC, which implies that a much larger area participates in the electrochemical process. On the other hand, if the mesopore area of powdered material (112 m^2/g) is taken into consideration, capacitances considerably lower than those quoted in the literature for carbon electrodes are obtained [31,200]. Furthermore, estimated capacitances related to the mass of powdered electrode materials (0.80 and 0.39 F/g for CWZ—NM and CWZ—Ox carbon, respectively) are significantly lower than the double layer capacity of

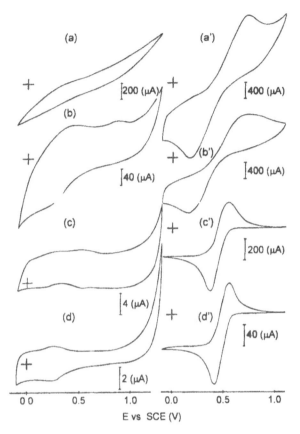

FIG. 35 CVs of carbon electrode materials investigated: a, a′, CWZ–NM; b, b′, CWZ–Ox; c, c′, Radelkis graphite electrode; d, d′, glassy carbon electrode. Curves recorded in 0.01 M HNO$_3$ in the absence (a–d) and in the presence (a′–d′) of 0.05 M Fe(NO$_3$)$_3$. Sweep rate $v = 5 \cdot 10^{-3}$ V/s. (From Ref. 227.)

active carbon materials [7,200]. The overall conclusion should be that probably only part of the electrode material or that an unknown proportion of the carbon particle surface (and macroporous surface area) participated in the studied electrochemical process.

Active carbon materials can be used for preconcentration (by adsorption) of diluted electroactive species (e.g., metal ions). Characterizing the electrochemical behavior of these systems is therefore important both for electroanalytical [248] and catalytic purposes [242–247,249]. On the other hand, how the adsorbed electroactive species interacts with the electrode material depends on its surface chemistry [247–256]. Whereas much attention has been given to the effects of surface treatment of graphite, glassy carbon, carbon fibers, carbon black, and pyrolytic carbon electrodes on electron transfer (ET) [250–256], the electrochemical properties of active carbon have been investigated only occasionally, despite their numerous applications (e.g., in fuel cells) [7,242].

Figure 36 shows the results of CV studies of modified D43/1 active carbon samples. CVs were recorded in 0.01 M $Fe(NO_3)_3$ solution for carbon samples with (c–c″) and without (curves b–b″) preadsorbed iron ions. In addition, CVs for carbon samples with preadsorbed iron were recorded in 0.01 M HNO_3 (curves a–a″). Comparing the electrochemical behavior of the oxidized samples (CWZ—Ox and D—Ox) shows the CVs obtained to be similar. The CV curves

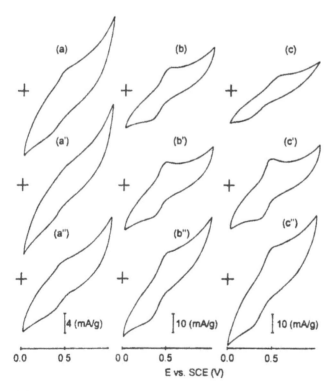

FIG. 36 Cyclic voltamograms of carbon electrodes with (a–a″, c–c″) and without (b–b″) preadsorbed Fe^{3+} ions recorded in different electrolyte systems: a–a″, 0.1 M $NaNO_3$ + 0.01 M HNO_3 (pH = 2.0); b–b″, 0.01 M $Fe(NO_3)_3$ + 0.1 M $NaNO_3$ (pH = 2.0); c–c″, 0.01 M $Fe(NO_3)_3$ + 0.1 M $NaNO_3$ (pH = 2.0); for D—H (a–c), D—Ox (a′–c′), and D—N (a″–c″). Sweep rate $\nu = 3.10^{-3}$V/s.

recorded for these carbons in the presence of iron (III) ions in solution (Figs. 35 and 36) display a pair of peaks ascribed to the Fe^{3+}/Fe^{2+} redox couple. The peak potentials depend on the kind of electrode used. In commercial carbon electrodes (RG and GC), they are similar to those quoted in the literature [254,256–258]. The peak currents also depend on the type of electrode materials used. The relationships between (a) anodic and (b) cathodic peak current densities and the depolarizer concentrations shown in Fig. 37 are linear over the concentration range studied. The linearity of these relationships for PACEs is not as good as that of commercial carbon electrodes. The linearity of the anodic peak current densities was inferior in all cases, which indicates that some interaction could have occurred between the electrogenerated Fe^{2+} ions and the electrode material surface. Hence both the porous structure of an electrode material and its surface chemical properties influence the charge transfer kinetics of this process. The formal redox potentials (E_f) of the Fe^{3+}/Fe^{2+} couple were estimated as average anodic and cathodic peak potentials at all the depolarizer concentrations and scan rates used. The respective charge transfer coefficients (n_α and n_β) and rate constants [$k_s(a)$ and $k_s(c)$] for the anodic and cathodic processes were calculated [259] from experimentally estimated peak and half-peak potentials and from diffusion coefficients of Fe^{3+} and Fe^{2+} ions gleaned from the literature ($0.35 \cdot 10^{-5}$ and $0.51 \cdot 10^{-5}$

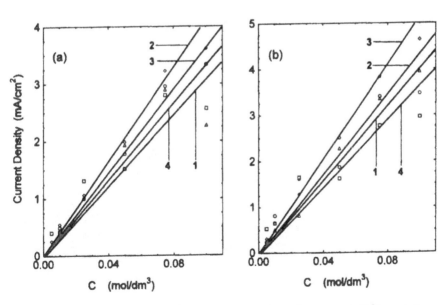

FIG. 37 Dependence of the anodic (a) and cathodic (b) peak currents on Fe^{3+} concentration in 0.01 M HNO_3: 1, CWZ—NM; 2, CWZ—Ox; 3, graphite electrode; 4, glassy carbon electrode. Sweep rate $v = 7.5 \cdot 10^{-3}$ V/s. (From Ref. 227.)

cm^2/s, respectively) [260]. The average values of these parameters at a selected depolarizer concentration ([Fe^{3+}] = 0.01 M) and several potential scan rates are given in Table 11. Additionally, the proportions between cathodic and anodic peak currents ($i_{p,c}/i_{p,a}$) in all the systems investigated have been recorded. As the differences between anodic and cathodic peak potentials for powdered electrodes are much higher than for commercial solid electrodes, charge-transfer processes are expected to the irreversible when the electrode surface becomes increasingly heterogeneous. The E_f values obtained for the commercial electrodes used are close to the literature data [260–262]. The kinetic parameters (n_α, n_β, k_s) quoted in Table 11 show that charge transfer is much slower at powdered electrodes, and that oxidative modification accelerates this process only slightly (approx. 20%) in relation to nonmodified carbon. Other authors [254,261] report very large increases (10- to 100-fold) in k_s for the Fe^{3+}/Fe^{2+} couple following oxidative modification of carbon electrodes. The differences observed for powdered carbons can be explained in that nonmodified carbon is partially covered by surface oxides (see IR and XPS spectra) and that the surfaces are additionally covered by adsorbed iron-containing ionic species. The adsorbed layer can play a significant role in charge-transfer processes [256,257], a fact that seems be confirmed by the long (24 h) equilibration (increase) of peak currents of the Fe^{3+}/Fe^{2+} couple. A similar period is required for equilibrium of cation adsorption from aqueous solution to be reached [224].

The CV curves for powdered active carbon (CWZ and D43/1) electrodes with previously adsorbed ionic iron species are shown in Figs. 36 (curves a-a", c-c") and 38. The presence of a pair of peaks, ascribed to the Fe(III)/Fe(II) redox couple ($E_{p,c}$ = 0.31 V, $E_{p,a}$ = 0.63 V), is very clear in the case of the oxidized carbon samples, which exhibit a higher adsorption capacity towards iron ions. By contrast, the nonmodified carbon sample produced only a scarcely visible cathodic peak. In view of the preadsorption conditions (acidic solution), these peaks should be ascribed to the iron ions bound to the carbon surface. The experimental variations in peak potential of Fe^{3+} ion reduction in solution (for CWN2—NM and CWN—Ox) and of preadsorbed iron ion (for CWZ—Ox carbon) vs. the logarithm of the sweep rate are given in Fig. 39 (lines 1–3, respectively) as an example. There is a more positive shift in reduction peak potentials for oxidized carbons, which may indicate greater irreversibility of the electrochemical process [255]. The relationship $E_{p,c}$ vs. log(v) for oxidized carbon is similar, both for the reduction of adsorbed iron (line 3) and for iron in bulk solution (line 2). On the other hand, the differences between nonmodified (line 1) and oxidized samples are clear. All this is indicative of the significance of surface functional groups and the adsorption process in the electrochemical reactions of powdered active carbons.

Oxidative modification of an active carbon surface raises considerably the amount of oxygen attached, particularly in the form of strongly acidic functional

TABLE 11 Redox Potentials ($E_{p,a}$, $E_{p,c}$, and E_f) and Kinetic Data of Fe^{3+}/Fe^{2+} Couple Obtained with Carbon Electrodes

Sample	$E_{p,a}$, V	$E_{p,c}$, V	E_f^a, V	n_α^b	n_α^b	$k_s(a)^c$, 10^{-5} cm/s	$k_s(c)^c$ 10^{-5} cm/s	$i_{p,a}$, mA	$i_{p,c}/i_{p,a}$
CWZ–NM	0.700	0.165	0.455	0.25	0.25	2.03	5.14	0.30	1.73
CWZ–Ox	0.640	0.170	0.445	0.30	0.26	2.48	5.40	0.32	1.47
RG	0.540	0.450	0.480	0.72	0.72	7.06	65.5	0.124	1.15
GC	0.530	0.450	0.480	0.72	0.78	9.36	63.5	0.0242	1.20

[a] Calculated as average values of anodic ($E_{p,a}$) and cathodic ($E_{p,c}$) peak potentials for all depolarizator concentrations and scan rates used.
[b] Charge transfer coefficients (n_α and n_β) for the anodic and cathodic processes, respectively.
[c] Rate constants [$k_s(a)$ and $k_s(c)$] for the anodic and cathodic processes, respectively.
Source: Ref. 227.

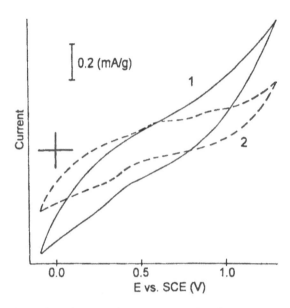

FIG. 38 Cyclic voltammograms of carbon electrodes (CWZ—NM, dashed line; CWZ—Ox, solid line) with preadsorbed Fe^{3+} ions recorded in 0.01 M HNO_3 solution. Sweep rate $v = 5 \cdot 10^{-3}$V/s. (From Ref. 227.)

groups, and significantly increases its capacity to adsorb water molecules and iron (III) ions. The spectral studies suggest that not only ion-exchange adsorption but also the formation of surface complexes with the participation of iron ionic species present in aqueous solution are possible. Oxidation of the carbon surface influences the charge transfer reaction in the Fe(III)/Fe(II) redox couple and, simultaneously, increases the irreversibility of this process.

B. Copper Ion Adsorption

The adsorption properties of modified active carbon samples with various oxygen- and/or nitrogen-containing surface functional groups towards copper ions have been studied [236]. Because surface modification procedures applied to the tested carbon alter its surface area only slightly (see Table 2), the nearly 10% differences in apparent surface area (S_{BET}) could not explain the marked differences in adsorption capacity (Fig. 40). The amount adsorbed is closely dependent on the pH of the solution and the modification procedure (Fig. 41). Cu^{2+} adsorption on AC samples was examined from external solutions with a pH range from 1 to 6. For greater pH values the possibility of copper being precipitated as $Cu(OH)_2$ is also increased [236]. The cation uptake increases linearly (0.07

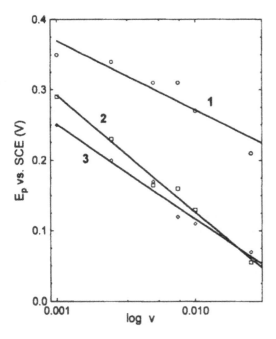

FIG. 39 $E_{p,c}$ vs. $\log(v)$ plots for Fe^{3+} ion reduction in bulk solution on CWZ—NM (1) and CWZ—Ox (2), and preadsorbed on CWZ—Ox (3) carbon electrodes. (From Ref. 227.)

mmol/pH unit) in this pH range for heat-treated sample (D—H), increases sharply between 1 and 3 in the case of the oxidized sample (D—Ox), while in the case of the ammonia-treated sample (D—N), it increases smoothly with increasing pH (0.03 mmol/pH unit). In acidic solutions (adjusted by the addition of sulfuric acid), adsorption increases in carbon samples modified with ammonia (D—N) and nitric acid (D—Ox) probably as result of the complexing action of surface functional groups. In a strongly acidic solution (pH = 1) Cu^{2+} adsorption is much better for the D—N modified sample; however, in a slightly acidic solution (pH = 3) it is the D—Ox modified sample that displays the greater adsorption. On the other hand, from nearly neutral solutions (adjusted by sodium hydroxide), adsorption is greatest on D—H modified active carbon. The jump in the adsorption found in the pH range 1 to 3 for the D—Ox sample might indicate that the point of zero charge of the oxidized AC lies between these two values, which is reasonable based on studies by Radovic and coworkers [8,233,263].

This seems to indicate that there is electrostatic repulsion between Cu^{2+} ions and the positively charged surface of carbons for the D—Ox sample at low pHs,

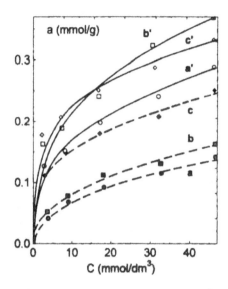

FIG. 40 The adsorption isotherms of Cu^{2+} ions on modified D43/1 carbons (a, a', D—H; b, b', D—Ox; c, c', D—N) from sulfate solution with various pH; a–c, 1.0; a'–c', 3.0. (From Ref. 236.)

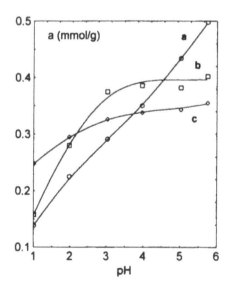

FIG. 41 The adsorption isotherms of Cu^{2+} ions on modified D43/1 carbons (a, D—H; b, D—Ox; c, D—N) from sulfate solution as a function of the pH. (From Ref. 236.)

as expected [233]. As the pH increases, the cations replace hydrogen ions from the carbon surface, and the extent of adsorption therefore increases rapidly [232]. The adsorption of Cu(II) ions by carbon is thus influenced principally by the presence of surface functional groups. Adsorption from copper sulfate solution (0.05 M, pH = 4.68) without additives alters the final (equilibrium) pH of the solution (Table 12). The reduction in the pH of the external solution for the D—Ox sample is caused by active hydrogen ions being released from the surface during ion exchange. The increase in pH for copper sulfate solutions with D—H and D—N samples is insignificant in relation to the changes observed for a neutral sodium inorganic salt solution (see Table 3). The surface group therefore influences the quantity of copper actions adsorbed both by direct interaction and by modification of the external solution's pH. The addition of any pH-adjusting electrolyte (acid or base) leads to competitive adsorption of the ions present in solution and influences the dissociative ionization of functional groups. In further studies the carbon samples were selected with copper adsorbed from 0.05 M CuSO$_4$. The quantities of adsorbed ions are given in Table 12.

The recorded differences in adsorption capability indicate a different mechanism of interaction between the carbon surface and the ionic metal species present in the aqueous solution (aqua and hydroxy complexes, hydroxide ions, and electronegative complexes). To discover the state of the adsorbed species, some independent measurements of the surface layer of adsorbent were carried out. The selected carbon samples were studied by the XPS method in powdered form following copper uptake (Figs. 42 and 43). Several peaks attributable to carbon, oxygen, nitrogen, and copper were present. The XPS survey spectra of the initial modified carbons (before adsorption) were discussed in the previous section. The surface elemental composition estimated from XPS data for modified D43/1 car-

TABLE 12 Cu^{2+} Adsorption Data and Surface Composition from XP Spectra

Modified sample	Cation uptake, mmol/g	pH[a]	C, %	O[b], %	N, %	Cu, %
D—H	—		96.88	2.72	0.40	—
D—H/Cu^{2+}	0.43	4.97	95.54	3.60	0.61	0.23
D—Ox	—		88.70	10.10	1.20	—
D—Ox/Cu^{2+}	0.25	2.45	87.20	10.30	0.90	0.11
D—N	—		90.89	6.20	2.81	—
D—N/Cu^{2+}	0.32	5.05	91.17	5.10	3.60	0.13

[a] Initial pH of 0.05 M CuSO$_4$—4.68.
[b] Oxygen from water excluded.
Source: Ref. 236.

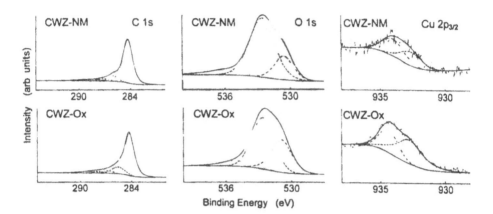

FIG. 42 High-resolution XPS spectra of C 1s, O 1s, and Cu 2p$_{3/2}$ regions of CWZ carbons with Cu^{2+} ions adsorbed from sulfate solution. (From Ref. 194.)

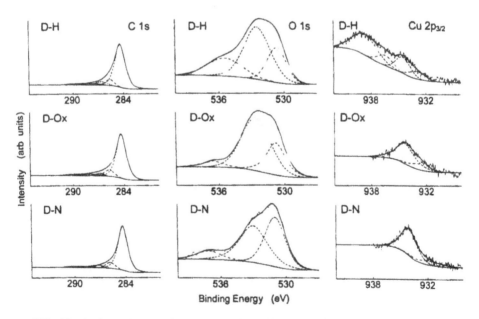

FIG. 43 High-resolution XPS spectra of C 1s, O 1s, and Cu 2p$_{3/2}$ regions of D43/1 carbons with Cu^{2+} ions adsorbed from sulfate solution.

bon sample is shown in Table 12. The amounts of surface copper were correlated with Cu^{2+} ion adsorption.

Figures 42 and 43 show the high resolution copper $2p_{3/2}$ spectra from the nonmodified and oxidized CWZ carbon samples as well as from modified D43/1 carbon (D—H, D—Ox, D—N) samples. There were marked differences between the experimental (dotted) and synthesized (continuous) lines in all the spectra. The position of the fitted peaks (dashed lines) were determined according to both the literature data [224,240,264–273] and empirically derived values. According to the literature data [240], the XPS Cu $2p_{3/2}$ lines calibrated with standard samples were found at binding energies (BE) of 932.4, 933.4, and 934.8 eV for copper in oxidation states 0, +1, and +2, respectively. Beyond the difference in energy shift, another important feature of the copper(II) XP spectrum is the presence, or absence, of a satellite peak at 942.5 eV. However, with the copper in the +1 and zero oxidation states, no such satellite peaks are observed. In other research [243], the binding energies of the Cu $2p_{3/2}$ transitions on copper-impregnated (5%) Cu activated carbon samples were reported as 933.4 and 934 eV and respectively interpreted as Cu_2O and CuO; the samples produced in this work were prepared in ammoniacal [269] or carbonate [243] solutions, dried, and heat-treated in air. This would result in a larger proportion of copper oxides. The BE near 935 eV was ascribed to the presence of $CuCO_3$ and $Cu(OH)_2$ species [270].

The main binding energy peaks recorded for D—Ox and D—N samples (near 80% relative peak area) in the 934.1–934.3 eV region can be ascribed to Cu(II) species adsorbed with participation of carbon surface functionalities. The second peak, with the lower binding energy, suggests the presence of nearly 20% of species with copper in a reduced form: Cu(I) and Cu(0) for D—Ox and D—N modified samples of active carbon, respectively. Unexpectedly, two other forms with higher binding energies are present (Fig. 43) for D—H carbon besides the Cu(I) (11.8%) and Cu(II) (23.7%) species. They can be assigned to $Cu(OH)_2$ species (936.4 eV, 15.8%) and Cu(III) (938.7 eV, 48.8%), which may be viewed as Cu(II) and a hole in the carbon matrix [271]. Additionaly, the positive shift in binding energy values could be explained as a consequence of the decreased relaxation in highly dispersed specimens of copper [272]. The influence of the chemical interaction of copper with surface functionalities (active adsorption centers) on binding energies cannot be ruled out either. The XPS results obtained for CWZ carbon samples (Fig. 42) generally confirmed the above discussion (especially for the oxidatively modified samples, CWZ—Ox and D—Ox).

Because the FTIR spectra of the carbon samples in question have already been discussed in detail (Section IV), Fig. 44 shows only the transmitance IR spectra of a number of carbons with preadsorbed copper ions. In the FTIR spectra of the carbon materials, the band of stretching OH vibrations (3600–3100 cm^{-1}) is due to the presence of surface hydroxylic groups, chemisorbed water, and adsorbed

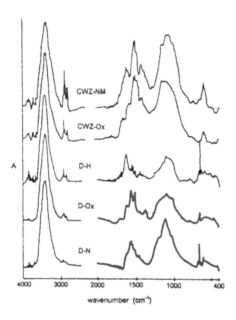

FIG. 44 FTIR spectra of CWZ (NM and Ox) and D43/1 (D—H, D—Ox, and D—N) carbons with Cu^{2+} ions adsorbed from sulfate solution.

copper hydroxy- and aqua-complexes [274]. For wavenumbers smaller than 2000 cm^{-1}, the shape of the FTIR spectrum depends on the kind of carbon (modification procedure), as discussed below.

For a heat-treated carbon sample (D—H), the presence of complex absorption bands in the 1650–1550 cm^{-1} region suggests that aromatic-ring-stretching and double-bond (C=C) vibration bands overlap the C=O stretching vibration and OH binding vibration bands. The presence of ion-radical structures with chemisorbed oxygen molecules is also possible [91,274]. Another broad band in the 1330–1000 cm^{-1} range consists of a series of overlapping absorption bands ascribable to etherlike (symmetrical stretching vibration) and OH bending modes. These results suggest that copper hydroxide species [$CuOH^+$, $Cu(OH)_2$] and copper (II) aqua complexes may be adsorbed on the carbon surface. The presence of bands below 800 cm^{-1} is characteristic of out-of-plane deformation vibrations of C—H moieties in aromatic structures. Additionally, the presence of narrow absorption peaks near 1110, 1000, and 670 cm^{-1} (the highest) points to the presence of sulfate ions (internal vibration of the SO_4^{2-}) [274] coadsorbed on the carbon surface as counterions. Furthermore, following the oxidation of active carbon (D—Ox and CWZ—Ox) and copper adsorption, there appears a band characteristic of carbonyl moieties in carboxylic acid (1720 cm^{-1}) and bands 1600 and 1580 cm^{-1} ascribable to metal chelates with carboxylate functionalities [274]. The presence of a peak at 1390 cm^{-1} characteristic of carboxylic ions is indicative of the partial ion-exchanging nature of Cu^{2+} adsorption on this carbon. For the D—N carbon sample the presence of overlapping peaks in the 1600–1500 cm^{-1} region and broad overlapping bands in the 1350–900 cm^{-1} region is seen following copper adsorption. An increase in the intensity of the bands in those regions is characteristic of pyridine-, pyridine-N-oxide-, and/or pyridonelike surface structures and related complexes with copper ions [274]. The peaks near 1100 and 670 cm^{-1} again indicate the presence of adsorbed SO$_4^{2-}$ ions. In accordance with the results of this spectroscopic study, the possible surface structures containing adsorbed copper can be shown schematically as follows:

For D—H carbon (Scheme 11 a–d),

$$>C—H + Cu^{2+} \rightarrow >C—H \cdot Cu^{2+} \quad \text{(dipole–dipole, } \pi\text{-}d \text{ interactions)} \quad (11a)$$

$$>C\bullet + Cu^{2+} \rightarrow >C \cdot Cu^{+} \rightarrow \oplus> C \cdot Cu^{0} \quad (11b)$$

$$>C—O\bullet + Cu^{2+} + H_2O \rightarrow >C—O—CuOH + H^{+} \quad (11c)$$

$$>C= O + Cu^{2+} + H_2O \rightarrow >C \cdot Cu(OH)_2 \quad (11d)$$

For D—Ox carbon (Scheme 12 a–c),

$$>C—COOH + Cu^{2+} \rightarrow >C—COOCu^{+} + H^{+} \quad (12a)$$

$$(>C—COOH)_2 + Cu^{2+} \rightarrow (>C—COO)Cu + 2H^{+} \quad (12b)$$

$$>C—OH + Cu^{2+} \rightarrow >C=O \cdot Cu^{+} + H^{+} \quad (12c)$$

For D—N carbon (Scheme 13 a–c),

$$>C—NH_2 + Cu^{2+} + H_2O \rightarrow >C—NH_2 \cdot Cu(OH)^{+} + H^{+} \quad (13a)$$

$$>N: + Cu^{2+} + H_2O \rightarrow >N—Cu(OH)^{+} + H^{+} \quad (13b)$$

$$>N: + Cu^{2+} + H_2O \rightarrow \oplus>NOH \cdot Cu^{0} + H^{+} \quad (13c)$$

where symbols $\bullet$ and $\oplus$ denote the ion-radical and positive holes in the carbon structure, respectively. In the interests of clarity, the coadsorbed counterions (mainly SO$_4^{2-}$) are not marked.

The possible interactions and surface structures presented above (Schemes 11–13) describing copper species sorbed on various modified active carbon samples have been deduced from the results obtained. It seems that the dominant mechanisms of copper adsorption on heat-treated active carbon (D—H sample) could be dipole–dipole (π-d) interactions between graphene layers and metal ionic species and the spontaneous electrochemical reduction of copper ions. For oxidized active carbon samples (D—Ox, CWZ—Ox), surface ionization and the ion-exchange mechanism can describe cation sorption from aqueous solutions.

Heat-treatment with ammonia (D—N sample) involves incorporation of nitrogen atoms into the graphene layers (preasumbly at their periphery), and these ad-atoms can play a dominant role, acting as ligands, when copper ions are adsorbed. This may additionally explain the low dependence of adsorption on pH.

The various forms of adsorbed copper can alter the electrochemical behavior of modified carbon samples used as electrode materials (powdered working elec-trodes in cyclic voltammetry). Figures 45 and 46 show cyclic voltammograms (CVs) for powdered electrodes prepared from selected active carbon samples with and without preadsorbed copper recorded in solution, which do or do not contain Cu^{2+} ions. An aqueous solution of 0.5 M Na_2SO_4 as background electro-lyte was employed. The CV curves recorded in the solution containing copper ions exhibit a pair of cathodic and anodic peaks, the potentials of which are dependent on the carbon modification procedure and the electrolyte's pH. The estimated peak potentials and the midpoint potentials [formal potentials, $E_f = (E_{p,a} - E_{p,c})/2$] are given in Table 13.

The electrochemistry of Cu (II) ions in aqueous solution is well documented [275–278] and can be summarized as follows:

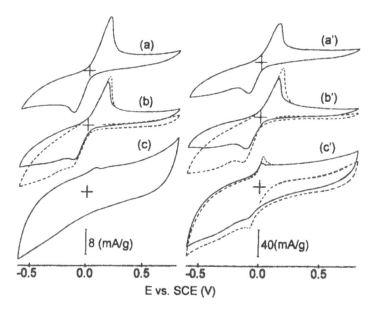

FIG. 45 Cyclic voltammograms of carbon electrodes with (b, b', c, c') and without (a, a') preadsorbed Cu^{2+} ions recorded in different electrolyte systems: a, a', b, b', 0.1 M $CuSO_4$ + 0.5 M Na_2SO_4 (pH = 4.46); c, c', 0.5 M Na_2SO_4 (pH = 6.02); for CWZ—NM (a–c) and CWZ—Ox (a'–c') ($v = 3\cdot10^{-3}$V/s). Dotted lines show the first cycles.

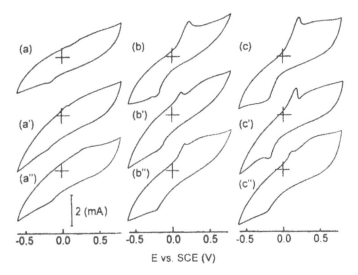

FIG. 46 Cyclic voltammograms of carbon electrodes with (a–a″, c–c″) and without (b, b′, b″) preadsorbed Cu^{2+} ions recorded in different electrolyte systems: a–a″, 0.5 M Na_2SO_4 (pH = 6.02); b–b″, 0.1 M $CuSO_4$ + 0.5 M Na_2SO_4 (pH = 4.46); c–c″, 0.1 M $CuSO_4$ + 0.5 M Na_2SO_4 (pH = 4.46) for D—H (a–c), D—Ox (a′–c′) and D—N (a″–c″) (v = 3·10^{-3}V/s). (From Ref. 236.)

$$Cu^{2+} + e \rightarrow Cu^+ \quad E^0 = -0.04 \text{ V} \tag{14}$$

$$Cu^+ + e \rightarrow Cu^0 \quad E^0 = +0.32 \text{ V} \tag{15}$$

$$Cu^{2+} + 2e \rightarrow Cu^+ \quad E^0 = +0.14 \text{ V} \tag{16}$$

In the absence of any complexing anions, Cu(I) is unstable and undergoes disproportionation. Complexing ligands stabilize Cu(I), and the Cu(II)/Cu(I) redox couple appears in cyclic voltammetry as a quasi-reversible wave [275].

The presence on the recorded CV curves peaks of the anodic response to the
one-electron reduction of Cu^{2+} ions and of formal potentials (E_f) close to E^0 for
the couple (14) (Table 13) suggests that both Cu(I) and Cu(II) surface-bound
species are present in these systems. For modified D-43/1 active carbon samples,
the CV curves recorded in the presence of Cu^{2+} in bulk solution (Fig. 46, curves
b–b″) display the redox peaks, which are the tallest for D—H samples; this corresponds well with the adsorption capacity towards copper ions (see Table 12).
The higher anodic peaks observed for D—H and D—N samples of modified
active carbons indicate that partial spontaneous reduction of adsorbed copper ions

TABLE 13 Characteristic Potentials of the Cu^+/Cu^{2+} Redox Couple in Various Systems ($v = 3 \cdot 10^{-3}$ V/s)

System studied	pH	$E_{p,a}$, V			$E_{p,c}$, V			E_f, V		
		D–H	D–Ox	D–N	D–H	D–Ox	D–N	D–H	D–Ox	D–N
in 0.1 M CuSO₄ + 0.5 M Na₂SO₄	4.46	+0.22	+0.12	+0.16	−0.22	−0.18	−0.21	0.00	−0.02	−0.02
in 0.1 M CuSO₄ + 0.5 M Na₂SO₄ + H₂SO₄	2.97	+0.25	+0.12	+0.20	−0.30	−0.19	−0.26	−0.03	−0.03	−0.03
ads. Cu from 0.1 M CuSO₄; in 0.5 M Na₂SO₄	6.02	+0.06	+0.06	+0.31	−0.16	−0.20	−0.15	−0.05	−0.06	−0.08
ads. Cu from 0.1 M CuSO₄; in 0.1 M CuSO₄ + 0.5 M Na₂SO₄	4.46	+0.21	+0.17	+0.12	−0.20	−0.21	−0.21	+0.01	−0.02	−0.04
ads. Cu from 0.1M CuSO₄; in 0.1 M CuSO₄ + 0.5 M Na₂SO₄ + H₂SO₄	2.97	+0.23	+0.18	+0.13	−0.23	−0.25	−0.23	−0.01	−0.02	−0.05

Source: Ref. 236.

(see Schemes 11b, 12c, 13c) and/or the disproportionation of reduced Cu(I) ions is occurring:

$$2 \, Cu^+ \rightarrow Cu^{2+} + Cu^0 \tag{17}$$

The CV curves obtained for carbons with preadsorbed copper shown in Figs. 45 (curves b, b′, c, c′) and 46 (a–a″)) exhibit only slight peaks of the Cu(II)/Cu(I) couple and broad waves due to the redox reaction of surface carbon functionalities (see Section IV). However, preadsorbed copper enhances the peaks of the redox process in bulk solution (especially the anodic peaks for D—H and D—Ox samples), as can be seen in Fig. 46 (curves c–c″). The low electrochemical activity of samples with preadsorbed copper species observed in neutral solution is the result of partial desorption (ion exchange with Na^+) of copper as well as the formation of an imperfect metalic layer (microcrystallites). Deactivation of the carbon electrode as a result of spontaneous reduction of metal ions (silver) was observed earlier [279,280]. The increase in anodic peaks for D—H and D—Ox modified samples with preadsorbed copper suggests that in spite of electrochemical inactivity, the surface copper species facilitate electron transfer reactions between the carbon electrode and the ionic form at the electrode–solution interface. The fact that the electrochemical activity of the D—N sample is lowest indicates the formation of strong complexes between adsorbed cations and surface nitrogen-containing functionalities (similar to porphyrin) [281]. Between −0.35 V and +0.80 V, copper (II) in the porphyrin complex (carbon electrode modifier) is not reduced, so there can be no reoxidation peak of copper (0) [281].

A general conclusion can be inferred from these observations: The number of copper ions adsorbed on active carbon depends on the nature and quantity of the surface acid–base and iono-radical functionalities as well as on the pH equilibria in the external solution. Possible dominant interactions between metal ions and various modified surfaces of active carbon could be considered:

Physical adsorption of ion (dipole–dipole interaction) and hydroxo-complexes as well as redox reactions with a change in metal valence (D—H sample)
Ion-exchange process with the participation of a strongly acidic surface group (D—Ox, CWZ—Ox samples)
Formation of surface complexes with nitrogen- and/or oxygen-containing surface groups (D—N sample)

The electrochemical behavior of the powdered active carbon electrode depends on the surface chemistry, and cyclic voltammetry can be used as a simple method of characterizing active carbon materials. A new heterogeneous copper catalyst was developed using highly porous active carbon as the catalyst support [282]. The advantages of a porous-medium supported catalyst are that the active phase could be kept in a dispersed but stable state, and that, as an example, the oxidized organic pollutant is adsorbed onto carbon, thereby enhancing its surface concentration; this favors the catalytic reaction. The state of adsorbed copper seems to have an effect on catalytic activity [249,273,282].

C. Silver Deposition

It has been shown that active carbon materials have not only excellent adsorption capabilities but also outstanding reduction properties. They are capable of reducing ions of higher standard potentials to adsorbed elemental metals or lower valence ions [283–306]. The contact of active carbon materials with an aqueous solution containing silver ion leads to cation adsorption and the formation of an imperfect metallic electrode sensitive to these ions. Silver films on platinum and carbon electrodes immersed in a solution of silver ions have already been observed [283]. The amount of silver taken up by active carbon depends closely on the immersion time [224,279], and the precipitation of metallic silver seems to predominate after the first two hours of contact. In order to measure the exchange capacity for Ag^+, a relatively short immersion time (2 h) was used, and the amounts of Ag^+ ions eluted with dilute nitric acid were determined (Table 14). The amount of silver deposited during the first 2 h of immersion depends on the chemical nature of the sorbent and is highest for D—H, intermediate for D—N, and nearly 2.5 times less for D—Ox carbon. Absorption of silver is accompanied by an increase in hydrogen ion concentration (in relation to the pH of carbon slurries in a neutral inorganic salt solution, see Tables 3 and 14) as a result of ion exchange and/or spontaneous oxidation of the carbon surface (with the participation of water molecules).

The cyclic voltammograms of Ag deposition/solution on the powdered active carbon electrodes (PACEs) in aqueous solutions with different Ag^+ concentrations are shown in Figs. 47 and 48. Figure 47 presents the CVs recorded in 0.05 M $NaNO_3$ + 0.01 M $AgNO_3$ solution. The cathodic wave for these electrodes occurs at higher potentials than for a platinum electrode; however, the positive shift is highest for the oxidized carbon (D—Ox) and lowest for the D—N carbon. This means that the presence of surface oxides facilitates reduction as opposed to nitrogen-containing functional groups (Brønsted bases). At low Ag^+ concentra-

TABLE 14 Ag$^+$ Ion Adsorption and Desorption Data

Carbon sample	Ag^+ uptake, mmol/g	pH	Ag^+ elution with 0.01 M HNO_3, mmol/g
D—H	1.120	3.60	0.180
D—Ox	0.484	2.45	0.182
D—N	0.872	3.66	0.242

Source: Ref. 279.

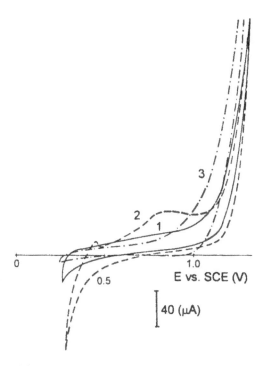

FIG. 47 Cyclic voltammograms of the PACEs studied (1, D—H; 2, D—Ox; 3, D—N) recorded after 2 h in 0.05 M $NaNO_3$ + 0.01 M $AgNO_3$ ($v = 10^{-2}$V/s). (From Ref. 279.)

tion, only the anodic response for the D—Ox carbon (curve 2, Fig. 47) is observed ($E_{p,a1} \cong +0.45$ V). Additionally, the presence of a broad anodic peak on this cyclic curve in the 0.8–0.9 V potential range ($E_{p,a2} \cong +0.85$ V) can be described by the oxidation of electroactive functional groups. The following reactions explain the observed peaks and waves:

$$\text{cathodic (waves): } Ag^+ + e = Ag^0 \qquad (E_{p,c1} < +0.3 \text{ V}) \qquad (18)$$

$$>C{=}O + H_3O^+ + e = {\equiv}C{-}OH + H_2O \qquad (E_{p,c2} < +0.3 \text{ V}) \qquad (19)$$

$$\text{anodic (peaks): } Ag^0 - e = Ag^+ \qquad (E_{p,a1} \cong +0.45 \text{ V}) \qquad (20)$$

$${\equiv}C{-}OH + H_2O - e = >C{=}O + H_3O^+ \qquad (E_{p,a2} \cong +0.85 \text{ V}) \qquad (21)$$

Reduction processes (18 and 19) create a cathodic wave whose half-wave potentials depend on the pH values of the electrolyte in the near-electrode-electrolyte layer.

The lack of an anodic response for nonoxidized carbons (D—H and D—N) and the weak response for oxidized carbon (D—Ox) suggests that silver ion re-

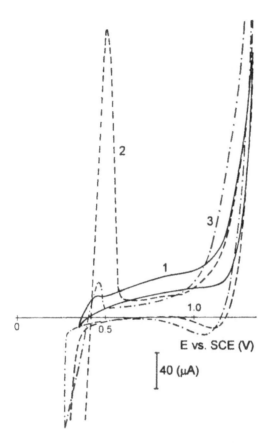

FIG. 48 Cyclic voltammograms of the PACEs studied (1, D—H; 2, D—Ox; 3, D—N) recorded after 2 h in 0.05 M NaNO$_3$ + 0.05 M AgNO$_3$ (v = 10^{-2}V/s). (From Ref. 279.)

duction is electrochemically irreversible. The shape of the cyclic curves in Fig. 47 indicates that for carbons with a basic surface (D—N, D—H) and a low silver ion concentration, all the ions in the near-electrode layer (volume of powdered electrode) are reduced to an electrochemically inactive metallic form. Furthermore, electro-oxidation of silver for these carbons is probably inhibited (behavior similar to metallic silver electrode).

This emerges clearly from Fig. 48, which shows CV curves recorded in a 0.05 M AgNO$_3$ + 0.05 M NaNO$_3$ solution. At higher Ag$^+$ ion concentration a strong anodic peak for oxidized carbon (D—Ox), very weak anodic peak for heat-treated carbon (D—H) and an intermediate peak for D—N carbon were recorded. The shape of the CV curves and hence the electrochemical equilibria for D—N and

D—H carbons are established very quickly; there is hardly any difference between the first and subsequent cycles. For D—Ox carbon the anodic peak during the first 2 h of cycling decreased in size (Fig. 49) and correlated well with the silver adsorption equilibrium on this carbon. Nearly 40% of the adsorbed silver can be removed from this carbon surface with diluted acid (Table 14), which indicates that the cation exchanger can be reactivated by reduction according to the scheme proposed by Jannakoudakis et al. for noble metal ions [303–306]:

$$\equiv C-COOH + Ag^+ \leftrightarrow \equiv C-COO \cdot Ag^+ + H^+ \tag{22}$$

$$\equiv C-COO^-Ag^+ + e + H^+ \leftrightarrow \equiv C-COOH \cdot Ag^0 \tag{23}$$

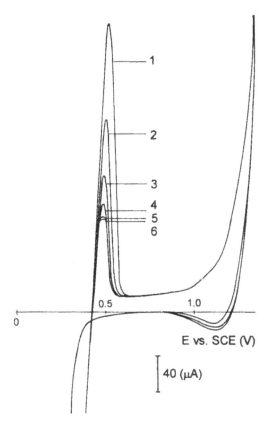

FIG. 49 Cyclic voltammograms of D—Ox carbon in 0.05 M NaNO$_3$ + 0.05 M AgNO$_3$ during cyclization: 1, 15 min; 2, 19 min; 3, 23 min; 4, 27 min; 5, 75 min; 6, 135 min (steady state) ($v = 10^{-2}$V/s). (From Ref. 279.)

The cyclic voltammograms obtained for powdered electrodes prepared from carbons with deposited silver are shown in Fig. 50–52. The CV recorded in neutral blank solution (0.05 M $NaNO_3$) (Fig. 50) for D—H carbon (curve 1) shows the presence of a pair of broad peaks ($E_{p,c} = +0.10$ V, $E_{p,a} = +0.65$ V) that can be described as surface oxide functional groups created as a result of spontaneous oxidative action of silver ions during deposition:

$$\equiv C\bullet + Ag^+ + H_2O = \equiv C-OH \cdot Ag^0 + H^+ \tag{24}$$

where $\equiv C\bullet$ is a Lewis base center.

The absence of silver oxidation and/or reduction peaks is evidence for the electrochemical inactivity of the silver deposited on this carbon (in the form of metallic crystallites). The cyclic voltammogram recorded for the D—Ox carbon (Fig. 50, curve 2) exhibits two anodic peaks ($E_{p,a2} = +0.27$ V, $E_{p,a2} = +0.77$ V) due to the oxidation of adsorbed silver and surface hydroquinone-like groups, respectively. A single cathodic peak ($E_{p,c1} = +0.16$ V) is due to the reduction of quinone-like surface groups according to Scheme 19. The large cathodic reduction wave confirms the presence of adsorbed silver cations and their reduction

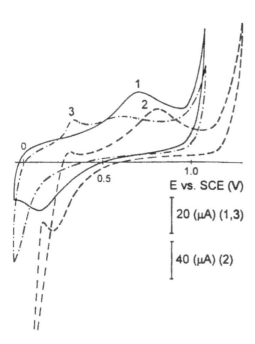

FIG. 50   CVs of the PACEs studied (1, D—H; 2, D—Ox; 3, D—N) with the preadsorbed silver recorded in 0.05 M $NaNO_3$ ($v = 10^{-2}$V/s). (From Ref. 279.)

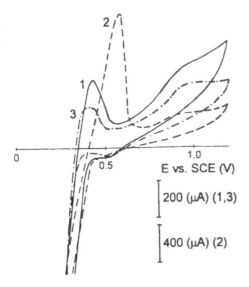

FIG. 51 CVs of the PACEs studied (1, D—H; 2, D—Ox; 3, D—N) with the preadsorbed silver recorded in 0.05 M NaNO₃ ($v = 10^{-2}$V/s). (From Ref. 279.)

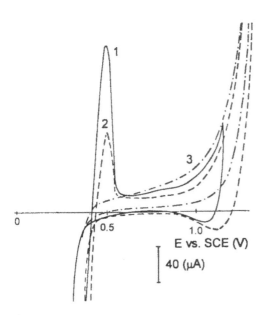

FIG. 52 CVs of the PACEs studied (1, D—H; 2, D—Ox; 3, D—N) with the preadsorbed silver recorded in 0.05 M NaNO₃ + 0.05 M AgNO₃ ($v = 10^{-2}$V/s). (From Ref. 279.)

according to Scheme 23. For D—N carbon (Fig. 50, curve 3) the anodic silver oxidation peak ($E_{p,a1} = +0.30$ V) is indicative of the presence of electroactive silver (e.g., isolated ion/atom couples) adsorbed on the surface. This can be explained by the presence of nitrogen atoms (Brønsted bases) in the surface structures of D—N carbon and their participation in adsorption as follows:

$$>N-H + Ag^+ = >N \cdot Ag^0 + H^+ \tag{25}$$

$$>N\bullet + Ag^+ = >N \cdot Ag^+ \tag{26}$$

Weakly marked and broad, the second anodic peak shows that the D—N carbon contains the smallest quantity of electroactive oxides, which indicates that it is resistant to oxidation during silver ion adsorption.

The cyclic voltammograms of all the carbons carrying preadsorbed silver, recorded in dilute nitric acid solution (Fig. 51), exhibit a Ag^+/Ag^0 couple (cathodic wave $< +0.4$ V and an anodic response in the $+0.40$–0.60 V potential range), as well as the electroactive quinone/hydroquinone-like surface system ($E_{p,c} \cong +0.50$ V; $E_{p,a2} \cong +0.90$ V). The presence of distinctly shaped anodic silver oxidation peaks indicates the partial solution of sorbed (deposited) metal. An almost sixfold higher anodic peak for D—Ox carbon confirms the partially ionic form of the adsorbed silver.

The cyclic voltammogram of Ag deposition/solution on carbons with preadsorbed silver in the 0.05 M NaNO₃ + 0.05 M AgNO₃ solution is shown in Fig. 52. In relation to the initial carbons (Fig. 48), a taller silver oxidation peak for the D—H carbon and a shorter one for the D—Ox are recorded; however, in the case of the D—N carbon, this peak almost completely disappears. This indicates that silver deposition on D—H carbon is easier if it has previously been covered with a metallic silver layer. This behavior is a typical example of the nucleation and growth during electrodeposition [302]. The different behavior of the D—N carbon is indicative of another kind of interaction (such as complex formation) between silver and surface functional groups.

All carbon samples were studied by the XPS method in powdered form before and after silver uptake. Some peaks attributable to carbon, oxygen, nitrogen, and silver (for samples after adsorption) are observed. The surface compositions estimated from XPS data for each sample are shown in Table 15. The amounts of surface silver correlated strongly with Ag^+ ion adsorption (Table 14). Figure 53 show the high-resolution spectra of the C 1s, O 1s, and Ag 3d regions. There were marked differences between the experimental (dotted) and synthesized (continuous) lines in all the spectra. The position of the fitted peaks (dashed lines) were determined according to the literature data [240,307]. Table 16 shows the results obtained by curve-fitting the Ag 3d$_{5/2}$ (364–370 eV region) spectra of silver deposited on the carbons. The binding energy (BE), the full width at half maximum (FWHM), and the relative peak area (r.p.a.) for separate peaks were

TABLE 15 Surface Composition from XPS Spectra (% at) of Active Carbons in Ground Form

Carbon sample	C	O[a]	N	Ag
D—H	96.88	2.72	0.40	—
D—H/Ag	94.60	3.77	0.47	1.15
D—Ox	88.70	10.10	1.20	—
D—Ox/Ag	88.98	9.42	1.12	0.48
D—N	90.81	6.20	2.91	—
D—N/Ag	90.18	638	2.76	0.67

[a] Oxygen from water excluded.
Source: Ref. 279.

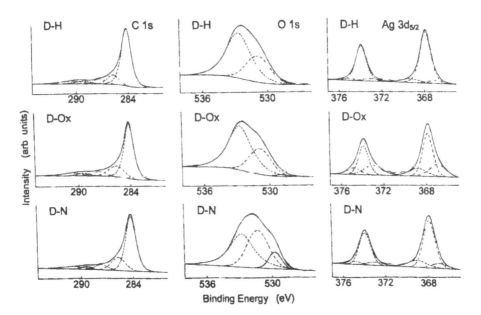

FIG. 53 High-resolution XPS spectra of C 1s, O 1s, and Ag 3d regions of D43/1 carbons with adsorbed Ag$^+$ ions from nitrate solution.

TABLE 16 Ag $3d_{5/2}$ Peak Parameters Deduced from
XPS Spectra

Carbon sample	BE, eV	r.p.a., %	FWHM, eV
	369.3	10.1	1.93
D—H	367.9	84.5	1.20
	367.0	5.4	1.62
	368.9	20.0	2.01
D—Ox	367.9	59.4	1.13
	366.9	21.6	1.80
	369.0	16.8	2.09
D—N	367.9	74.5	1.30
	367.0	8.7	1.71

Source: Ref. 279.

estimated and collected. The results show that the main peak (BE = 367.9 eV),
due to the deposition of metallic silver, is the largest for D—H carbon. Comparison of these peaks with the $3d_{5/2}$ peaks obtained from metallic Ag and solid AgO/
Ag_2O samples [307] shows them to be correlated to some extent. The BE shift
of the $3d_{5/2}$ peak from Ag metal (BE = 368 eV) to AgO (BE = 367.3 eV) is
negative, which is the opposite of what is expected between a metal and its oxides
[240,245]. This effect was explained by differences between Ag^0 and AgO other
than electronegativity, such as lattice potential, work function changes, and extra-
atomic relaxation energies [307]. Another difference between the $3d_{5/2}$ spectrum
obtained from metallic silver and silver oxides is that the oxide peak is much
broader (FWHM = 1.9 eV) than that due to Ag^0 (FWHM = 1.15 eV). Therefore
the broad peak obtained from the silver absorbed on the carbons (BE = 367.0
eV) may be due to the ionic form of the metal. The next broad peak with BE =
369.3 eV is probably caused by the presence of isolated zero-valence silver atoms
deposited on the carbon surface.

The XPS results presented in Table 16 confirm the adsorption/desorption data
(silver ions on carbon) as well as the electrochemical deposition/dissolution (silver on the electrodes) proposal. The smallest amount of adsorbed metallic silver
(in lattice form) and the highest quantity of ionic silver on the oxidized carbon
surface confirms the partial ion-exchange nature of the adsorption process for
this material. Moreover, the highest number of isolated silver atoms on this carbon (r.p.a. = 20%) shows that the reduction of metal ions took place after sorption
on surface functional groups. The ionic and isolated atomic forms of adsorbed
silver exhibit electrochemical activity (see Figs. 48 and 51) and can be eluted
with dilute nitric acid (see Table 14). The relatively high participation of metallic

silver (both latticed and isolated) in the total XPS signal for D—N carbon is indicative of the different nature of the interaction between nitrogen-containing surface functional groups and adsorbed silver species. Coordination of silver ions with nitrogen probably disperses the silver on the carbon surface, which prevents large crystallite formation.

Generally speaking, the surface chemistry of the electrode material strongly influences the Ag^+/Ag^0 redox process. First, spontaneous and rapid reduction of silver on the D—H carbon (Lewis-base) causes silver crystallites to be deposited on the carbon surface, which results in electrode deactivation. Second, the interaction of silver ions with surface oxygen- or nitrogen-containing functional groups (on the D—Ox and D—N carbons) gives rise to an anodic response, a peak due to the oxidation of isolated silver atoms. The preadsorption of silver on carbons alters the electrochemical behavior of the electrodes in the following way: (1) electroactive redox surface centers appear on the D—H carbon as a result of oxidation with Ag^+ ions (probably quinone or hydroquinonelike surface forms); (2) the anodic response (the silver oxidation peak) occurs in acid solution for all the active carbon materials studied.

VI. CONCLUSIONS

The properties of active carbon render it a difficult material to use as an electrode. Electrochemical processes occur more often in the inner cavities (pore structure) of active carbon particles than on their outer, planar surface. For this reason, three-dimensional electrochemical activity is observed rather than the planar responses characteristic of solid carbon electrodes. It is generally assumed that the total area of the internal structure of the porous carbon electrode is completely wetted by electrolyte, although this may not be the case with high-surface-area carbons containing micropores inaccessible to electrolyte. The main difficulty is estimating the electrochemically active part of the total surface area of the active carbon electrode material.

Therefore the electrochemical response with porous electrodes prepared from powdered active carbons is much increased over that obtained when solid electrodes are used. Cyclic voltammetry used with PACE is a sensitive tool for investigating surface chemistry and solid–electrolyte solution interface phenomena. The large electrochemically active surface area enhances double layer charging currents, which tend to obscure faradic current features. For small sweep rates the CV results confirmed the presence of electroactive oxygen functional groups on the active carbon surface. With peak potentials linearly dependent on the pH of aqueous electrolyte solutions and the Nernst slope close to the theoretical value, it seems that equal numbers of electrons and protons are transferred.

Cyclic voltammetry can also be used for investigating the influence of surface chemistry on the adsorption of selected heavy metal ions from aqueous solution on active carbons subsequently utilized as electrode material. The nature of the carbon materials and the kind of metal ion are highly significant. The measurements enabled the adsorbed metal species to be identified and their surface distribution to be estimated.

REFERENCES

1. M Smisek, S Cerny. *Active Carbon—Manufacture, Properties and Applications.* New York: Elsevier, 1970.
2. JS Mattson, HB Mark Jr. *Activated Carbon.* New York: Marcel Dekker, 1971.
3. RP Bansal, JB Donnet, F Stoeckli. *Active Carbon.* New York: Marcel Dekker, 1988.
4. F Rodriguez-Reinoso, A Linares-Solano, in PA Thrower, ed. *Chemistry and Physics of Carbon.* Vol 21. New York: Marcel Dekker, 1993, pp 1–146.
5. H Jankowska, A Świątkowski, J Choma. *Active Carbon.* Chichester: Ellis Horwood, 1991.
6. F Rodriguez-Reinoso, in H Marsh, EA Heintz, F Rodriguez-Reinoso, eds. *Introduction to Carbon Technologies.* Alicante: Universidad, 1997, pp 35–101.
7. K Kinoshita. *Carbon—Electrochemical and Physicochemical Properties.* New York: John Wiley, 1988.
8. CA Leon y Leon, LR Radovic, in PA Thrower, ed. *Chemistry and Physics of Carbon.* Vol 24. New York: Marcel Dekker, 1994, pp 213–310.
9. K Kinoshita, JAS Bett. *Carbon 11*:403–411, 1973.
10. K Kinoshita, JAS Bett. *Carbon 12*:523–533, 1974.
11. J Schreurs, J van der Berg, A Wonders, E. Barendrecht. *Recl Trav Chim Pays-Bas 103*:251–259, 1984.
12. A Surya, NA Murthy. *Electroanalysis 5*:265–268, 1993.
13. D Laser, M Ariel. *J Electroanal Chem 52*:291–303, 1974.
14. Y Yang, Z-G Lin. *J Electroanal Chem 364*:23–30, 1994.
15. JH Marsh, SW Orchard. *Carbon 30*:895–901, 1992.
16. KF Blurton. *Electrochim Acta 18*:869–875, 1973.
17. F Régisser, M-A Lavoie, GY Champagne, D Bélanger. *J Electroanal Chem 415*: 47–54, 1996.
18. AD Jannokoudakis, PB Jannokoudakis, E Theodoridou, JO Besenhard. *J Appl Electrochem 20*:619–624, 1990.
19. S Neffe. *Carbon 25*:761–767, 1987.
20. C Kozlowski, PMA Sherwood. *J Chem Soc Faraday Trans I 80*:2099–2107, 1984.
21. S Biniak, M Pakuła, A Świątkowski. *Monatsh Chem 125*:1365–1370, 1994.
22. BD Epstein, E Dalle-Molle, JS Mattson. *Carbon 9*:609–615, 1971.
23. AL Beilby, A Carlsson. *J Electroanal Chem 248*:283–304, 1988.
24. DR Lowde, JO Williams, PA Attwood, RJ Bird, B Nicol, RT Short. *J Chem Soc Faraday Trans I 77*:2312–2324, 1979.
25. P Żołtowski. *J Power Sources 1*:285–298, 1976/77.
26. JD Kaplan, JH Marsh, SW Orchard. *Electroanalysis 5*:509–516, 1993.
27. M Pakuła, A Świątkowski, S Biniak. *J Appl Electrochem 25*:1038–1044, 1995.

28. A Świątkowski, M Pakuła, S Biniak. *Electrochim Acta* 42:1441–1447, 1997.
29. M Yaniv, A Soffer. *J Electrochem Soc* 132:506–511, 1976.
30. JS Newman, CW Tobias. *J Chem Soc* 109:1183–1191, 1962.
31. JC Card, G Valentin, A Storck. *J Electrochem Soc* 137:2736–2745, 1990.
32. GM Jenkins, K Kawamura. *Polymeric Carbons—Carbon Fibre, Glass and Char*. Cambridge: Cambridge University Press, 1976.
33. IJ Spain, in PL Walker Jr, PA Thrower, eds. *Chemistry and Physics of Carbon*. Vol 16. New York: Marcel Dekker, 1981, pp 119–310.
34. G Dresselhaus, in S Sarangapani, JR Arkide, B Schumm, eds. *Electrochemistry of Carbon*. Proceedings of the Workshop. Electrochemical Society. Vol. 84–5, Pennington, NY, 1984, pp 5–60.
35. R Schlögl, in G Ertl, H Knözinger, J Weitkamp, eds. *Preparation of Solid Catalysts*. Weinheim, NY: Wiley-VCH, 1999, pp 150–240.
36. MM Dubinin, in PL Walker Jr, ed. *Chemistry and Physics of Carbon*. Vol 2. New York: Marcel Dekker, 1966, pp 51–120.
37. MM Dubinin, in JF Danielli, MD Rosenberg, D Cadenhead, eds. *Progress in Surface Membrane Science*. Vol. 9. New York: Academic Press, 1975, pp 1–70.
38. KSW Sing, DH Everett, RAW Haul, L Moscou, RA Pierotti, J Rouquerol, T Siemieniewska. *Pure Appl Chem* 57:603–619, 1985.
39. SJ Gregg, KSW Sing. *Adsorption, Surface Area and Porosity*. London: Academic Press, 1982.
40. T Paryjczak. *Gas Chromatography in Adsorption and Catalysis*. Chichester: Ellis Harwood, 1986.
41. JW Patrick, ed. *Porosity in Carbons*. London: Edward Arnold, 1995.
42. AJ Groszek. *Carbon* 29:821–830, 1991.
43. J Goworek, W Stefański, A Świątkowski. *Ads Sci Techn* 11:105–112, 1994.
44. DJ O'Connor, BA Sexton, RSC Smart, eds. *Surface Analysis Methods in Material Science*. Berlin: Springer-Verlag, 1992.
45. G Horvath, K Kawazoe. *J Chem Eng Japan* 16:470–475, 1983.
46. C Lastoskie, KE Gubbins, N Quirke. *Langmuir* 9:2693–2702, 1993.
47. SR Tennison, N Quirke. *Carbon* 34:1281–1286, 1996.
48. C Lastoskie, KE Gubbins, N Quirke. *J Phys Chem* 97:4786–4796, 1993.
49. CM Lastoskie, N Quirke, KE Gubbins, in W Rudzinski, WA Steele, G Zagrablich, eds. *Equilibria and Dynamics of Gas Adsorption on Heterogeneous Solid Surfaces*. (B Delmon, JT Yates, eds. *Studies in Surface Science and Catalysis*. Vol 120.) Amsterdam: Elsevier, 1999, pp 745–775.
50. S Scaife, P Kluson, N Quirke. *Exten Abstr 24th Biennial Conference on Carbon, Charleston, 1999*, pp 460–461.
51. Special issue. Fundamental aspects of active carbon. *Carbon* 36:1417–1554, 1998.
52. D Avnir, D Farin, P Pfeifer. *J Chem Phys* 79:3566–3571, 1983.
53. P Pfeifer, YJ Wu, MW Cole, J Krim. *Phys Rev Lett* 62:1997–2000, 1989.
54. P Pfeifer, K-Y Liu, in W Rudzinski, WA Steele, G Zagrablich, eds. *Equilibria and Dynamics of Gas Adsorption on Heterogeneous Solid Surfaces*. (B Delmon, JT Yates, eds. *Studies in Surface Science and Catalysis*. Vol 120.) Amsterdam: Elsevier, 1999, pp 625–677.
55. AT Terzyk, PA Gauden, G Rychlicki, R Wojsz. *Langmuir* 15:285–288, 1999.

56. AT Terzyk, PA Gauden, G Rychlicki, R Wojsz. *Colloids Surf A Physicochem Engin Aspects 152*:293–313, 1999.
57. RL McCreery, in A. J. Bard, ed. *Electroanalytical Chemistry.* Vol. 17. New York: Marcel Dekker, 1991, pp 221–374.
58. MR Tarasevich, EI Khrushcheva, in BE Conway, JO'M Bockris, KN Reddy, eds. *Modern Aspects of Electrochemistry.* Vol 19. New York: Plenum Press, 1989, pp 295–372.
59. MR Tarasevich. *Elektrokhimiya uglerodnych materiolov.* Moscow: Izdatelstvo Nauka, 1984.
60. LR McCreery, KK Cline, CA McDermott, MT McDermott. *Colloids Surf A 93*: 211–219, 1994.
61. LR McCreery, KK Claine, in PT Kissinger, WR Heineman, eds. *Laboratory Techniques in Electroanalytical Chemistry.* 2d. ed. New York: Marcel Dekker, 1996, pp 293–332.
62. E Yeager, JA Molla, S Gupta, in S Sarangapani, JR Akridge, B Schumm, eds. *Proceedings of the Workshop on the Electrochemistry of Carbon.* Electrochemical Society. Vol. 84–85. Pennington, NY, 1984, pp 123–157.
63. HP Boehm, E Diehl, W Heck, R Sappok. *Angew Chem 76*:742–751, 1964.
64. BR Puri, in PL Walker Jr, ed. *Chemistry and Physics of Carbon.* Vol 6. New York: Marcel Dekker, 1970, pp 191–282.
65. JB Donnet. *Carbon 20*:266–282, 1982.
66. HP Boehm. *Carbon 32*:759–769, 1994.
67. JT Fabish, DE Schleifer. *Carbon 22*:19–38, 1984.
68. JA Menendez, J Phillips, B Xia, LR Radovic. *Langmuir 12*:4404–4410, 1996.
69. CA Leon y Leon, JM Solar, V Calemma, LR Radovic. *Carbon 30*:797–811, 1992.
70. A Contescu, M Vass, C Contescu, K Putyera, JA Schwarz. *Carbon 36*:247–258, 1998.
71. MA Montes-Moran, JA Menendez, E Fuente, D Suarez. *J Phys Chem B 102*:5595–5601, 1998.
72. VA Garten, DE Weiss. *Austral J Chem 10*:309–321, 1957.
73. M Voll, HP Boehm. *Carbon 9*:481–488, 1971.
74. E Papirer, S Li, J-B Donnet. *Carbon 25*:243–247, 1987.
75. JA Menendez, D Suarez, E Fuente, MA Montes-Moran. *Carbon 37*:1002–1006, 1999.
76. D Suarez, JA Menendez, E Fuente, MA Montes-Moran. *Langmuir 15*:3897–3904, 1999.
77. P Davini. *Carbon 28*:565–571, 1990.
78. A Bismarc, C Wuertz, J Spinger. *Carbon 37*:1019–1027, 1999.
79. JA Menendez, MJ Illan-Gomez, CA Leon y Leon, LR Radovic. *Carbon 33*:1655–1659, 1995.
80. MV Lopez-Ramon, F Stoeckli, C Moreno-Castilla, F Carrasco-Marin. *Carbon 37*: 1215–1221, 1999.
81. E Papirer, J Dentzer, S Li, J-B Donnet. *Carbon 29*:69–72, 1991.
82. M Voll, HP Boehm. *Carbon 9*:473–480, 1971.
83. B Steenberg. *Adsorption and Exchange of Ions on Activated Charcoal.* Uppsala: Almquist and Wiksell, 1944.

84. HP Boehm, M Voll. *Carbon 8*:227–233, 1970.
85. M Voll, HP Boehm. *Carbon 8*:741–752, 1970.
86. J Zawadzki, S Biniak. *Polish J Chem 62*:195–202, 1988.
87. JS Noh, J Schwarz. *Carbon 28*:675–682, 1990.
88. JB Tomlinson, JJ Freeman, CR Theocharis. *Carbon 31*:13–20, 1993.
89. S Biniak, GS Szymański, J Siedlewski, A Świątkowski. *Carbon 35*:1799–1810, 1997.
90. P Vinke, M van der Eijk, M Verbee, AF Voskamp, H van Bekkum. *Carbon 32*: 675–686, 1994.
91. J Zawadzki, in PA Thrower, ed. *Chemistry and Physics of Carbon*. Vol 21. New York: Marcel Dekker, 1988, pp 147–380.
92. IA Tarkovskaja. *Okisliennyj ugol.* Kijev: Naukova Dumka, 1991.
93. T van der Plas, in BG Linsen, ed. *Physical and Chemical Aspects of Adsorbents and Catalysts.* London: Academic Press, 1970.
94. C Ishizaki, I Marti. *Carbon 19*:409–412, 1981.
95. MO Corapcioglu, CP Huang. *Carbon 25*:569–578, 1987.
96. SS Barton, J Dacey, MJB Evans. *Colloid Polymer Sci 260*:726–731, 1982.
97. JT Cookson, in PN Cheremisinoff, F Ellerbusch, eds. *Carbon Adsorption Handbook.* Ann Arbor: Science Publication, 1978, pp 241–280.
98. BR Puri, DD Shing, LR Sharma. *J Phys Chem 62*:756–758, 1958.
99. J Zawadzki, S Biniak, J Siedlewski. *Polish J Chem 57*:207–215, 1983.
100. J Zawadzki. *Carbon 19*:19–25, 1981.
101. PE Fanning, MA Vannice. *Carbon 31*:721–730, 1993.
102. H Marsh, AD Ford, JS Mattson, JM Thomas, EL Evans. *J Colloid Intrface Sci 49*: 368–382, 1974.
103. J Zawadzki, J Siedlewski, S Biniak. *Polish J Chem 55*:1575–1583, 1981.
104. BR Puri, RC Bansal. *Carbon 1*:457–464, 1964.
105. S Neffe. *Carbon 25*:441–443, 1987.
106. A Contescu, CC Contescu, K Putyera, JA Schwarz. *Carbon 35*:83–94, 1997.
107. AS Arico, V Antonucci, M Minutoli, N Giordano. *Carbon 27*:337–347, 1989.
108. BE Reed, MR Matsumoto. *Carbon 29*:1191–1201, 1991.
109. TJ Bandosz, J Jagiello, C Contescu, JA Schwarz. *Carbon 31*:1193–1202, 1993.
110. MM Dubinin, VV Serpinsky. *Doklady AN SSSR 285*:1151–1154, 1981.
111. J Kazmierczak, S Biniak, A Świątkowski, K-H Radeke. *J Chem Soc Faraday Trans 87*:3557–3561, 1991.
112. F Stoeckli, L Currit, A Leaderach, TA Centeno. *J Chem Soc Faraday Trans 90*: 3689–3691, 1994.
113. SS Barton, MJB Evans, E Halliop, JAF MacDonald. *Carbon 35*:1361–1366, 1997.
114. AP Terzyk, PA Gauden, G Rychlicki. *Colloids Surfaces A 148*:271–281, 1999.
115. F Carrasco-Martin, J Rivera-Utrilla, J-P Joly, C Moreno-Castilla. *J Chem Soc Faraday Trans 92*:2779–2782, 1996.
116. Y Otake, RG Jenkins. *Carbon 31*:109–121, 1993.
117. S Haydar, JP Jolly. *J Thermal Anal 52*:345–353, 1998.
118. JL Figueiredo, MFR Pereira, MMA Freitas, JJM Orfao. *Carbon 37*:1379–1389, 1999.
119. JV Hallum, MV Druschel. *J Phys Chem 62*:110–117, 1958.

120. RA Fridel, LIE Hofer. *J Phys Chem* 74:291–292, 1970.
121. C Moreno-Castilla, MA Ferro-Garcia, JP Joly, I Bautista-Toledo, F Carrasco-Marin, J Rivera-Utrilla. *Langmuir* 11:4386–4392, 1995.
122. M Acedo-Ramos, V Gomez-Serrano, C Valenzuela-Calahorro, AJ Lopez-Peinado. *Spectroscopy Letters* 26:1117–1137, 1993.
123. MJD Low, C Morterra, AG Severida, M Lacroix. *Appl Surface Sci* 13:429–440, 1982.
124. AS Glass, MJD Low. *Spectroscopy Letters* 19:397–404, 1986.
125. JJ Venter, MA Vannice. *Carbon* 26:889–896, 1998.
126. BJ Meldrum, CH Rochester. *J Chem Soc Faraday Trans* 86:1881–1884, 1990.
127. BJ Meldrum, CH Rochester. *Fuel* 70:57–63, 1991.
128. A Dandekar, RTK Baker, MA Vannice. *Carbon* 36:1821–1831, 1998.
129. PA Christensen, A Hamnett, in RG Compton, A Hamnett, eds. *Chemical Kinetics*. Vol 29. Oxford: Elsevier, 1989, pp 1–78.
130. M Datta, JJ Freeman, REW Jansson. *Spectroscopy Letters* 18:273–282, 1985.
131. DM Anjo, S Brown, L Wang. *Anal Chem* 65:317–319, 1993.
132. Y Yang, Z-G Lin. *J Appl Electrochem* 25:259–266, 1995.
133. M Polovina, B Babic, B Kaluderovic, A Dekanski. *Carbon* 35:1047–1052, 1997.
134. F Kapteijn, JA Moulijn, S Matzner, H-P Boehm. *Carbon* 37:1143–1150, 1999.
135. P Alberts, K Deller, BM Despeyroux, G Prescher, A Schafer, K Seibold. *J Catal* 150:368–375, 1994.
136. LR Radovic, IF Silva, JI Ume, JA Menendez, CA Leon y Leon, AW Scaroni. *Carbon* 35:1339–1348, 1997.
137. B Buczek, S Biniak, A Świątkowski. *Fuel* 78:1443–1448, 1999.
138. S Mrozowski. *Carbon* 9:97–109, 1971.
139. A Espinola, PM Miguel, MR Salles, AR Pinto. *Carbon* 24:337–341, 1986.
140. K-H Radeke, KO Backhaus, A Swiatkowski. *Carbon* 29:122–123, 1991.
141. SS Barton, JE Koresh. *Carbon* 22:481–486, 1984.
142. B Kastening, M Hahn, B Rabanus, M Heins, U zum Felde. *Electrochim Acta* 42:2789–2800, 1997.
143. B Kastening. *Ber Bunsenges Phys Chem* 102:229–237, 1998.
144. Z Szeglowski, A Świątkowski. *Extended Abstracts*. 37th Meeting Internat Soc Electrochem, Vilnus, 1986, Vol 3, pp 341–343.
145. Z Szeglowski, A Świątkowski. *Mat Sci Forum* 25/26:431–433, 1988.
146. H Jankowska, A Świątkowski, Z Szeglowski, JK Garbacz. *Przemysl Chemiczny* 68:507–510, 1989 (in Polish).
147. LA Fokina, HA Shurmovskaia, RH Burshtein. *Kinetik Katal* 4:143–148, 1963 (in Russian).
148. B Kastening, W Schiel, M Henschel. *J Electroanal Chem* 191:311–328, 1985.
149. B Kastening, S Spinzing. *J Electroanal Chem* 214:295–302, 1986.
150. M Muller, B Kastening. *J Electroanal Chem* 374:149–158, 1994.
151. A Soffer, M Folman. *J Electroanal Chem* 38:25–43, 1972.
152. J Koresh, A Soffer. *J Electrochem Soc* 124:1379–1385, 1977.
153. B Kastening, M Hahn, J Kremeskotter. *J Electroanal Chem* 374:159–166, 1994.
154. Y Oren, A Soffer. *J Electroanal Chem* 186:63–77, 1985.
155. Y Oren, A Soffer. *J Electroanal Chem* 206:101–114, 1986.

156. D Golub, Y Oren, A Soffer. *J Electroanal Chem* 227:41–53, 1987.
157. JP Randin, E Yeager. *J Electroanal Chem* 36:257–276, 1972.
158. C Urbaniczky, K Lundstrom. *J Electroanal Chem* 176:169–182, 1984.
159. H Tobias, A Soffer. *J Electroanal Chem* 148:221–232, 1983.
160. V Majer, J Vesely, K Stulik. *J Electroanal Chem* 45:113–125, 1973.
161. H Jankowska, S Neffe, A Świątkowski. *Electrochim Acta* 26:1861–1866, 1981.
162. A Frumkin. *Usp Khim* 18:9–21, 1949 (in Russian).
163. A Frumkin, E Ponomarenko, R Burstein. *Dokl Akad Nauk SSSR* 149:1123–1126, 1963 (in Russian).
164. E Ponomarenko, A Frumkin, R Burstein. *Izv Akad Nauk SSSR Otdel Khim Nauk* 9:1549–1555, 1963 (in Russian).
165. VA Garten, DE Weiss. *Rev Pure Appl Chem* 7:69–122, 1957.
166. JA Korver. *Chem Week* 46:301–302, 1950.
167. A Świątkowski, A Deryło-Marczewska, J Goworek, S Biniak. *J Colloid Interface Sci* 218:480–487, 1999.
168. MM Dubinin. *Carbon* 19:321–324, 1981.
169. K Tsutsumi, Y Matsushima, A Matsumoto. *Langmuir* 9:2665–2669, 1993.
170. W Kopycki, D Fraise, J Binkowski. *Chem Anal* 25:829–839, 1980 (in Polish).
171. MM Dubinin, VV Serpinsky. *Carbon* 19:402–403, 1981.
172. G Rychlicki, in *Materiały Seminarium. Węgiel-sorbenty węglowe.* Kraków, 1998, pp 35–36 (in Polish).
173. BK Pradhan, NK Sandle. *Carbon* 37:1323–1332, 1999.
174. CA Toles, WE Marshall, MM Johns. *Carbon* 37:1207–1214, 1999.
175. E Yeager. *Electrochim Acta* 29:1527–1537, 1984.
176. B Stöhr, HP Boehm, R Schlögl. *Carbon* 29:707–720, 1991.
177. D Mang, HP Boehm, K Stanczyk, H Marsh. *Carbon* 30:391–398, 1992.
178. HP Boehm, G Mair, T Stöhr, AR de Rincon, B Tereczki. *Fuel* 63:1061–1063, 1984.
179. RJJ Jansen, H van Bekkum. *Carbon* 32:1507–1516, 1994.
180. SS Barton, MJB Evans, JAF MacDonald. *Adsorption Sci Technol* 10:75–84, 1993.
181. BJ Meldrum, JC Orr, CH Rochester. *J Chem Soc Chem Comun* 1985:1176–1177, 1985.
182. M Sakogushi, K Uematsu, A Sakata, Y Sato, M Sato. *Electrochim Acta* 34:625–630, 1989.
183. DJ Suh, T-J Park, SK Ihm. *Carbon* 31:427–436, 1993.
184. A Macias-Garcia, C Valenzuela-Calahorro, V Gomez-Serrano, V Espinosa-Mansilla. *Carbon* 31:1249–1256, 1993.
185. J Zawadzki. *Carbon* 16:491–497, 1978.
186. V Gomez-Serrano, J Pastor-Villegas, A Perez-Florindo, C Durran-Valle, C Valenzuela-Calahorro. *J Anal Appl Pyrolysis* 36:71–80, 1996.
187. C Ishizaki, in *Fundam Adsorpt Proc Eng Found Conf 1983* (Pub 1984), pp 229–238.
188. F Atamy, J Blöcker, A Dübotzky, H Kurt, O Tipe, G Loose, W Mahdi, R Schlögl. *Molecular Phys* 76:851–886, 1992.
189. T Grzybek. *Polish J Chem* 69:1649–1657, 1994.
190. WH Lee, PJ Reucroft. *Carbon* 37:7–14, 1999.

191. V Alderucci, L Pino, PL Antonucci, W Roh, J Cho, H Kim, DL Cocke, V Antonucci. *Materials Chem Phys 41*:9–14, 1995.
192. D Briggs, MP Seah, eds. *Practical Surface Analysis by Auger and X-Ray Photoelectron Spectroscopy.* New York: John Wiley, 1983.
193. J Lahaye, G Nanse, A Bagreev, V Strelko. *Carbon 37*:585–590, 1999.
194. M Pakuła. PhD thesis. Warsaw: Military Technical Academy, 1999 (in Polish).
195. OS Ksenzhek. *Zh Fiz Khim 37*:2007–2011, 1963 (in Russian).
196. K Mund, F v Sturm. *Electrochim Acta 20*:463–467, 1975.
197. A Srikumar, TG Stanford, JW Weidner. *J Electronal Chem 458*:161–173, 1998.
198. CA Frysz, X-P Shui, DDL Chung. *Carbon 35*:893–916, 1997.
199. A Świątkowski, J Rubaszkiewicz, E Bednarkiewicz. *J Electroanal Chem 239*:91–105, 1988.
200. T Momma, X-J Liu, T Osaka, Y Ushio, Y Sawada. *J Power Sources 60*:249–253, 1996.
201. ML Bowers, BA Yenser. *Anal Chim Acta 243*:43–53, 1991.
202. FE Woodart, DE McMackins, REW Jansson. *J Electroanal Chem 214*:303–330, 1986.
203. F Mohammdi, P Timbrel, S Zhong, C Pedeste, M Skyllas-Kazacos. *J Power Sources 52*:61–68, 1994.
204. I Tanahashi, A Yoshida, A Nishino. *J Appl Electrochem 21*:28–31, 1991.
205. H-P Dai, K-K Shiu. *J Electronal Chem 419*:7–14, 1996.
206. SI Bailey, IM Ritchie. *Electrochim Acta 30*:3–12, 1985.
207. MR Tarasevich, SI Sysov, WA Bogdanovskaya. *Elektrokhimia 20*:1202–1210, 1984 (in Russian).
208. MR Tarasevich, WA Bogdanovskaya, NM Zagudaeva. *J Electroanal Chem 223*: 161–169, 1987.
209. KT Kawagoe, PA Garris, RM Wightman. *J Electroanal Chem 259*:193–207, 1993.
210. G Gun, M Tsionsky, O Lev. *Anal Chim Acta 294*:261–270, 1994.
211. T Nagaoka, T Yoshino. *Anal Chem 58*:1037–1042, 1986.
212. JO Oliveira, EL Barish, RO Allen. *Anal. Chem 56*:1747–1749, 1986.
213. N Motta, AR Guadelupe. *Anal Chem 66*:566–571, 1994.
214. J Mattusch, G Werner. *Z Chem 30*:345–351, 1990 (in German).
215. RC Engstrom. *Anal Chem 54*:2310–2314, 1982.
216. D Jürgen, E Steckhan. *J Electroanal Chem 333*:177–193, 1992.
217. HA Fishman, AG Ewing. *Electroanal 3*:899–907, 1991.
218. MR Deakin, RM Wightman. *J Electroanal Chem 206*:167–177, 1986.
219. SA Pietrova, MW Kolobiazhnyi, OS Ksenzhek. *Elektrokhimia 25*:1493–1497, 1989 (in Russian).
220. F Pariente, F Tobalina, M Darder, E Lorenzo, HD Abruna. *Anal Chem 68*:3135–3142, 1996.
221. C Fernandez, E Chico, P Yanez-Sedeno, JM Pingarron, LM Polo. *Analyst 117*: 1919–1923, 1992.
222. C Barbero, JJ Silber, L Sereno. *J Electroanal Chem 248*:321–340, 1988.
223. MO Corapcioglu, CP Huang. *Wat Res 21*:1031–1044, 1987.
224. A Świątkowski, GS Szymański, S Biniak, in MD LeVan, ed. *Fundamentals of Adsorption.* Boston: Kluwer, 1996, pp 913–919.

225. CP Huang, in PN Cheremisinoff, F Ellerbusch, eds. *Carbon Adsorption Handbook.* Ann Arbor: Science Publishers, 1978, pp 281–330.
226. A Gierak. *Adsorption Sci Techn 14*:47–57, 1996.
227. M Pakuła, S Biniak, A Świątkowski. *Langmuir 14*:3082–3089, 1998.
228. M Goyal, VK Rattan, RC Bansal. Extended Abstracts of the "Eurocarbon" Conference, Strasbourg, 1998, p 269.
229. A Seco, P Marzal, C Gabalón, J Ferrer. *J Chem Tech Biotechnol 68*:3–12, 1997.
230. N Petrov, T Budinova, I Khavesov. *Carbon 30*:135–139, 1992.
231. MA Kahn, YI Khattak. *Carbon 30*:957–960, 1992.
232. MA Ferro-Garcia, J Rivera-Utrilla, J Rodriguez-Gordillo, I Bautista-Toledo. *Carbon 26*:363–374, 1988.
233. LR Radovic, F Rodriguez-Reinoso, in PA Thrower, ed. *Chemistry and Physics of Carbon.* Vol 25. New York: Marcel Dekker, 1997, pp 243–358.
234. JS Johnson Jr, CG Westmoreland, FH Sweeton, KA Kraus, EW Hagmann, WP Eatherly, HR Child. *J Chromatogr 354*:231–248, 1986.
235. RE Panzer, PJ Elving. *Electrochim Acta 20*, 635–647, 1975.
236. S Biniak, M Pakuła, GS Szymański, A Świątkowski. *Langmuir 15*:6117–6122, 1999.
237. PJM Carrott, MML Ribeiro Carrott, JMV Nabais, JP Prates Ramalho. *Carbon 35*: 403–410, 1997.
238. BE Reed, MR Matsumoto. *Separ Sci Techn 28*:2179–2195, 1993.
239. S Biniak, GS Szymański, A Świątkowski, in A Ziegler, KH van Heek, J Klein, W Wanzl, eds. *Proceedings 9th ICCS.* Vol 2. Essen, 1997, pp 1835–1838.
240. JF Moulder, WF Stickle, PE Sobol, DK Bomben, in J Chastain, ed. *Handbook of X-Ray Photoelectron Spectroscopy.* Eden Praire: Perkin-Elmer Corporation, 1992.
241. PL Gai, BHM Billinge, AM Brown. *Carbon 27*:41–53, 1989.
242. J Fournier, G Lalande, R Côté, D Guay, J-P Dodelet. *J Electrochem Soc 144*:218–226, 1996.
243. SH Park, S McClain, ZR Tian, SL Suib, C Karwacki. *Chem Mater 9*:176–183, 1997.
244. T Grzybek, H Papp. *Appl Catal B Environ 1*:271–283, 1992.
245. NS McIntyre, TC Chan, in D Briggs, MP Seah, eds. *Practical Surface Analysis.* Vol. 1. *Auger and X-Ray Photoelectron Spectroscopy.* New York: John Wiley, 1994, p 485.
246. MJ Illan-Gomez, E Raymondo-Piñero, A Garcia-Garcia, A Linares-Solano, C Salinas-Martinez de Lecea. *Appl Catal B Environ 20*:267–275, 1999.
247. A Guerrero-Ruiz, A Sepulveda-Escribano, I Rodriguez-Ramos, A Lopez-Agudo, JLG Fierro. *Fuel 74*:279–283, 1995.
248. GA Rivas, PI Ortiz. *Anal Lett 27*:751–753, 1994.
249. A Dandekar, RTK Baker, MA Vannice. *J Catal 183*:131–154, 1999.
250. GE Cabaniss, AA Diamantis, WR Murphy Jr, RW Linton, TJ Meyer. *J Am Chem Soc 107*:1845–1853, 1985.
251. K Ravichandran, RP Baldwin. *Anal Chem 56*:1744–1747, 1984.
252. CA Frysz, DDL Chung. *Carbon 35*:1111–1127, 1997.
253. X Chu, K Kinoshita. *Mater Sci Eng B49*:53–60, 1997.

254. RI Taylor, AA Hamffray. *Electroanal Chem Interfac Electrochem* 42:347–354, 1973.
255. RM Wightman, MR Deakin, PM Kovach, WG Kuhr, KJ Stutts. *J Electrochem Soc* 131:1578–1583, 1984.
256. P Chen, RL McCreery. *Anal Chem* 68:3958–3965, 1996.
257. P Chen, MA Fryling, RL McCreery. *Anal Chem* 67:3115–3122, 1995.
258. E Hollax, DS Cheng. *Carbon* 23:655–664, 1985.
259. RS Nicholson. *Anal Chem* 37:1351–1355, 1965.
260. ME Rice, Z Galus, RN Adams. *J Electroanal Chem* 143:89–102, 1983.
261. CA McDermott, KR Kneten, RL McCreery. *J Electrochem Soc* 140:2593–2599, 1993.
262. CA Frysz, X-P Shui, DDL Chung. *Carbon* 32:1499–1508, 1994.
263. JM Solar, CA Leon y Leon, K Osseo-Asare, LR Radovic. *Carbon* 28:369–375, 1990.
264. MC Biesinger, NS McIntyre, I Bello, S Liang. *Carbon* 35:475–482, 1997.
265. L Zhiqiang, Z Mingrong, C Kuixue. *Carbon* 31:1179–1184, 1993.
266. J Rossin, E Petersen, D Tevault, R Lamontagne. *Carbon* 29:197–205, 1991.
267. JA Rossin. *Carbon* 27:611–613, 1989.
268. RH Bradley. *Appl Surface Sci* 90:271–276, 1995.
269. JA Rossin, RW Morrison. *Carbon* 31:657–659, 1993.
270. T Grzybek, J Klinik, B Buczek. *Sur Interface Anal* 23:815–822, 1995.
271. J Lin, ATS Wee, KL Tan, KG Neoh, WK Teo. *Inorg Chem* 32:5522–5527, 1993.
272. I Jirka, J Dubsky. *Appl Surface Sci* 40:135–143, 1989.
273. H-W Chen, JM White, JG Ekerolt. *J Catal* 99:293–303, 1986.
274. G Socrates. *Infrared Characteristics Group Frequencies*, 2d ed. Chichester: John Wiley, 1994.
275. AM Bond. *Modern Polarographic Methods in Analytical Chemistry*. New York: Marcel Dekker, 1980.
276. DT Napp, DC Johnson, S Bruckenstein. *Anal Chem* 39:481–485, 1967.
277. S Basak, PS Zacharias, K Rajeshawar. *J Electroanal Chem* 319:111–123, 1991.
278. K-K Shiu, K Shi. *Electroanalysis* 14:959–964, 1998.
279. S Biniak, M Pakula, A Swiatkowski. *J Appl Electrochem* 29:481–487, 1999.
280. V Majer, J Vesely, K Stulik. *Electroanal Chem Interfac Electrochem* 45:113–125, 1973.
281. H Sugawara, F Yamamoto, S Tanaka, H Nakamura. *J Electroanal Chem* 394:263–265, 1995.
282. X Hu, L Lei, H Ping, PL Yue. *Carbon* 37:631–637, 1999.
283. BU Yoon, K Cho, H Kim. *Anal Sci* 12:321–326, 1996.
284. YA Tarasenko, AA Bagreev, BB Yatsenko. *Russ J Phys Chem* 67:2099–3103, 1993.
285. J Rivera-Utrilla, MA Ferro-Garcia. *Ads Sci Techn* 3:293–302, 1986.
286. R Fu, H Zeng, Y Lu. *Carbon* 31:1089–1094, 1993.
287. R Fu, H Zeng, Y Lu. *Carbon* 32:593–598, 1994.
288. NS Marinkovic, A Dekanski, Z Laušević, B Vucurovic, M Laušević, J Stevanovic. *Vacuum* 40:95–97, 1990.

289. A Dekanski, NS Marinkovic, J Stevanovic, VM Jovanovic, Z Lauševic, M Lauševic. *Vacuum 41*:1772–1775, 1990.

290. GA Ragoisha, VM Jovanovic, MA Avramov-Ivic, RT Atanasoski, WH Smyrl. *J Electroanal Chem 319*:373–379, 1991.

291. PG Hall, PM Gittins, JM Winn, J Robertson. *Carbon 23*:353–271, 1985.

292. A Oya, S Yoshida, J Alcañiz-Monge, A Linares-Solano. *Carbon 34*:53–57, 1996.

293. CY Li, YZ Wan, J Wang, YL Wang, XQ Jiang, LM Han. *Carbon 36*:61–65, 1998.

294. L Lakov, P Vassileva, O Peshev. *Carbon 37*:1655–1657, 1999.

295. ZR Yue, W Jiang, L Wang, SD Gardner, CU Pittman Jr. *Carbon 37*:1785–1796, 1999.

296. SK Ryu, SY Kim, N Gallego, DD Edie. *Carbon 37*:1619–1625, 1999.

297. CU Pittman Jr, W Jiang, ZR Yue, S Gardner, L Wang, H Toghiani, CA Leon y Leon. *Carbon 37*:1797–1807, 1999.

298. ZR Yue, W Jiang, L Wang, H Toghiani, SD Gardner, CU Pittman Jr. *Carbon 37*:1607–1618, 1999.

299. YL Wang, YZ Wan, XH Dong, GX Cheng, HM Tao, TY Wen. *Carbon 36*:1567–1571, 1998.

300. A Oya, T Wakahara, S Yoshida. *Carbon 31*:1243–1247, 1993.

301. YZ Wan, YL Wang, TY Wen. *Carbon 37*:351–358, 1999.

302. JP Sousa, S Pons, M Fleischmann. *J Chem Soc Faraday Trans 90*:1923–1929, 1994.

303. AD Jannakoudakis, PD Jannakoudakis, N Pagalos, E Theodoridou. *J Appl Electrochem 23*:1162–1168, 1993.

304. AD Jannakoudakis, PD Jannakoudakis, N Pagalos, E Theodoridou. *Electrochim Acta 39*:1881–1885, 1994.

305. E Theodoridou, AD Jannakoudakis JO Besenhard, RF Sauter. *Extended Abstracts of the International Conference Carbon '86, Baden-Baden 1986*, pp 623–625.

306. E Theodoridou, AD Jannakoudakis, PD Jannakoudakis, N Pagalos, JO Besenhard, CI Donner, M Wicher. *Electrochim Acta 38*:793–798, 1993.

307. JF Weaver, GB Hoflund. *J Phys Chem 98*:8519–8524, 1994.

4

Carbon Materials as Adsorbents in Aqueous Solutions

Ljubisa R. Radovic

The Pennsylvania State University, University Park, Pennsylvania

Carlos Moreno-Castilla and José Rivera-Utrilla

University of Granada, Granada, Spain

I. INTRODUCTION

On July 7, 1855, Michael Faraday wrote to *The Times* of London to complain
that the river Thames was a "real sewer" and that the "whole of the river was
an opaque pale brown fluid." He argued that "[i]f we neglect this subject, we
cannot expect to do so with impunity; nor ought we to be surprised if, ere many
years are over, a hot season give us sad proof of the folly of our carelessness"
(http://dbhs.wvusd.k12.ca.us/Chem-History/Faraday-Letter.html, viewed 4/19/
98). Society has made much progress since then, but many concerns remain. The
use of charcoals and activated carbons in water treatment is probably one of the
oldest chemical technologies, and a vast literature has accumulated on this subject
[1–6].

The steady interest in the effects of the chemistry and physics of the carbon
surface on pollutant removal from waters has been ignited by the U.S. Clean
Water Act (enacted in 1972, amended as the Water Quality Act in 1987). The
most recent interest stems from the Safe Drinking Water Act Amendment of
1996. Activated carbon adsorption has been cited by the U.S. Environmental
Protection Agency (www.epa.gov) as one of the best available control technologies. Furthermore, the most recent efforts to understand the adsorption of the
same pollutants by soils [7,8] can benefit from comparisons of similarities and
differences with respect to the behavior of activated carbons.

The level of fundamental understanding of liquid-phase adsorption [9] is well
below that of gas- or vapor-phase adsorption. For example, it was recently concluded by Radeke et al. [10] that "it is not possible to extrapolate straightforward
from the gas-phase adsorption of organics (e.g., phenol) in the presence of water
vapor to the separation of organics from liquid water." In adsorption of vapors
a suitable normalization parameter in the comparison of isotherms for different
adsorbates is the saturation vapor pressure. The analogous parameter for liquid-
phase isotherms is the solubility of the adsorbate. However, even approaches
such as quantitative structure–activity relationships (QSARs) [11] do not always
"factor out" this parameter before attempting to attribute the remaining differences in uptakes of different adsorbates on a given adsorbent to other, more subtle
causes. And when they do, important differences remain, and these need to be
related to the nature of the adsorbent surface.

Adsorption of both organic and inorganic solutes from the aqueous phase has
been a very important application of activated carbons. With current increasing
emphasis on the more thorough removal of pollutants from potable and waste
waters, the use of carbons and the demands placed on their performance are
expected to increase. Many buyers of activated carbon will not be able to afford
its underutilization or inefficient use. A similar situation, "greatly underutilized"
carbon adsorbents, exists in liquid chromatography applications, and it has been

attributed to "our limited understanding of solute interactions with the solid surface" [12].

This comprehensive review is thus thought to be timely from a practical point of view. It is also thought to be timely in a fundamental sense: significant progress has been achieved in recent years in our understanding of the chemistry of the carbon surface, and, as will be shown here, this is crucial for the understanding of adsorption of both organic and inorganic solutes. Brief annual reviews of the more practical aspects of the issues to be discussed here are available elsewhere [13–17].

Relatively few critical reviews of this topic are available in the scientific literature. For example, the only one published in this series is that of Zettlemoyer and Narayan [18], on adsorption by graphite surfaces, in which no attempt is made to discuss the acknowledged role of carbon surface chemistry. The most important early review of adsorption of organic solutes is the book by Mattson and Mark [6]. In addition to discussing adsorption of electrolytes (with 44 references from the period 1919–1969) and nonelectrolytes and weak electrolytes (with 79 references from the period 1938–1971), it includes an extensive review of the surface chemistry of carbons. In an earlier study, Weber [19] has reviewed "sorption from solution by porous carbon" based on "what [was then] known of the properties of the adsorbent and of variables observed to influence [this] process." In particular, among the factors influencing adsorption on carbon he discussed the effects of adsorbent surface area, pore size distribution, particle size, and, very briefly, those of the chemical nature of the carbon surface. In subsequent publications by the same principal author, it is concluded that "fundamental investigations are required for further delineation of the underlying principles" [20], that "the heterogeneous nature of natural surfaces and engineered adsorbents, such as activated carbon, necessarily precludes any thorough evaluation of surface and [that the] potential sorbent-sorbate chemical interactions remain largely speculative" [21].

The present review attempts to offer precisely such a delineation and evaluation, as well as to illustrate how a judicial characterization of carbon surface chemistry has made this possible. We shall limit our discussion to equilibrium adsorption of both electrolytes and nonelectrolytes from dilute aqueous solutions at or around room temperature. We do not cover desorption, an area of increasing importance because of the interest in regeneration (or reactivation) of activated carbons; nor do we discuss adsorption kinetics, acknowledging that in some studies true equilibrium may not have been reached, especially for large molecules such as those found in natural organic matter. In discussing these issues we shall take a historical approach and take full advantage of the discerning benefit of hindsight. (For a similar approach in the review of carbon materials as catalyst supports and catalysts, see Ref. 22.) We make every attempt to give credit where credit is really due, to highlight progress (or lack thereof), and to emphasize the key unanswered questions. Before proceeding, however, we briefly summarize some of the key conclusions of other relevant reviews.

Cookson [23] has emphasized the importance of oxygen functional groups on the carbon surface, either those deliberately produced or formed as a consequence of adsorbent aging. In particular, the author discussed the mechanism of adsorption of organics, endorsed the proposal of Mattson et al. [24] regarding the key role of carbonyl groups on the surface (see Section VI), but concluded that "[a]lthough the carbonyl groups enhance the adsorption of aromatics, this is not the sole means for adsorption."

Cini et al. [25] reviewed "those aspects of activated carbon which are often forgotten when applied to potable water systems." Among the adsorbate/adsorbent interactions they mentioned "dispersion forces, polarization forces and electrostatic forces" but did not provide examples of their relative importance.

The monograph by Perrich [26] lists typical uptakes of a large number of adsorbates. The following factors that influence adsorption were discussed: molecular structure, solubility, and ionization of the adsorbate. The following properties of the adsorbent were also briefly discussed: total surface area, apparent density, particle size distribution, iodine number, ash levels, molasses number, abrasion number, butane number, and adsorptive capacity. Regarding this last parameter, the rather disheartening statements are made that the "best measure of adsorptive capacity is the effectiveness of the carbon in removing the critical constituents . . . from the wastewater in question" and that "feasibility tests should always be run to determine which carbon would be best for a given application." The pH of the solution is mentioned only in the following context: "[I]n some applications, for example, organic acids and bases, it might be beneficial to adjust the pH. In such applications, a shift in pH towards a less soluble, more adsorbable species is desirable."

As late as 1987, McDougall and Fleming [27], whose interests rested primarily in the extraction of gold by activated carbons, commented that "the adsorption of organic and inorganic species onto carbon may occur by several mechanisms, the unraveling of which, especially in the extraction of certain metal complexes from solution, is extremely difficult."

Derylo-Marczewska and Jaroniec [28] have reviewed the adsorption of organic solutes from dilute solutions and have provided a useful compilation of published experimental data for both single- and multisolute adsorption isotherms on carbonaceous adsorbents. They also presented a survey of theoretical approaches used to describe the solute adsorption equilibria, including the Polanyi adsorption model, the solvophobic interaction model, the Langmuir adsorption theory, the vacancy solution model, as well as considerations based on the energetic heterogeneity of the adsorbent. In particular, these authors emphasize the complexities of the interplay between the chemical properties of solute, solvent, and adsorbent and conclude that "further studies in this field are required."

Najm et al. [29] presented a critical review of the uses of powdered activated carbon (PAC), in contrast to the more conventional granular activated carbon (GAC), with special emphasis on practical aspects such as its point of addition in the water treatment plant. The dependence of pollutant removal efficiency on the properties of the adsorbate was mentioned; the role of adsorbent properties was not discussed.

In two recent reviews that form part of a comprehensive monograph on the quality and treatment of drinking water, Haist-Gulde and coworkers [30,31] discuss both the experimental and the theoretical aspects of micropollutant removal by adsorption on activated carbon. The authors acknowledge that the experience of their institute is mainly presented and their emphasis is clearly on practical aspects, including the importance of competitive adsorption in the presence of natural organic matter, but excluding any discussion of the chemistry or physics of the carbon surface.

In a recently updated monograph on the chemistry of water treatment, Faust and Aly [32] devote one of 11 chapters to the removal of organics and inorganics by activated carbon. Their emphasis is overwhelmingly on the organic adsorbates and the effectiveness of removal of the various pollutants. Very little attention is paid to the role and the nature of the carbon surface. Thus, for example, in a very brief paragraph on the chemistry of the surface, they emphasize both detrimental and beneficial effects. As an illustration of the former, they state that the "surface oxides consisting of acidic functional groups reduce the capacity of carbon for adsorption of many organic solutes" (e.g., oxalic and succinic acids and trihalomethanes) and attribute this effect to surface blockage by preferential adsorption of water. As an illustration of the latter, they support the argument of Mattson and coworkers [24,6] (see Section VI) regarding the beneficial "interaction of aromatic ring π electrons with the carbonyl groups by a donor–acceptor mechanism involving the carbonyl oxygen as the electron donor and the aromatic ring as the acceptor." Of their 173 references only 17 are from the 1990s, and the most recent ones are from 1991. In the earlier version of their important monograph [33], the same authors offer a more balanced, albeit an even more utilitarian, discussion of the same topics; thus, for example, a detailed analysis of the important inorganic water pollutants (As, Cd, Cr, Cu, etc.) is provided.

Several general monographs on carbonaceous adsorbents have been published in the last decade. The ones on activated carbons devote much more space to gas-phase adsorption than to liquid-phase adsorption, and they discuss surface physics in more detail than surface chemistry. Thus, for example, Jankowska et al. [34] devote less than 10% of their book to adsorption from the liquid phase, but they do mention that the "process is also affected by the nature of the adsorbent, i.e., its surface area, porous structure, and surface properties;" their discussion of this last issue is quite selective, but the authors do make the point that previous studies "reveal the importance of changes of the carbon surface charge with pH for the adsorption equilibrium." On the other hand, Bansal et al. [35] ignore this issue completely (see Section IV.A.1). The book by Kinoshita [36] contains a much more detailed discussion of both physicochemical and electrochemical properties of carbons (mostly carbon blacks), but the brief review of liquid-phase adsorption has only 11 references, the most recent ones being from 1973.

The literature reviewed here is summarized in Fig. 1. Clearly, substantially more work has been done on the removal of organic compounds and phenols in particular. There is a good reason for this, which we shall attempt to clarify in what follows: adsorption of organics is more difficult to understand, and the role of the carbon surface in determining adsorption uptakes is much more subtle. Of course, even though a reasonably thorough and extensive literature search has been carried out, we claim to have analyzed only the most representative studies.

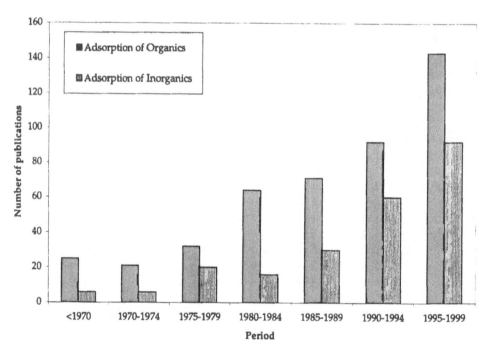

FIG. 1 Statistical summary of the publications cited in this review.

The versatility of carbon is indeed overwhelming: the publications that were scrutinized for this review are disseminated in as many as 120 different journals.

The main focus in the present review is on the relationship between the chemical nature of the activated carbon adsorbent and its adsorptive capacity for single solutes (i.e., excluding the effect of competitive adsorption). It is argued that a large surface area and a suitable pore size distribution are necessary but not sufficient attributes of an optimized adsorbent for both inorganic and organic solutes. The rule of thumb—simple to enunciate but sometimes difficult to realize in practice—is to use an activated carbon of the highest possible surface area (i.e., narrowest pore size distribution) that avoids at the same time the always latent problems of accessibility of the solute to the entire surface. While the importance of carbon surface chemistry has been recognized for a long time, the exact nature of this importance and the desirable chemical surface characteristics have been quite controversial and often misunderstood. In particular, the literature is reviewed here for clues regarding the relative importance of electrostatic (coulombic) and dispersive interactions in the adsorption equilibria. The former are governed by the interplay between solution and surface chemistry, while the latter depend on both surface physics and surface chemistry. For a detailed review of the surface chemistry of carbons, the reader is referred to the extensive chapter by Leon y Leon and Radovic [37] and the authoritative brief review by Boehm [38]. For an authoritative recent review of carbon surface physics, see the recent chapter by Rodríguez-Reinoso and Linares-Solano [39] in this series and the more recent monograph edited by Patrick [40].

II. AMPHOTERIC CHARACTER OF CARBONS

It is not our intention to present an exhaustive review of this important subject. Up-to-date reviews are provided elsewhere [38,37]. The classic review by Garten and Weiss [41] offers an excellent historical perspective. We do need to summarize here the issues that are essential for understanding the aqueous-phase adsorption phenomena. The main features of carbon surface chemistry are presented first and the consequent acid/base behavior of carbons is briefly discussed to illustrate their amphoteric character. In Section III it is shown that these phenomena often govern the adsorption of most inorganic compounds. In Section IV we argue that these phenomena can be dominant in the adsorption of organic compounds as well, but they are more often only a part of the whole story.

The essential features of the surface chemistry of carbons, relevant to their behavior in aqueous solution, are summarized in Figs. 2 and 3. These are discussed in more detail elsewhere [37], and the reader will appreciate from such discussions that the features presented here are perhaps gross (over)simplifications; we shall argue here that they are also very convenient and useful simplifi-

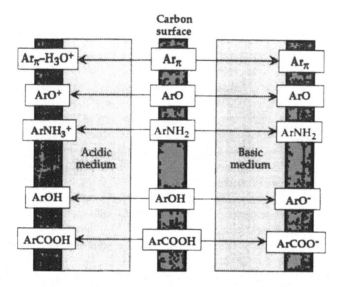

FIG. 2 Macroscopic representation of the features of carbon surface chemistry that are thought to be sufficient for understanding aqueous-phase adsorption phenomena.

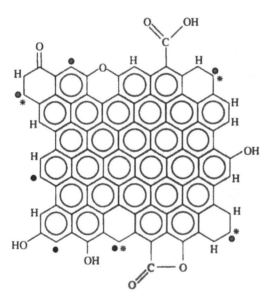

FIG. 3 Microscopic representation of the features of carbon surface chemistry that are thought to be sufficient for understanding aqueous-phase adsorption phenomena. ● represents an unpaired sigma electron; ●* represents an in-plane sigma pair with * being a localized π electron.

cations. Figure 2 illustrates the macroscopic chemical features of the carbon surfaces, both internal and external. In response to pH changes, the surface develops coexisting electric charges of opposite sign, whose prevalence depends on the chemistry of the solution. Therefore attractive or repulsive electrostatic interactions between the adsorbate and the adsorbent must be taken into consideration. Two points in this figure deserve special attention: (1) the origin of the positive charge, and (2) the relative contributions of functional groups vs. graphene layers to the development of surface charge.

Three possible sources of the positive charge are highlighted: (1) basic oxygen-containing functional groups (e.g., pyrones or chromenes); (2) protonated amino groups (for nitrogen-rich carbons); (3) graphene layers acting as Lewis bases and forming electron donor–acceptor (EDA) complexes with H_2O molecules. When N-containing functional groups are not abundant, which is the case in most practical situations, the relative contributions of basic oxygen and graphene layers to the positive charge are the key issue. Such information is not readily available in the literature, even though it has important practical implications: electrostatic repulsion of adsorbates from the graphene layers, where most of the surface area resides, is thought to be much more detrimental for adsorbent effectiveness than repulsion from the basic oxygen functional groups.

Electrophoretic studies on activated carbon and carbon black particles, and their comparison with the results for amphoteric oxides, contain very revealing clues about the contributions of functional groups and graphene layers to the development of surface charge. Figure 4 illustrates a typical set of results. Electrophoretic mobilities of particles are measured as a function of pH and are seen to be an order of magnitude smaller than those of hydrogen ions under comparable conditions. They are commonly converted to either zeta potentials or surface charges using standard electrical double layer (EDL) equations. In theory, each graph represents "a titration of surface charged groups and as such is strictly comparable with a potentiometric titration" [42]. The so-called H-carbons [41] have a high isoelectric point (e.g., $pH_{IEP} > 7$), while the so called L-carbons [41] have a low isoelectric point (e.g., $pH_{IEP} < 7$). A common observation in the surface charge characterization of carbon materials is that the absolute value of the positive charge is often much smaller than that of the negative charge [43–54]. This has been obvious for more than half a century [55]; for example, from the data of Steenberg, as reproduced by Garten and Weiss [41], the uptake of NaOH on an L-carbon (sugar-derived char activated at 400°C) was ~700 µmol/g, while the uptake of HCl on an H-type carbon (same char activated at 800°C) was ~300 µmol/g. In contrast, for carbon blacks, either the situation can be reversed [41] or the electrophoretic mobility vs. pH diagrams tend to be much more "symmetric" [56]. More recently, for example, Dobrowolski et al. [47] report the charges for a "basic" carbon (outgassed in argon at 1400 K) and an "acidic" carbon (oxidized in H_2O_2) of 15 mC/m² (at pH = 2.4) and −60 mC/

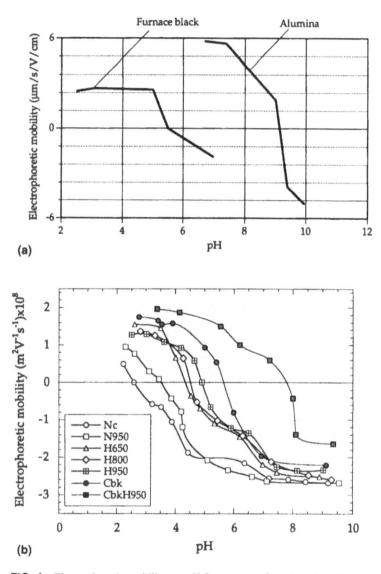

FIG. 4 Electrophoretic mobility vs. pH for a range of amphoteric solids, including alumina [777], a furnace carbon black [56] and chemically modified carbon blacks and activated carbons [82].

m^2 (at pH = 8.4), respectively. To put these numbers in proper perspective, an activated carbon ($\sim$1000 m^2/g) which is heavily oxidized (e.g., with HNO_3) may contain as much as 10 wt% oxygen; if all this oxygen is in the form of carboxyl groups, this is equivalent to $\sim$3 mmol COO^- per gram, or 300 mC/m^2 ($\sim$1.5 V).

The comparison with amphoteric oxides [57–59] is also instructive. In an early review, Snoeyink and Weber [60] compared the surface functional groups on carbons and silicas but failed to point out the resulting differences in the "symmetries" of their electrokinetic behavior. For amphoteric oxides, the symmetry (see Fig. 4a) is a consequence of the following equilibrium [57,61–66]:

$$RO^- + H^+ + H_2O = ROH_2^+ + OH^-$$

and at pH = pH_{IEP}, $[ROH_2^+]$ = $[RO^-]$.

Taken together, these results strongly suggest the following: (1) there is an important contribution of the graphene layers to the positive surface charge of carbon adsorbents, and (2) the widely held belief (especially reflected in the literature reviewed in Section III) that surface charge develops as a consequence of deprotonation and protonation of oxygen functional groups should be discarded. Protonation of aromatic carboxyl groups in aqueous solutions is highly unlikely. (Thus, for example, only acids whose strength exceeds $pK_a \sim -7$, e.g., HI ($pK_a = -10$), can protonate aromatic carboxyl groups, while $ArCOOH_2^+$ is of course a much stronger acid than H_3O^+ ($pK_a = -1.74$), and $ArCOOH$ is the dominant species in water.) Instead, protonation of the basal plane [67] is considered to be much more important for the applications of interest here.

The use of titrations is commonly considered to be an alternative method to electrophoresis for the estimation of surface charge. Actually, in the case of solid carbons, and especially in the case of high-surface-area activated carbons, this turns out to be a complementary method [52], and failure to make the distinction can be misleading. For convenience we shall refer to the titration-derived parameter as the point of zero charge (pH_{PZC}), while the electrokinetically derived parameter will be referred to as the isoelectric point (pH_{IEP}). We acknowledge that a more fundamental distinction between these two parameters [66] requires a discussion of specific adsorption (in an ideal situation, the two should be identical in the absence of specific adsorption), but such discussion is typically not included in most adsorption studies, especially of organic compounds. In most cases of practical interest, $pH_{IEP} < pH_{PZC}$ because of preferential (diffusion-controlled) ambient-air-induced oxidation (e.g., carboxyl group generation) on the external surfaces of carbon particles. Hence, the smaller the difference between the two values, the more homogeneous the distribution of surface charges. This is illustrated in Table 1: short-contact-time oxidation (e.g., in concentrated nitric acid or in air, especially at elevated temperature) introduces acidic surface groups primarily on the external surface of the particles. A dramatic example of this difference between the pH_{PZC} and pH_{IEP} has been shown (albeit implicitly) by

TABLE 1 Differences Between Point of Zero Charge (pH_{PZC}) and
Isoelectric Point (pH_{IEP}) Values for a Series of Chemically Treated
Activated Carbons

Carbon	pH_{PZC}	PH_{IEP}	$pH_{PZC} - pH_{IEP}$
Nc (Commercial Norit C)	2.5	2.6	−0.1
NcA	2.2	2.2	0.0
NcN	5.2	3.5	1.7
NcNA	4.2	2.9	1.3
NcH	9.0	4.9	4.1
NcH-1ox	8.9	4.2	4.7
NcH-7ox	9.0	4.2	4.8
NcH-20ox	8.6	3.7	4.9
NcH-30ox	8.4	3.3	5.1
NcH-HNO3d16h	3.9	3.5	0.4
NcH-HNO3c0.25h	3.3	1.6	1.7
NcH-HNO3c0.30h	2.8	1.4	1.4

A = treatment in air at 250°C for 3 h.
N = treatment in N_2 at 950°C for 3 h.
NA = same as N but subsequently oxidized in air at 250°C for 7 min.
H = treatment in H_2 at 950°C for 3 h.
H-Xox = room temperature exposure of sample H to ambient air for X days.
H-HNO3xYh = treatment of sample H with concentrated (x=c) or dilute (x=d)
HNO_3 for Y h.
Source: Ref. 52.

Dixon et al. [68] for the case of a commercial activated carbon, ~10.2 vs. ~3.5;
differences of ca. 5–6 pH units are indeed rather common [69].

Table 2 is a compilation of pH_{PZC} and pH_{IEP} values for several commercial
activated carbons. It is seen, for example, that the values for Filtrasorb 400, a
commonly used adsorbent in water treatment, vary by several pH units. This
reinforces the point that was clearly illustrated in the important paper of Lau et
al. [70]: because of the large affinity of carbon for oxygen, the storage and thermal
history of the adsorbent often has a large influence on its chemical surface properties (see also Fig. 25).

Figure 3 is an attempt to synthesize and highlight the key molecular features
of the edges and the basal planes of graphene layers. Only some of the most
prevalent surface functional groups are depicted [37]. Special attention should
be paid to the delocalized π electron system that acts as a Lewis base in aqueous
solution [71–73]:

$$C_\pi + 2H_2O = C_\pi\text{—}H_3O^+ + OH^-$$

TABLE 2 Compilation of Literature Data of pH_{PZC} and pH_{IEP} Values for Commercially Available Activated Carbon Materials

Adsorbent	pH_{PZC}	pH_{IEP}	Reference
Filtrasorb 400	10.4	~2.0	171
Filtrasorb 300	9.8	—	171
Nuchar SA	4.0	~2.0	171
Nuchar WVL	7.9	~2.0	171
Darco G60	6.2	~2.0	171
Darco (Aldrich)	6.1	—	172
Nuchar SA	3.8	—	44
Filtrasorb 400	7.1	—	44
Nuchar SA	8.3	—	206
Nuchar WVL	4.3	—	206
Darco G60	2.2	—	167
ACC (cloth)	2.7	—	222
Norit PK 3–5	8.2	—	186
Hydraffin B10/2	6.7	—	524
Norit ROW	5.8	—	524
Norit HD4000	~9	—	237
Filtrasorb 400	~6.5	—	a
Filtrasorb 300	~6.0	3.2	713
Hydrodarco	~8	—	a

a Bjelopavlic et al., *J. Colloid Interf. Sci.* 210:271–280, 1999.

In addition to giving rise to a positive surface charge, the graphene layers participate in π–π interactions [74–77] with aromatic adsorbates. These are known to be sensitive to substituent effects on the aromatic adsorbates [78–80,74,81]. The substituent effects on the carbon adsorbents are discussed in detail in Sections IV and V. Suffice it to note here that there are at least three ways in which the π electron density of a graphene layer can decrease [82–84]: (1) deliberate "decoration" of the edges with oxygen functional groups, by controlled oxidative treatments; (2) accidental (and more or less inevitable) decoration of the edges with oxygen functional groups upon adsorbent exposure to ambient air; and (3) additional localization of π electrons at the edges by forming "in-plane sigma pairs" with the unpaired sigma electrons (remnants of high-temperature treatment of the material). The degree to which these processes take place can be fine-tuned by carbon pretreatment in inert or reactive gases, as illustrated in Fig. 5 (see also Fig. 4). The dramatic decrease in both the O_2 uptake (at room temperature) and the heat of O_2 adsorption upon H_2 treatment at 950°C (Fig. 5a)

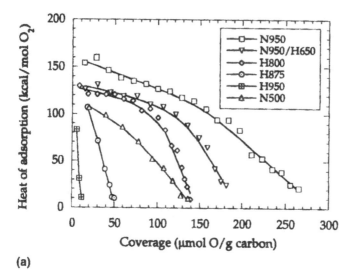

(a)

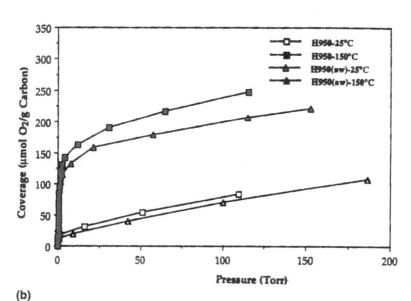

(b)

FIG. 5 Effects of treatments in H_2 at 650–950°C (H950, etc.) and N_2 at 500–950°C (N500, etc.) on the surface chemistry of a commercial activated carbon (Norit C extra). (a) differential heats of adsorption of O_2 at 25°C; (b) adsorption isotherms of O_2 at 25 and 150°C (aw = acid-washed) [82–84].

is not a consequence of the creation of stable C—H groups at the edges of the graphene layers; it is achieved by selective gasification of the most reactive carbon atoms. The resulting structure is less reactive upon subsequent ambient air exposure, even though it contains many unsaturated sites [85].

Interaction of π systems with water, as well as with cations, has recently been a topic of considerable fundamental [86–89] and practical [90] interest. Even though solid carbons are generally considered to have a hydrophobic character (which is certainly true in comparison with adsorbents such as zeolites and clays), they can be very effective in removing both organic and inorganic acids and bases, especially from aqueous solutions [91]. A critical review of the carbon–water interactions has long been overdue, but space constraints do not allow us to offer it here. Suffice it to emphasize the perhaps obvious but often neglected trends: progressive incorporation of acidic oxygen functional groups can transform a very hydrophobic carbon into a very hydrophilic one [92–94], while treatment at an optimum temperature in H_2 can result in both hydrophobic (basic) and stable carbon surfaces [95,96,82–84].

III. ADSORPTION OF INORGANIC SOLUTES

This is a topic of great practical interest because of water treatment and metal recovery applications. Its fundamental aspects are also important for the preparation of carbon-supported catalysts [22], where the catalyst precursor is typically dissolved in water prior to its loading onto the porous support.

A convenient historical point of reference is the chapter by Huang [97]. Even though there is some confusion in this review between activated carbons and carbon blacks, it contains a very useful compilation of studies on the removal of cadmium, chromium, mercury, copper, iron, vanadium, and cyanide species. It also contains a summary of four models of adsorption of inorganics, those of Gouy–Chapman Stern–Grahame and James–Healy, as well as the ion exchange model and the surface complex formation model. The author noted that "most of the applications of and research effort on activated carbon in the water and wastewater industries are oriented towards organics removal" and that "research efforts on inorganics removal by activated carbon, specifically metallic ions, are markedly limited." Statistical analysis of the papers reviewed in this chapter, summarized in Fig. 1, shows that this is the case even today. The later reviews by Faust and Aly [33,32] provide a wealth of information, but their focus is on issues of practical interest, with illustrative cases whose general validity has not been demonstrated. The review presented here is an attempt to rescue some generally valid trends from the now numerous, sometimes contradictory, and too often reciprocally oblivious publications.

A. Phenomenological Aspects

In this section we first briefly summarize the speciation characteristics of inorganic compounds present in typical residual waters (see also the appendix) and then review the factors that affect their removal by adsorption on activated carbons. The primary focus is on heavy metals whose removal from waste waters is an environmental issue of pressing concern. The landmark reviews on this topic [97,33] discuss the removal of barium, iron, vanadium, cyanide, fluoride, chlorine compounds, hydrogen sulfide, and selenium. More recently, studies on the use of carbonaceous adsorbents for the removal of sulfides [98], halides [99,100], cyanides [101,102], nitrates [103], chlorites and chlorates [104], bromates [105], boric acid and borax [106], aqueous SO_2 [107], heteropolyacids [108], rhodium chloride complexes [109], lithium [110,111], cesium [112], strontium [113], iron [114,115], iodine [116], and dysprosium [117] have been published. Ammonia removal has been of long-standing interest [118–120] and the importance of carbon surface chemistry is well known here; one of the few recent studies [121] has emphasized the usefulness of ammonia adsorption as a tool for characterizing the surface of porous and nonporous carbons. We do not review these studies here, even though they do exhibit many intriguing effects of the nature of the carbon surface.

1. Chromium

Chromium exists in aqueous solution primarily as Cr(II) (quite unstable and easily oxidized), Cr(III) (stable over a wide range of conditions), or Cr(VI) (stable in strongly oxidizing conditions). In natural water streams the principal trivalent Cr species are $CrOH^{2+}$ and $Cr(OH)_3$ [122], while the principal hexavalent species are $HCrO_4^-$ and CrO_4^{2-}. Trivalent Cr has a propensity to form the mononuclear $Cr(OH)^{2+}$, $Cr(OH)_2^+$, and $Cr(OH)_4^-$ species as well as the neutral $Cr(OH)_3(aq)$. At room temperature, the rate of formation of the polynuclear species $Cr_2(OH)_4^{4+}$ and $Cr_3(OH)_4^{5+}$ is quite low [123]. The speciation diagrams of both trivalent and hexavalent chromium have been reproduced recently by Leyva-Ramos and coworkers [124,125]; see also Figure A.3.

Although essential for life's metabolic processes, chromium in high concentrations can cause serious illnesses. In particular, Cr(VI) accumulated in waste waters from steelwork, electroplating, leather tanning and chemical manufacturing plants can be carcinogenic. A substantial literature thus exists on the removal of chromium by adsorption on carbonaceous adsorbents [126,127,97,128–130,33, 131–139,124,140–143,125,144–156].

Adsorption of both Cr(III) and Cr(VI) species is known to be pH-dependent. From Huang's review not many general conclusions can be drawn except that "Cr(VI) can be readily reduced to Cr(III) at acidic conditions" and that intermediate pH favors the removal of both Cr(III) and Cr(VI) [97]. No specific reference has been made to the key role of surface chemistry of the carbon except to note that "by heating the activated carbon in 1M HNO_3 solution for 30 minutes . . . part of the H-carbon (was converted) into the L-form which has greater affinity towards the reduced Cr(III) species than the H-carbon (at pH = 2.5)." The review by Faust and Aly [33] cites quite a few specific cases, including the relevant Freundlich and Langmuir parameters, but here too no attempt is made to synthesize the phenomena observed.

Golub and Oren [132] used electrochemical methods to reduce Cr(VI) to Cr(III) on a graphite felt electrode surface. Conflicting pH requirements were discussed: low pH is needed for efficient reduction, while high pH is needed for the precipitation of the hydroxide. Furthermore, the adhesion of the negatively charged hydroxide particles is affected by the pH of the solution: below 8.5 ["zero zeta potential" of $Cr(OH)_3$] it was expected that there would be electrostatic attraction between positively charged particles and the (presumably) negatively charged surface. More recently, Farmer et al. [150] performed similar experiments using a high-surface-area carbon aerogel. They concluded that the "mechanism for Cr(VI) separation involves chemisorption on the carbon aerogel anode" and that "Cr(VI) removal is not based upon simple double-layer charging" but they did not clarify the role of pH or surface chemistry.

Moreno-Castilla and coworkers [139,140] did clarify the relationship between carbon surface chemistry and chromium removal. Table 3 summarizes some of the key results. Upon oxidation of carbon M in nitric acid (sample MO), the surface has become much more hydrophilic and more acidic, and the uptakes increased despite a decrease in total surface area. The enhancement in Cr(III) uptake was attributed to electrostatic attraction between the cations and the negatively charged surface. The enhancement in Cr(VI) uptake (at both levels of salt concentration) was attributed to its partial reduction on the surface of carbon MO (perhaps due to the presence of phenolic or hydroquinone groups), which is favored by the lower pH. The increase in uptake on carbon MO with increasing NaCl concentration is consistent with this explanation, from a straightforward analysis of the Debye–Hückel and Nernst equations; the decrease in uptake on carbon M was attributed to the competition of specifically adsorbed Cl^- and CrO_4^{2-} ions on the positively charged surface.

Perez-Candela et al. [145] performed a similar study using a wide range of activated carbons, but they focused on the role of surface physics. Like many other investigators [128,130,134,138,143,144,156], they confirmed the large increases in Cr(VI) uptake as pH decreased and attributed them to its reduction to Cr(III). They did not characterize the surface chemistry of the adsorbents but did note a rough correlation between Cr(VI) removal efficiency at pH = 3 and "the pH of carbons;" they did not clarify the nature of this correlation, however, even though Bowers and Huang [128] (not cited by the authors) clearly stated that the uptake decrease "between pH 2.5 and 7.1 [was] primarily due to the decreasing electrostatic attraction between the positively charged carbon surface and the anionic Cr(VI) species in solution.''

Leyva-Ramos and coworkers [124,125] reported similar results after examining the role of pH in the removal of both Cr(III) and Cr(VI) by adsorption on a commercial activated carbon (pH$_{PZC}$ = 4.9). As the pH decreased from 10 to 6 there was a marked increase in Cr(VI) uptake, but the authors related this only to ''the different complexes that Cr(VI) can form in aqueous solution.'' Regarding the effects observed for Cr(III) (increase in uptake from pH = 2 to pH = 5, especially between 4 and 5, and a drastic decrease at pH = 6), the authors were almost as vague [''Cr(III) can form different complexes in aqueous solution''], but they did mention the key point: at pH < 4.9, the surface carries a net positive charge thus leading to electrostatic repulsion of Cr^{3+} and $CrOH^{2+}$ cations. The uptake decrease at pH = 6, at which such repulsion is not invoked, was left unexplained, however. Interestingly, the authors do not cite the study of Jayson et al. [137], which reports very similar results: a monotonic increase in Cr(VI) uptake with decreasing pH from 13 to 1 and a maximum uptake of Cr(III) at pH between 4 and 5. Jayson et al. [137] explain their results both in terms of the microporous structure of the activated carbon cloth (ACC, pH$_{PZC}$ = 2.7) and as a function of ''electrostatic attraction or repulsion between the charcoal surface and the ions in solution.'' In particular, they invoke ''repulsion between negative charges'' at pH > 5 due to the replacement of ''associated water molecules around the chromic ion . . . by hydroxyl ions.'' The interesting finding that, at comparable concentrations and at any and all pH values, the uptake of Cr(VI) was greater than that of Cr(III) was attributed not to solubility or surface chemistry effects but to the physical inaccessibility of Cr(III) species to the narrow micropores in ACC.

Tereshkova [148] has arrived at a rather unique conclusion about the adsorption of Cr^{3+} and $Cr_2O_7^{2-}$ species on fibrous carbonaceous sorbents: she states that the ''mechanism of sorption and the activity of sorbents depend on both the sample preparation and the state of the surface.'' This rather general conclusion is not (apparently) based on careful consideration of any of the studies discussed above (most of the relevant literature is not cited by the author); rather, it is based on puzzling speculations regarding a ''proton migration mechanism.'' Lalvani et al. [153] did cite some of the relevant literature, but only very superficially (all 15 references are cited in the Introduction; and only one of them, from 1977 [127], is mentioned again in the Discussion). They compared the Cr(VI) and Cr(III) uptake capacity of a commercial activated carbon with that of a ''selective and novel carbon adsorbent,'' produced (in the form of soot) by graphite electrode arcing. The latter carbon selectively removed Cr(VI) anions from solution, whereas, depending on the solution pH, a very small or negligible uptake of cations was observed. In contrast, the activated carbon ''showed great affinity for cations of lead, zinc and trivalent chromium, but none for the anion of hexavalent chromium.'' They identified ion exchange as the mechanism responsible for the metal uptake and, intriguingly, attributed this high selectivity to "positive charges'' on the carbon surface; nevertheless, they did not characterize the chemistry of the carbon surfaces nor did they suggest which positive charges may be responsible for anion exchange (see Section II).

In the very recent study by Bello et al. [155] the important role of surface chemistry (which was characterized by desorption experiments), in addition to porosity development upon carbon activation in H_2O or CO_2, is mentioned, but the authors also fail to compare their results, or their (rather vague) conclusions, with much of the available literature. In a more detailed and more enlightening study, Aggarwal et al. [154] evaluated both granular and fibrous activated carbons for their uptakes of Cr(III) and Cr(VI) and concluded that "the presence of acidic surface groups enhances the adsorption of Cr(III) cations'' and "suppresses the adsorption of Cr(VI) anions.'' They attributed the behavior of Cr(III) cations, which is in agreement with the results shown in Table 3, to "electrostatic attractive interactions between the carbon surface and the chromium ions'' but did not explain why "on degassing at 950°C, there is little or no adsorption of Cr(III) ions.'' Their explanation for the more complex behavior of Cr(VI) anions (maximum uptake at intermediate carbon degassing temperature) was quite convoluted and oblivious to earlier proposals available in the literature; it is complicated not only by the reduction of Cr(VI) to Cr(III) but by the lack of control of ionic strength and pH in their experiments. Support (albeit unacknowledged) for the trend exhibited in Table 3 by the Cr(VI) species can be found in a recent study of the effect of anodic oxidation of activated carbon fibers (ACF): Park et al. [157] observed "that the amount of adsorption and the adsorption rate of Cr(VI) increase with increases in the surface oxide groups of ACFs;'' the authors do not mention the relevance of Cr(VI) reduction, however, and simply state that the beneficial effect of carbon oxidation is due "to a larger content of the surface functional groups on ACFs.''

2. Molybdenum

The interest in adsorption of molybdenum stems primarily from its use as a carbon-supported catalyst [22]. The use of activated carbon for the removal of Mo-99 (used in nuclear medicine) has also been reported [158]. Only hexavalent Mo is stable under a wide pH range and in the absence of other complexing agents. At pH > 8, the dominant species is MoO_4^{2-}, while at very low pH there is precipitation of the hydrated oxide; between these two extremes polymeric anions are formed [123,159].

The dominant role of electrostatic adsorption has been obvious since the study of Dun et al. [160], who were surprised to find that there was an 18-fold decrease in the uptake of Mo(VI) anions when the pH was raised from 2.1 to 9.4. Solar et al. [60] clarified this trend and explained their own adsorption results based

TABLE 3 Effect of Activated Carbon Oxidation on the Adsorption of Cr(III) and Cr(VI)

Carbon	% oxygen[a]	Surface area[b] (m²/g)	V_{H2O}[c] (cm³/g)	pH_{pzc}	Maximum adsorption capacity		
					mg Cr(III)/g[d]	mg Cr(VI)/g[e]	mg Cr(VI)/g[f]
M	2.0	1089	0.654	7.0	2.7	6.9[g]	3.4[g]
MO	23.0	164	0.371	2.6	25.3	15.5[h]	23.4[h]

[a] From the amount of CO and CO_2 evolved during temperature-programmed desorption to 1200 K.
[b] From BET equation.
[c] Pore volume accessible to water.
[d] Equilibrium pH = 3.5–4.0.
[e] [NaCl] = 5 × 10⁻⁴.
[f] [NaCl] = 5 × 10⁻².
[g] Equilibrium pH = 6.5.
[h] Equilibrium pH = 3.5.
Source: Ref. 140.

on electrostatic arguments. This is illustrated in Table 4. All the carbon supports/ adsorbents shown in the table had pH_{IEP} values less than 8.5; at a slurry pH of 8.5, therefore, they have an overall negatively charged surface. Since both the oxalate and (presumably) the acetylacetonate species are anionic upon dissociation in solution, the low Mo uptakes were indeed to be expected. The higher Mo uptakes at $pH_{slurry} < pH_{IEP}$ were also explained by invoking the same electrostatic arguments. These findings and interpretations were confirmed by Rondon and coworkers [161,162] who also noted that "at pH 2 and 5 the major Mo surface species is octahedrally coordinated" [161]. Such coordination had also been postulated by Cruywagen and de Wet [159] in a detailed study of Mo(VI) adsorption on a commercial activated carbon.

3. Cobalt

Relatively few studies are available on cobalt removal from waters even though it is an important industrial material and may be toxic (though not as much as some of the other heavy metals). The dominant species in aqueous solutions is Co^{2+} [123]. The mononuclear hydrolysis products of Co(II) from $CoOH^+$ to

TABLE 4 Influence of Carbon Surface Chemistry (given as pH_{IEP}) on the Adsorption of Mo(VI) Anions

Adsorbent[a]	pH_{IEP}	Precursor[b]	pH_{slurry}	Uptake (wt% Mo)
M	6.2	MoOx	8.5	0.06
		MoOx	2.6	1.39
		MoAcAc	8.5	0.03
		MoAcAc	3.9	2.62
M-HNO₃	1.3	MoOx	8.5	0.02
		MoOx	1.8	0.85
M-2500-HNO₃	3.0	MoOx	8.5	0.02
		MoOx	1.9	0.52
CPG-1	1.2[c]	AHM	1.4[d]	16
CPG-2	2.0[c]	AHM	2.1[d]	18
CPG-3	5.2[c]	AHM	5.5[d]	11
CPG-4	9.0[c]	AHM	9.4[d]	1

[a] M is a commercial carbon black (Monarch 700, Cabot) treated in HNO₃ (M-HNO₃) or first heat-treated at 2500°C in inert atmosphere and then soaked in HNO₃ (M-2500-HNO₃); CPG is a commercial activated carbon (CPG, Calgon Carbon Corp.).
[b] MoOx = molybdenum oxalate, $H_2(MoO_3C_2O_4)$; MoAcAc = molybdenum acetylacetonate, $MoO_2(C_5H_7O_2)_2$; AHM = ammonium heptamolybdate.
[c] Initial pH of the slurry.
[d] Final pH of the slurry.
Source: Refs. 160 and 50.

$Co(OH)_4^{2-}$ have all been well established. There is also evidence for formation of $Co_2(OH)_3^+$ and $Co_4(OH)_4^{4+}$ at relatively high concentration of Co(II).

Rivera-Utrilla et al. [163] studied the adsorption of both radioactive (Co-60) and nonradioactive Co^{2+} species on activated carbons derived from almond shells. The typically observed increases in the uptake of Co-60 in the presence of anions (at uncontrolled pH) were interpreted using electrostatic arguments: it was postulated that the anions can be adsorbed on the carbon surface (whose pH was >6.5), thus increasing the density of negative charge on it and therefore its capacity to attract cations [163]. The role of pH and carbon surface chemistry was addressed [164], with somewhat inconclusive results, and then clarified [165]; the latter results are shown in Fig. 6. In all cases there was an increase in uptake with increasing pH because of a reduced repulsion between the positively charged surface and Co^{2+} cations. Furthermore, the lower the slurry pH of the carbon (roughly equivalent to its point of zero charge), the greater the uptake (at the same solution pH). Large uptake differences for the three "basic" carbons (A-5, A-8, and A-24) remained unexplained, however; they may be due to large differences in their physical surface properties.

Netzer and Hughes [166] analyzed the behavior of a large number of commercial activated carbons but did not characterize their surface chemistry. They also concluded that the "solution pH was the single most important parameter affecting adsorption" of cobalt (as well as copper and lead) and that "a pH of 4 was determined to be the lowest pH for maximum adsorption." Surprisingly, how-

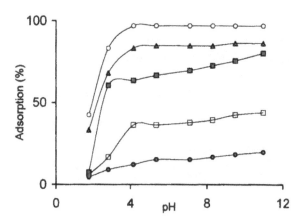

FIG. 6 Effect of pH on the uptake of cobalt ions by a commercial activated carbon (M) and activated carbons derived from almond shells: ●, A-5 (BET SA = 747 m²/g; pH = 8.42); □, A-8 (837, 8.56); ■, A-24 (1381, 9.05); ▲, M (1114, 6.66); ○, A-5HNO₃ (668, 4.06). (From Ref. 165.)

ever, electrostatic interactions were not invoked; rather, these results were interpreted in terms of hydrolysis and precipitation phenomena.

Huang et al. [167] used one H-carbon and one L-carbon (see Section II), with "pH as the master variable" and interpreted similar results in terms of the "formation of surface complexes." They showed that "Co(II) was removed by adsorption mechanism rather than precipitation," especially for the L-carbons. They proceeded to postulate the "formation of specific multidentated Co(II) complexes with surface hydroxo groups as ligands" and analyzed the inhibiting or enhancing effect of several organic ligands.

More recently, several largely phenomenological studies on the behavior of commercial and experimental carbons have been published [168,169]. In particular, the effects of pH were analyzed and the reported results (monotonic increase in uptake with increasing pH) were consistent with the earlier studies summarized above (but not cited by the authors). It is thus puzzling why Paajanen et al. [168] conclude that "the probable uptake mechanism is ion exchange of cobalt ions, but *more experimental work needs to be done* to verify this" (emphasis added).

A brief reassessment of the relationship between surface acidity and metal ion adsorption by C. P. Huang [170], a pioneer in this research field, is worth mentioning here. The authors studied the uptakes of Co, Cd, Cu, Pb, Ni, and Zn (from perchlorate solutions) by an "acidic" carbon ($pH_{PZC} = 4.0$) and a "basic" carbon ($pH_{PZC} = 10.4$) of comparable surface areas. In both cases a monotonic increase in uptake was observed as pH increased from 2 to 11. However, there was an important subtle difference: while the steepest rise for the low-pH_{PZC} carbon was in the acidic range (pH = 2–6) for all adsorbates, the steepest rise for the high-pH_{PZC} carbon was in the approximate range of 6–9 for Co(II), 8–10 for Cd(II), 6–8 for Zn(II) and Ni(II), and 3–6 for Cu(II) and Pb(II). The authors felt it necessary to reiterate the more obvious (albeit very important) conclusion: "Information on the surface acidity of an activated carbon is crucial to the satisfactory achievement of metal removal from aqueous solution. One should carefully consider the type of carbon (L-type or H-type carbon) to be used in the removal process as well as the chemical characteristics of the metals." Explanation (and confirmation) of the subtle differences mentioned above remains an interesting challenge from both a fundamental and a practical point of view.

4. Nickel

The only mononuclear hydrolysis product of Ni(II) whose stability is known well is $NiOH^+$. The other mononuclear species, $Ni(OH)_2(aq)$ to $Ni(OH)_4^{2-}$, are less well known. Small amounts of polynuclear species $Ni_4(OH)_4^{4+}$ form rapidly at high Ni(II) concentration before precipitation of $Ni(OH)_2$ occurs.

In most respects, removal of Ni(II) from water follows the same patterns as that of cobalt. Thus, for example, both Corapcioglu and Huang [171] and Seco et al. [172] show significant uptake increases (for most adsorbents) with increasing pH. The former authors conclude that the "deprotonated surface functional groups, thereby, favor the attachment of cationic metal ions;" the latter authors invoke not only "the decrease in positive surface charge, which results in a lower electrostatic repulsion of the sorbing metal," but also "a decrease in competition between proton and metal species for the surface sites."

Jevtitch and Bhattacharyya [45] noted that the adsorption capacity (at pH = 7.5–8.0) of a commercial activated carbon (pH_{PZC} = 9.0) was slightly lower for Ni^{2+} than for Cd^{2+} in the presence of an ethylenediaminetetraacetate (EDTA) complexing agent, which was opposite to the results obtained in the absence of EDTA [173]. They commented that the "charge distribution of the various species in solution is an important factor which can affect the extent of adsorption of metal-ligands on the activated carbon surface" and concluded that "the charge barrier between the carbon surface and solute species plays a predominant role during adsorption." In a subsequent study of the same adsorbent, performed at pH = 5.0, the same group [173] reported that the uptake of both Ni^{2+} and Cd^{2+} increased in the presence of EDTA and reached a maximum value of its adsorption capacity. They also offered a straightforward explanation for this trend: "At this pH, both Cd(II) and Ni(II) form negatively charged complexes . . . Since activated carbon . . . is positively charged . . . the adsorption of Cd(II) and Ni(II) complexes is favorable." In a similar study, using both organic and inorganic ligands and three commercial activated carbons, Reed and Nonavinakere [174] do not cite the work of Bhattacharyya and coworkers [45,173] but offer essentially the same electrostatic arguments for the results reproduced in Fig. 7: "The

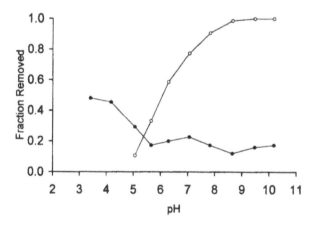

FIG. 7 Effect of pH on the uptake of nickel by a commercial activated carbon (F-400, Calgon Carbon Corp.): comparison of ligand-free system (O) to that in which the Ni/EDTA molar ratio is 1/10 (●). (From Ref. 174.)

pH_{PZC} for the F400 sample was reported to be 10.4 [171]. Below a pH of 10.4 the surface has a net positive charge, and above a pH of 10.4 the surface has a net negative charge. The EDTA [complex of nickel carries] a negative charge for the pH investigated in this study. Thus, at lower pH values the electrostatic force is attractive. As the pH increases, this attractive force decreases until it becomes repulsive." The authors cautioned, however, that "metal adsorption cannot be explained solely by electrostatic theory," but did not elaborate which additional factors must be considered.

More recently, Brennsteiner et al. [175] noted that the electrochemical removal efficiency for nickel is dependent on the pH of the contaminant solution. Maximum efficiency was achieved at pH = 7.0, but only when the carbon electrode was preplated with a layer of copper; the role of surface chemistry was not investigated. Seco et al. [172] did characterize the surface chemistry of a commercial activated carbon (pH_{PZC} = 6.1) and studied its uptake of heavy metals (Ni, Cu, Cd, Zn), as well as of some binary systems. They interpreted the monotonic uptake increase with pH to be consistent with the surface complexation model: "a decrease in competition between proton and metal species for the surface sites" and "a decrease in positive surface charge, which results in a lower coulombic repulsion of the sorbing metal." In the binary uptake studies, they concluded that Ni (as well as Cd and Zn) is not as "strongly attracted to the sorbent" as Cu.

5. Copper

In contrast to the situation of a decade ago [33], a substantial literature has now accumulated on copper removal by activated carbons. This is not only because of metal recovery from acid mine wastes [176] and acidic corrosion of pipes [33] but also because of increasing industrial contamination of water streams [177–182]. In particular, many wastewaters contain complexing ions such as ethylenediaminetetraacetate (EDTA) and the removal of EDTA-chelated copper (and other) ions has been a special focus of attention [45,173,183–186].

The dominant oxidation state in solution is Cu(II), while the dominant hydrolysis product is the dimer $Cu_2(OH)_2^{2+}$. A series of mononuclear species, $CuOH^+$ to $Cu(OH)_4^{2-}$, is formed as well, the first three only in very dilute solutions between pH 8 and 12 and the last only in more alkaline solutions.

Huang [97] noted that H-type carbons are positively charged at pH = 4–5 and that "electrostatic repulsion tends to play its role in repelling the predominant Cu(II) (positively charged species) from the carbon surface." He also emphasized "insignificant" uptakes of Cu(II) by these carbons even "under the optimal pH conditions and Cu(II) concentration. Slight improvement was achieved by "treating the H-carbons with organic chelates." Subsequent work by the same research group [171] confirmed—as did many other investigators [166,187–190,184, 191,192,179,193,172,194,195]—the important effect of pH ("[a]s expected, the

metal removal efficiency of the activated carbons increases as pH increases")
but concluded, intriguingly, that "[e]lectrostatic interaction plays an insignificant
role in adsorption reaction" (see Section III.B). The authors note very high removal efficiencies even for carbons with high pH_{PZC} values (i.e., H-type carbons)
and dismiss precipitation as a major removal mechanism; intriguingly, however,
they do not comment on the earlier conclusion about insignificant uptakes even
under optimal pH conditions.

Ferro-García et al. [187] noted that EDTA complexation (see the section on
nickel above) may not have the beneficial effect of eliminating the electrostatic
repulsion between a positively charged surface and Cu^{2+} (as well as Zn^{2+} and
Cd^{2+}) cations. This is illustrated in Table 5 and was rationalized by postulating
that, in contrast to the smaller inorganic complexes, the Cu-EDTA complex may
be excluded from a large fraction of pores in the olive-stone-derived microporous
carbon used (with 60% of pores less than 7.5 nm in diameter).

Chang and Ku [184] revisited this issue by studying adsorption and desorption
of EDTA-chelated copper on a commercial activated carbon. Their results were
in agreement with those discussed previously for nickel (see above), and they
invoked the same electrostatic arguments to explain them; additionally, the authors concluded that "[c]helated metal ions can be adsorbed on an activated carbon surface with either the metal or the ligand end bonding directly to the surface." In a more recent study of the same issue [186], the authors still ignore
the study of Ferro-García et al. [187], even though their discussion of breakthrough curves focuses again on electrostatic interactions. They also do not make
it clear to what extent they agree (or disagree) with the studies of Bhattacharyya
and coworkers [45,173]—which they do cite—and arrive at the unnecessarily
cautious conclusion that the "adsorption of chelated copper on activated carbon
[varies] over the entire solution pH range, *possibly* [emphasis added] because the
electrostatic interactions between the activated carbon surface and the dominant
species is [sic] highly pH dependent."

TABLE 5   Influence of Anions on the Adsorption (%) of Zn, Cd, and
Cu Cations on an Olive-Stone-Derived Carbon

Adsorbate	No salt	Cl^-	CN^-	SCN^-	EDTA
Zn^{2+}	35.5	56.2	36.0	96.9	4.5
Cd^{2+}	10.0	41.1	42.7	88.1	2.7
Cu^{2+}	79.0	85.0	100.0	95.0	7.5

Note: Cations were added as nitrates (2×10^{-5} M for Zn and Cd; 2×10^{-4} M
for Cu); anions were added as sodium salts (10^{-3} M); the pH was adjusted to 5.
Source: Ref. 187.

The combined importance of the chemical nature and the porous texture of a carbonaceous adsorbent was emphasized in the study of Petrov et al. [189], who examined the uptakes of copper, zinc, cadmium, and lead on as-received and oxidized anthracite. A subsequent study by the same group [190], using the same adsorbates but different adsorbents, ignores the porous texture effects, reemphasizes the "great significance" of the "chemical nature of the carbon surface," and also highlights the different effects observed for different metals by switching from one adsorbent to another.

In recent studies of the use of peat, lignite, and activated chars to remove copper, other metals, and dyes from wastewaters [196,197], Allen and coworkers show wide variations in uptakes, attributed in part to "the heterogeneous nature of the adsorbents used" [197], but they do not consider the role of surface chemistry. In contrast, Johns et al. [198] do show, of course, that even in the case of activated carbons derived from a wide range of agricultural by-products, judicious tailoring of the surface chemistry (e.g., O_2 treatment) can be very beneficial for adsorbing copper. Exactly which features of carbon surface chemistry are responsible for this effect has been addressed in the detailed and important study of Biniak et al. [194]. Their uptake and surface characterization results—the latter obtained using both wet chemistry and surface spectroscopy—are reproduced in Table 6. Because there was no correlation between uptake and surface area, in agreement with many previous studies, the authors concluded that the "number of adsorbed ions depends on the nature and quantity of surface acid–base functionalities and on the pH equilibrium in the aqueous solution." The authors invoke variations in electrostatic interactions as a function of pH and surface treatment (albeit without acknowledging previous studies in this area), but they also consider the interaction with the graphene layers, especially in the case of oxygen-free carbons. The summary of their proposed adsorption schemes (see also pp 201–202) is as follows (symbols ◆ and ⊕ represent the ion radical and positive holes in the carbon structure):

-carbon D—H (annealed in vacuum at 1000 K)

$$>C—H + Cu^{2+} \rightarrow >C—H—Cu^{2+} \quad \text{(dipole–dipole, π-d interactions)}$$
$$>C^{\bullet} + Cu^{2+} \rightarrow >C—Cu^{+} \rightarrow \oplus >C—Cu^{0}$$
$$>C—O^{\bullet} + Cu^{2+} + H_2O \rightarrow >C—O—CuOH + H^{+}$$
$$>C{=}O + Cu^{2+} + H_2O \rightarrow >C—Cu(OH)_2$$

-carbon D—O (oxidized with concentrated nitric acid)

$$>C—COOH + Cu^{2+} \rightarrow >C—COOCu^{+} + H^{+}$$
$$(>C—COOH)_2 + Cu^{2+} \rightarrow (>C—COO)_2\,Cu + 2H$$
$$>C—OH + Cu^{2+} \rightarrow >C{=}O—Cu\ + H^{+}$$

TABLE 6 Effect of pH and Carbon Surface Chemistry on the Adsorption of Cu(II) Cations

Sample[b]	pH[c]	Functional group concentration (meq/g)[a]					Total O/N (wt%)	pH[d]	Uptake (mmol/g)
		—COOH	—COO—	>C—OH	>C=O	Basic			
D—H	10.1	0.00	0.01	0.12	0.09	0.42	0.6/0.2	4.97	0.43
D—O	3.08	0.72	0.38	0.56	0.39	0.13	10.8/0.6	2.45	0.25
D—N	10.4	0.00	0.04	0.10	0.21	0.62	0.4/1.9	5.05	0.32

[a] Determined using Boehm's titration method (Ref. 38).
[b] Commercial Carbo-Tech activated carbon (AC) treated in vacuum at 1000 K (D—H), in conc. HNO$_3$ (D—O) or ammonia at 1170 K (D—N).
[c] 1 g of AC in 100 cm^3 of 0.1 M Na$_2$SO$_4$ (pH = 6.42).
[d] Initial pH of 0.05 CuSO$_4$.
Source: Adapted from Ref. 194.

-carbon D—N (treated in ammonia at 1170 K)

$$>C-NH_2 + Cu^{2+} + H_2O \rightarrow >C-NH_2-Cu(OH)^+ + H^+$$
$$>N: + Cu^{2+} + H_2O \rightarrow >N-Cu(OH)^+ + H^+$$
$$>N: + Cu^{2+} + H_2O \rightarrow \oplus>NOH-Cu^0 + H^+$$

6. Zinc

In contrast to inorganics such as mercury, copper, arsenic, lead, and cadmium, there have been fewer studies that focus only on zinc removal by carbonaceous adsorbents [199–203]. The presence of zinc in water appears to be due primarily to corrosion of galvanized metals [33]. The dominant oxidation state in aqueous solution is Zn(II), while the dominant species (see Fig. A1) are Zn^{2+} at pH < 9 and $Zn(OH)_3^-$ at pH > 9 [202].

Corapcioglu and Huang [171] studied the performance of a wide range of activated carbons as a function of pH. Vast differences were observed in acidic solution; for example, at pH = 3 the efficiency of Zn^{2+} removal for one commercial activated carbon ($pH_{PZC} = 8.2$) was negligible, while for another ($pH_{PZC} = 4.0$) it was about 70%. Removal efficiencies higher than 80% and much smaller differences were observed at pH > 7, as expected based on electrostatic arguments. Two of the carbons exhibited removal maxima at intermediate pH; the authors speculated that the decrease at high pH was due to the presence of phosphorus impurities in these adsorbents, which presumably led to the formation of "phosphoryl compounds on the carbon surface." Ferro-García et al. [187] also found the same trends with pH. In addition they analyzed the very often neglected issue of the fraction of adsorbent surface occupied by the adsorbed species, as well as the temperature effects on the uptake of zinc and other cations (see Section III.B).

Budinova et al. [190], as well as Petrov et al. [189], confirmed the higher uptakes by three H-type activated carbons in alkaline solutions and commented that "[a]t the higher pH values metal ions will replace hydrogen from the carbon surface." From the "jump in adsorption capacity in the pH range 3–5," they concluded that "the pH corresponding to the zero point of charge of those carbons lies between these values" even though they reported pH values of the adsorbents in the range 8–10; intriguingly, they supported the former statement by citing a reference to pH_{PZC} values of insoluble oxides. In a rather confusing paper on zinc uptakes by a commercial charcoal, whose abstract states that the "adsorption kinetics [sic] followed a Freundlich adsorption isotherm," Mishra and Chaudhury [199] confirmed higher uptakes as the pH increased from 3 to 5, but in their discussion of the "mechanism of adsorption" they make no reference to the role of surface chemistry. Allen and Brown [196] also emphasize the isotherm analyses for single- and multicomponent metal sorption (onto lignite, as "a possible alternative to the use of more expensive activated carbons'') but they ignore the importance of chemical surface characterization.

Marzal et al. [201] do cite some of the key papers on this topic and they also confirm the greater effectiveness of a commercial activated carbon for Zn (and Cd) removal at higher pH. In addition to determining the pH_{PZC} of the carbon, they postulated the existence of a single diprotic acid surface with pK_a values of 2.80 and 9.40, but they did not discuss its physical significance. Carrott et al. [202] studied in detail the surface chemistry of five commercial activated carbons as well as the influence of surface ionization on Zn uptake. They agreed with many other investigators of cation adsorption: on most carbons the adsorption of Zn^{2+} occurs on ''ionized acid sites;'' in contrast, for one carbon which contained ''virtually no acid sites, but a high concentration of strong basic sites''—see Table 7—the uptake was apparently ''due to adsorption of a negatively charged hydroxy complex on protonated basic sites.'' The nature of these protonated sites was not discussed. Interestingly, however, the calculated adsorption equilibrium constants in the latter case were one order of magnitude higher, in apparent support of the adsorption site asymmetry argument discussed in Section II. The authors concluded that ''equilibrium constants (and not just site concentrations) [are important] in controlling the adsorption of ionic species by activated carbons.''

7. Cadmium

The aqueous chemistry of Cd is very similar to that of Zn, i.e., dominated by Cd^{2+} [33,204], but the literature on Cd(II) adsorption is much more abundant (see below). The hydrolysis of Cd(II) starts at pH > 7. The species $Cd_2(OH)_3^+$,

TABLE 7 Relationship Between Surface Chemical Properties and Zn(II) Adsorption Equilibrium Constants for a Series of Commercial Activated Carbons

Carbon	C_{acidic} (mmol/g)[a]	C_{basic} (mmol/g)[a]	pH_{PZC}[b]	$pK_{ads,a}$[c]	$pK_{ads,b}$[c]
S51	0.133	0.254	8.3	−7	—
AZO	0.040	1.158	11.8	—	−8
SX+	0.150	0.372	7.5	−7	—
AX24	0.477	0.381	6.9	−6	—
AX21	0.378	0.167	6.2	−6	—

[a] Experimentally determined using 0.01 M NaOH or 0.01 M HNO_3.
[b] Experimentally determined using mass titration.
[c] Adjustable parameters in a surface ionization/adsorption model.
Source: Ref. 202.

$Cd(OH)^+$, $Cd(OH)_2$, and $Cd(OH)_4^{2-}$ seem to be clearly established; $Cd_4(OH)_4^{4-}$ and $Cd(OH)_3^-$ very likely exist as well but not in appreciable amounts.

Huang's review [97] emphasized the "increasing removal efficiency with pH," the advantages of using L-type carbons at low pH, and the advantages of using chelating agents such as EDTA. Regarding this last point, the improvement in % Cd removed was attributed to the formation of anionic complexes (such as $CdEDTA^{2-}$) "which may be electrostatically attracted by Filtrasorb 400, whose surface charge is positive at pH < 7" and to subsequent "association of the Cd^{2+} cations with the adsorbed anions." In a subsequent study whose objective was to identify the physicochemical factors that affect the removal of cadmium, Huang and Smith [44] used four commercial activated carbons with pH_{PZC} ranging from 3.8 to 7.1. Uptake increases with pH were confirmed, and carbons with lower pH_{PZC} were found to be much more effective, especially at intermediate pH. The authors dismissed the difference in surface areas as an important factor by noting that the fraction of surface covered by the adsorbed species is quite small and concluded that "generally, powdered activated carbon has low pH_{PZC} and excellent adsorption capacity for cationic metal ions," while "[granular] activated carbon, having high pH_{PZC}, is rather poor for metal ion adsorption." (In Section II we discuss why powdered carbons may have lower pH_{PZC} values than granular carbons, but clearly this generalization appears to be incorrect.) Perhaps paradoxically, however, they concluded that these uptake differences are due "to the presence of certain functional groups on the surface of the powdered carbons which have a greater chemical affinity for Cd(II) species in solution." The possibility of ion exchange was not mentioned.

Dobrowolski et al. [47] explored the modification of a commercial activated charcoal with the aim "to obtain an adsorbent with a well-defined chemical nature of the surface" and study its effect on Cd(II) adsorption. The isotherms were obtained and compared at unadjusted pH, however, and the discussion of trends for one "basic" and two "acidic" carbons was made quite complicated. The authors concluded that the adsorption mechanism on an acidic carbon was "physisorption in micropores at low pH values and exchange adsorption at high pH values." Similar surface chemical modifications were performed more recently by Polovina et al. [205]. In addition to exposure to Ar at 1000°C and H_2O_2 treatment, an activated carbon cloth was subjected to air and HNO_3 oxidation. The corresponding chemical surface properties and effects on cadmium adsorption (at pH = 5.2) are summarized in Table 8. The authors concluded that "[o]bviously the amount of cadmium adsorbed is in agreement with the carboxylic and lactone group concentration on the activated cloth indicating that ion exchange is the dominant adsorption mechanism." The former conclusion is not that obvious, however: (1) a small fraction of the acidic functional groups had been exchanged; (2) sample Ar-1000°C had a very low acidity, yet its uptake was not suppressed; (3) the uptake was much lower than the ion exchange capacity of the carbons.

TABLE 8 Effect of Surface Chemistry on the Adsorption of Cd(II) by an Activated Carbon Cloth

Sample	Surface area (m^2/g)	Acidity[a] $(mmol/g)$	Carboxyls + lactones[a] $(mmol/g)$	Uptake[b] $(\mu mol/g)$	Surface coverage[c] $(\%)$
As-received	1156	2.27	1.52	10.5	0.3
Ar-1000°C	1013	0.50	0.50	10.8	0.4
Air-oxidized	805	5.07	4.92	34.7	1.5
H_2O_2-oxidized	1104	1.64	0.45	14.3	0.4
HNO_3-oxidized	729	7.29	4.86	35.6	1.7

[a] Determined using Boehm's method (see Ref. 38).
[b] At 10 mg/L of Cd(II) in solution at pH = 5.2.
[c] Based on BET surface area.
Source: Ref. 205.

The authors did not characterize further the surface chemistry (and pore size distribution) of their carbons in order to clarify some of the apparent contradictions.

In an earlier study, Rubin and Mercer [206] wrote, without justification (see Section II), that activated carbon has "little net surface charge and is thus ineffective for adsorbing free hydrated metal ions" and thus proposed, because of its great effectiveness for removing organics, to complex "the metal with an organic molecule prior to contacting the carbon." In particular, these authors analyzed the removal of Cd, both free and complexed. Tables 9 and 10 summarize some of their key results. The role of the adsorbent was clarified by invoking electrostatic arguments (increasing repulsion of cations as pH_{PZC} increases). Surprisingly, however, the effect of EDTA, which was found to be dependent on the EDTA/Cd ratio, was not discussed in the same terms. The authors hastily concluded that "the usefulness of EDTA to enhance ca[d]mium adsorption to activated carbon is quite limited at best." Nevertheless, simple electrostatic arguments show that they selected the "wrong" carbon to test such usefulness: at pH = 7.1 the "acidic" carbon used was negatively charged and there was no Cd^{2+} repulsion. As shown by Bhattacharyya and coworkers [45,173] and Reed and Nonavinakere [174]—see the section on nickel above—when a "basic" carbon is used, the beneficial chelating effect is clearly observed (even at high EDTA/Cd ratios).

In a study mentioned already (see the section on nickel above), Bhattacharyya and coworkers [45,173] studied in some detail the beneficial effect of chelating agents in the adsorption of Cd^{2+} on a commercial activated carbon whose only reported chemical property is pH_{PZC}. These authors provide very dubious evidence for the existence of diprotic acid functional groups on the carbon surface

TABLE 9 Summary of Freundlich Parameters for Cd Adsorption[a] on a Commercial Activated Carbon as a Function of pH and EDTA Chelation

Adsorbate	pH	K	$1/n$	EDTA/Cd
Cadmium	5.7	1.6	0.34	0
	7.1	3.7	0.54	0
	8.1	8.4	0.78	0
EDTA	7.1	3.3	0.41	—
Cd-EDTA	7.1	3.2	0.50	0.1
	7.1	3.7	0.33	0.5
	7.1	1.8	0.24	1.0

[a] Using 50–500 mg/L of commercial Nuchar WV-L carbon (1000 m²/g; $pH_{PZC} = 4.3$), 5–50 μM Cd(II), and 1–50 μM EDTA.
[b] $X = KC^{1/n}$, with X in μmol/g and K in μM.
Source: Ref. 206.

and go on to postulate the following adsorption processes in the absence and presence of EDTA (as well as other ligands):

$$-COH + Cd^{2+} = (-CO^- - Cd^{2+}) + H^+$$
$$-COH + H^+ + (Cd-EDTA)^{2-} = (-COH_2^+ - Cd-EDTA^{2-})$$

They acknowledge that the "exact mechanism of the adsorption of heavy metal chelates is quite complex" but do not hesitate to propose "the formation of an electron donor–acceptor complex of the chelate and the active sites" (e.g., carbonyl groups) and possible beneficial effect of "hydrogen bonding between the

TABLE 10 Effects of pH and Surface Chemistry on the Adsorption of Cd(II) by Four Commercial Activated Carbons

Adsorbent	Surface area (m²/g)	pH_{PZC}[a]	Uptake of Cd (μmol/g)[b] pH = 6.5	pH = 8.0
Darco HDC	650	3.8	3.2	178
Nuchar WV-L	1000	4.3	3.2	160
Aqua-Nuchar	1000	6.2	2.0	125
Nuchar S-A	1500	8.3	<0.3	<10

[a] Determined by titration (the authors refer to it as the isoelectric point).
[b] Maximum surface coverage obtained from the Langmuir isotherm.
Source: Ref. 206.

surface groups and the chelated metals or free ligands'' [45]. Yet the key arguments that they use to explain the observed experimental results are primarily electrostatic. Indeed, Bhattacharyya and Cheng [173] conclude that "electrostatic binding is important in the adsorption of metal ions and metal-ligand complexes."

As mentioned in the discussion on nickel, Reed and Nonavinakere [174] reached similar conclusions. These authors also illustrated the occurrence of three different effects produced when a complexing agent is added to Cd-containing aqueous solution. Figure 8 illustrates the effect discussed above ("the ligand adsorbs on the solid and the metal ion reacts with the solid–ligand complex"). In addition to this effect, the authors reproduced (albeit implicitly) the case discussed above in the context of the study by Rubin and Mercer [206] (for an "acidic" carbon adsorption was suppressed by EDTA addition at all but the lowest pH values). They also showed that, upon increasing EDTA loading, the general shape of the pH-adsorption edge (uptake vs. pH curve) remained unchanged, except that the adsorption edge (pH at which the uptake exhibits the highest slope) was shifted to a higher pH region; one reason for this effect is competitive adsorption of the complexing agent, whose impact depends on the surface charge variations with pH.

Periasamy and Namasivayam [207] claimed that the "adsorption capacity [of a peanut hull carbon] is much superior to commercial activated carbon." The relevant uptakes achieved their maximum and constant values, of ~20 and ~1.8 mg/g, respectively, at pH > 3. The corresponding surface areas, slurry pH values, and ion exchange capacities were, respectively, 208 and 354 m^2/g, 6.7 and 8.2, and 0.49 and ~0.0 meq/g. The authors discuss these results in terms of electrostatic forces, but they conclude that (unspecified) "specific chemical interaction also plays an important role in Cd(II) adsorption."

Macías-García et al. [208] did not cite any of these studies and thus came to very confusing conclusions, but they did analyze an intriguing set of chemically modified activated carbon samples (exposure to N_2 at 900°C, exposure to H_2S at 900°C, exposure to SO_2 and then H_2S at 30°C followed by treatment in N_2 at 200°C). Their results are summarized in Table 11. The characterization of the surface chemistry after various treatments is limited to statements about "the formation of surface sulfur" and "S=O groups [which] may belong to the species SO_2 or SO_4^{2-}." The authors then simply conclude that "the most effective [method] to increase the adsorption capacity of AC [is] the heat treatment in N_2," which they attributed to "the development of porosity in this sample" (even though their own data—see Table 11—do not really support such an explanation). The authors do not comment on the fact that the uptake at low pH is electrostatically unfavorable and is expected to be even more so after the removal of acidic surface groups (unless these are reconstituted upon exposure of the sample treated in N_2 at 900°C to room-temperature air; see Refs. 82 and 84). Finally, a

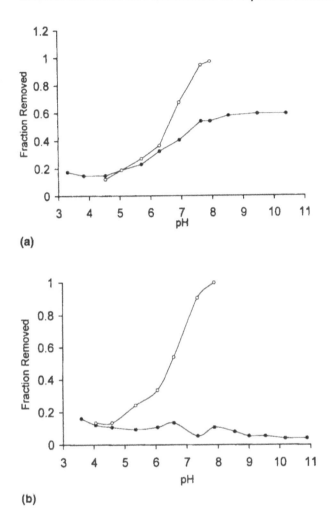

(a)

(b)

FIG. 8   Effect of pH on the uptake of cadmium by a commercial activated carbon (Darco KB): comparison of ligand-free system (○) to that in which the Cd/EDTA molar ratio is (a) 1/1 (●) or (b) 1/10 (●). (From Ref. 174.)

detailed comparison of these (unusually high?) uptakes (e.g., in terms of surface coverage) with the abundantly available literature data (e.g., Refs. 44, 209, 207, 205, 197, and 203) would have been and would still be desirable.

Brennsteiner et al. [175] also offer no arguments about the role of surface chemistry in their promising electrochemical method of heavy metal removal, even though they note that the "removal efficiency [was] dependent upon the

TABLE 11 Effects of Heat Treatment and Sulfurization of an Activated Carbon on the Adsorption of Cd(II)

Sample	V_{mu} (cc/g)	V_T (cc/g)	$X_{25,un}$ (mmol/g)	$X_{25,2.0}$ (mmol/g)	$X_{45,un}$ (mmol/g)
AC	0.37	0.83	0.193	0.190	0.102
AC-N2-900	0.37	0.87	0.320	0.299	0.124
AC-H2S-900	0.31	0.78	0.241	0.242	0.110
AC-SO2-H2S-200	0.30	0.72	0.291	0.255	0.128

Note: V_{mu} = micropore volume; V_T = total pore volume; X is the uptake of Cd^{2+} at either 25 or 45°C and at either unadjusted pH (un) or at pH = 2.0.
Source: Ref. 208.

pH of the contaminant solution." Several "novel carbon materials"—including vapor-grown carbon fibers, a coal-derived foam, and carbon nanofibers—were used to remove Cd (as well as Pb and Ni). The significant differences reported in the amounts of metal removed by these materials, from 0 to 91%, were left unexplained: the authors simply made a vague reference to their "high surface areas and good electrical conductivity."

Allen et al. [197] compared the effectiveness of an activated bone char, a peat-derived char, a lignite char, and a commercial activated carbon; in addition to invoking surface area effects, they attributed the highest uptake by the bone char "largely to the presence of chemical species in these materials." Intriguingly, and rather unconvincingly, these "species" were claimed to form a "phosphorus-based hydroxy-apatite/metal association." Mostafa [210] also argued that the chemistry of the carbon surface is "a prominent factor which determines the removal capacity" and not "the extent of the surface." Jia and Thomas [211] introduced various types of oxygen functional groups onto the surface of a coco-nut-shell-derived activated carbon and confirmed a dramatic enhancement in the uptake of Cd^{2+} "indicating cation exchange was involved in the process of adsorption." In addition to this irreversible uptake, they reported that "reversible adsorption probably involved physisorption of the partially hydrated cadmium ion."

8. Mercury

Mercury is one of the most toxic water contaminants, and its discharges from chloralkali, paper and pulp, oil refining, plastic, and battery manufacturing industries [212] need to be controlled. It can exist in aqueous solution as the free metal, as Hg(I), or as Hg(II) [32], the latter being favored in well-aerated water (see Fig. A2).

The review by Huang [97] highlights the findings that Hg adsorption capacity increases as pH decreases [213] and that reduction of Hg(II) occurs on the carbon surface at high pH, as well as the observations that different carbons prepared by different activation methods have widely varying capacities. Here we review the post-1975 literature with two principal objectives: (1) confirmation and explanation of the former points, and (2) more useful guidelines regarding the latter point. In particular, we search for clues that might clarify one of the interesting but unexplained conclusions in the early work of Yoshida et al. [214]: "Contents of acidic and basic surface oxides of activated carbons also affect the adsorption characteristics of Hg(II) on activated carbons." A subsequent investigation by the same group [215] did not report any chemical properties of the activated carbons studied, but the following explanations were offered: "Mechanisms of adsorption of Hg(II) in an acidic (HCl) and an alkaline (NaOH) medium differ from one another." In the former medium, "the $HgCl_4^{2-}$-complex is reversibly adsorbed;" in the latter, adsorption is irreversible and "Hg(II) is considered to be adsorbed . . . in the form of $Hg(OH)_2$ and Hg metal which is generated by the reduction of Hg(II) on the surface." The authors do not cite the interesting work by Humenick and Schnoor [216] in which more detailed (and controversial) mechanistic explanations (e.g., for the beneficial effect of low pH) were offered after analyzing the behavior of an H-type commercial activated carbon: "[T]he pronounced pH effect may be explained in terms of the surface charge on the carbon surface." At low pH, "the charges of the surface oxides could be neutralized by hydrated protons" and "pore diffusion of the mercury-chloride complexes were [sic] facilitated." At high pH, "[s]ince most of the mercury ions were in the neutral to negative form, the basic surface charges of the carbon may have repelled the mercury complexes." The authors did not give examples of such repulsive interactions, however; they did quote the study by Rivin [217] in which the possibility was mentioned that the "π-electrons of the aromatic basal planes of the carbon may be available for reduction."

López-González et al. [218] also failed to cite the study by Humenick and Schnoor [216], but they analyzed in some detail the effect of carbon–oxygen and carbon–sulfur surface complexes on the uptake of mercuric chloride, which is very weakly ionized in aqueous solution. (The effectiveness of sulphurized carbons in removing mercury from air or water streams had been demonstrated earlier by Sinha and Walker [219], Humenick and Schnoor [216], and more recently by Gómez-Serrano and coworkers [208].) Their key results are summarized in Fig. 9. There was a noticeable uptake decrease when the activated carbons were oxidized with H_2O_2 (AO). This decrease was not due to a reduction in the surface area and is contrary to the behavior of cationic metallic species, whose uptake is typically enhanced as a consequence of a lower pH_{PZC} of the oxidized carbon. Upon subsequent heat treatment in helium at 873 K (A-873), the adsorption ca-

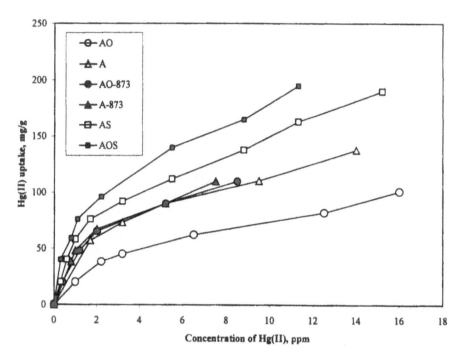

FIG. 9 Effects of oxidation and sulfurization on the uptake of mercury by activated carbon. (Adapted from Ref. 218.)

pacity was recovered. Evidence was presented for the reduction of Hg(II) to Hg(I) on the residual phenolic and hydroquinonic surface groups:

$$2(-OH) + 2HgCl_2 \rightarrow 2(=O) + Hg_2Cl_2 + 2HCl$$

It was thus proposed that the uptake of Hg(II) by activated carbons occurs by the parallel mechanisms of adsorption of molecular $HgCl_2$ and reduction. Upon H_2O_2 oxidation, the beneficial phenolic and hydroquinone groups were hypothesized to be lost and converted to carboxyl groups. In contrast to the earlier report by Sinha and Walker [219], significant uptake increase was observed upon sulphurization (saturation with CS_2 at 288 K and subsequent heating to 773 K) of the as-received (AS), and especially so, the oxidized carbon (AOS). This effect was attributed to the analogous reduction reaction:

$$2(-SH) + 2HgCl_2 \rightarrow 2(=S) + Hg_2Cl_2 + 2HCl$$

Finally, and in apparent contradiction with both the results of Yoshida et al. [215] and the report by Huang [97], the efficiency of removal was in all cases lower in acidic than in neutral solutions. This was explained by noting that enhanced reduction is favored at higher pH, as shown in the above equations. More recently, Adams [220] performed electron microscopic analyses of an activated carbon surface that was contacted with $HgCl_2$ solution and confirmed the presence of both Hg and Cl, thus endorsing the importance of the reduction mechanism.

Another detailed study of adsorption of Hg(II), on the surface of a range of commercial activated carbons, was carried out by Huang and Blankenship [221]. In the adsorbent screening stage of their study, they noted that maximum uptakes were typically observed at pH = 4–5. A more extensive analysis of the contrasting behavior of two L-type carbons and one H-type carbon led them to echo the conclusion of López-González et al. [218], "that total Hg(II) removal was brought by two mechanisms: adsorption and reduction." Thus for example they noted that, even though the "two acidic activated carbons [used] have definitely superior Hg(II) removal capacity over a wide pH range 3–11," the main reason for the superior performance of the H-carbon, "at least in the acidic pH region," was its greater Hg(II) reduction capacity.

It is clear from these uncommitting (and apparently contradicting) references to the nature of Hg(II) species in solution that the challenge for the late 1980s and the 1990s was to describe more clearly the nature of adsorbent/adsorbate interactions in the absence of significant reduction. Are they predominantly electrostatic or dispersive? And how exactly are they affected by variations in both pH and carbon surface chemistry? A study that would offer the answers to these questions has yet to be performed, however. The emphasis in the available literature has been directed elsewhere, for the most part (see below). The studies by Jayson et al. [222] and Gomez-Serrano et al. [223] are notable exceptions. The former authors studied the adsorption and electrosorption of Hg(II) acetate onto an activated charcoal cloth (pH_{IEP} = 2.7) and noted that the "saturation adsorption increases slightly as the pH increases from 3 to 5." They attributed this trend to "a continual increase in the electro negative [sic] character of this activated charcoal surface which increases its attraction for the positively charged mercury(II) ions." Gomez-Serrano et al. [223] revisited the issue of Hg removal using a sulphurized activated carbon (treated in H_2S at 900°C), by comparing its adsorption behavior to that of a sample heat-treated in N_2 at 900°C. They confirmed a drastic reduction of mercury uptake in acidic solution (pH = 2) and an enhancement upon both carbon sulphurization and heat treatment in N_2. These results, combined with the fact that the uptake of $HgCl_2$ and $(HgCl_2)_2$ (at pH = 4.4) was much greater than that of Cd^{2+} (at pH = 6.2) and Pb^{2+} (at pH = 5.4), led them to postulate an acid–base mechanism of adsorbate/adsorbent interaction following Pearson's HSAB principle [224], but the role of the chemical surface properties (in particular, the beneficial effect of the removal of surface oxygen) remained unclear.

Kaneko [225] demonstrated the enhancement of Hg(II) uptake in the presence of well dispersed FeOOH species on the surface of activated carbon fibers. Youssef et al. [226] were interested in exploiting the virtues of an adsorbent derived from a subbituminous coal. Polcaro et al. [227] studied the removal of "mercury ions from concentrated sulphuric acid" using electrodeposition on carbon or graphite electrodes. Namasivayam and Periasamy [212] developed an adsorbent from bicarbonate-treated peanut hulls. They did examine the effect of pH and their results are shown in Fig. 10. The very different pH ranges in which the two carbons are effective are striking. The authors argued that the decrease at pH < 4.0 in both cases is due to the formation of $HgCl_2$ which "has been found to decrease the Hg(II) sorption onto a commercial [activated carbon]." The decrease at pH > 4.0 for one of the adsorbents was tentatively attributed to the "formation of soluble hydroxy complexes of mercury, $Hg(OH)_2$ (aq)." The reason the same process does not reduce the uptake on the other adsorbent was, apparently, the "retention of the $Hg(OH)_2$ species in the micropores." The authors provide no experimental evidence for this statement; instead they note that supporting evidence can be found in the studies of Humenick and Schnoor [216] and Ma et al. [228], but this assertion is far from clear. The former authors do not mention $Hg(OH)_2$ at all. Ma et al. [228], on the other hand, show a very different trend with pH: a maximum at intermediate pH which, contrary to the assertion of the authors, was not consistent with the study of López-González et al. [218], who found an increase in uptake from pH = 2 to pH = 5.

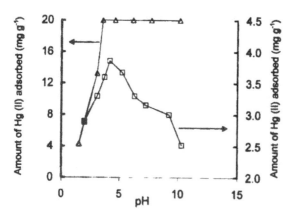

FIG. 10 Effect of pH on the uptake of mercury by a commercial activated carbon (CAC, □) and a bicarbonate-treated peanut hull carbon (BPHC, △). (Adapted from Ref. 212.)

More recently, Peräniemi et al. [229] advocated the use of zirconium-loaded activated charcoal as an effective adsorbent for mercury (and especially for arsenic and selenium), the rationale being that the "presence of active metal on an impregnated charcoal surface can greatly affect the adsorption affinity." They compared the pH effects on the uptakes by both a loaded and an unloaded commercial charcoal powder and concluded that the adsorption mechanism of mercury differs from that of the anionic arsenic and selenium species. They also noted the "highly complicated" behavior of mercury in aqueous solutions and did not attempt to explain the apparent absence of pH dependence of the uptakes.

Adsorption of mercury from a cyanide solution is of interest because of its relevance to the debate regarding the mechanism of gold cyanide adsorption. Adams [220] has investigated this issue and concluded that "[m]ercury adsorbs on to activated carbon in the form of $Hg(CN)_2$" while the anionic "$Hg(CN)_4^{2-}$ species adsorbs to a negligible extent." The author interpreted this finding to be consistent "with the theory that aurocyanide adsorbs on to carbon as an ion pair" (see section on gold below). Recent studies of Hg(II) removal in the presence of Br^- ions [230] indicate that the "[a]dsorption ability of [a coal- and a coconut-shell-derived] activated carbons for $[HgBr_2]$ and $[HgBr_3]$- at pH 7 and for $[HgBr_3]$- at pH 1.4 resembled each other." The authors concluded "that activated carbon reduces Hg^{2+} at pH 7 even if Br^-ion is present" and that the "adsorption species varies with Br^-ion concentration."

The important recent study by Carrott et al. [231] comes close to answering the key fundamental questions; it parallels the one on zinc adsorption discussed above [202]. The same carbons were used (see Table 7) for the removal of Hg chlorocomplexes. The authors found that "whereas the adsorption of neutral $HgCl_2$ or positive Hg^{2+} was very low, significant quantities of . . . $HgCl_4^{2-}$ were adsorbed" and concluded that "surface charge is important in controlling the adsorption of mercury species from dilute solutions." For example, in agreement with the electrostatic arguments discussed throughout this review, the *low-coverage* uptake of anions on the basic (H-type) AZO carbon decreased with increasing pH, while the uptake of cations was very low at pH = 0.2; at higher concentrations, which are of course less relevant for water treatment applications, the authors note that "the electrostatic approximation begins to break down and the adsorption mechanism changes."

9. Lead

This metal becomes a pollutant when released into aqueous streams from metal treatment and recovery plants (e.g., lead-using battery manufacturers). In industrialized nations, until the 1970s, a major source of air pollution was lead emission from the tetraethyllead antiknock additives to gasoline; this is still a problem in many underdeveloped nations. Once ingested, lead readily forms complexes with enzymatic oxo groups in the organism, thus interfering with all the steps in haemo synthesis and in the porphyrin-based metabolism [232]. Hydrolysis of Pb(II) is well documented [123]. Lead is also readily oxidized in acidic media: $2Pb + O_2 + 4H^+ \rightarrow 2Pb^{2+} + 2H_2O$.

The adsorption of aqueous Pb(II) has been studied extensively. The following important factors have been studied: solution pH [233,234,190,235–239], type of adsorbent [166,171,233,234,190,236,198] and chemical surface modification [210,223,240]. As in the case of many metallic cations, Pb^{2+} uptake increases with increasing aqueous solution pH, with a sharp increase ("adsorption edge") being observed in a narrow pH range, typically between 3 and 6 [171], depending on the pH_{PZC} of the carbon used. Adsorption of Pb(II) as a function of solution pH for different initial concentrations is illustrated in Fig. 11. As the pH increases further, there is surface precipitation of the products of hydrolysis of Pb^{2+} (see Table A1 in the Appendix).

It was proposed by Corapcioglu and Huang [171] that Pb^{2+} adsorption obeys the surface complex formation model (see Section III.B). Accordingly, the chemical bonding energy ("probably hydrogen bonding") was assumed to have a more important role than the energy of electrostatic interaction. On the basis of such analysis, these authors have calculated surface densities of the various adsorbed species, e.g., $(CO^-)_2 Pb^{2+}$ being dominant at pH < 7 and the uptake at higher pH being due to several species adsorbing simultaneously. In an earlier study, ignored by these authors, Netzer and Hughes [166] had suggested that the role of pH is simply related to the onset of metal hydrolysis and precipitation and that "no single carbon property appears to be dominant in determining its metal adsorptive characteristics from aqueous solution."

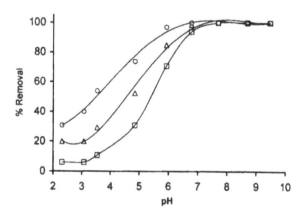

FIG. 11 Effect of pH and surface loading on the removal of lead by a commercial activated carbon (Filtrasorb 400, Calgon Carbon Corp.): 2.0×10^{-5} M (O); 1.0×10^{-4} M (△); 8.0×10^{-4} M (□). (Adapted from Ref. 171.)

Cheng et al. [234] tackled these issues head on but arrived at very few clear conclusions; they did note several complications due to experimental artifacts (e.g., use of CO_2-free solutions). They followed the earlier proposal of Corapcioglu and Huang [171] and proposed the existence of protonated oxygen functional groups on the carbon surface (see Section II). A contemporary paper by Taylor and Kuennen [241] also claimed that the "importance of the surface chemistry of the carbon and the relationship with the pH of the water for lead reduction was demonstrated." They showed first that the solubility of lead decreased from 100 to 30% as the pH increased from 5 to 9, with a rapid decline between 7 and 8. In minicolumn tests they then showed that an L-type carbon had a much later breakthrough than an H-type carbon; however, when both the L-type and the H-type carbons were deoxygenated at 600°C in N_2, their breakthrough times were dramatically reduced at the same pH (6.5). The authors thus concluded that "the oxides on the surface of the carbon are extremely important for the removal of lead" but did not clarify the nature of this importance. Budinova et al. [190] also concluded that the "chemical nature of the carbon surface [has] great significance for the adsorption process" and proceeded to characterize the surface chemistry of three activated carbons. Ironically, by citing a study of Huang [97] in which the concept of pH_{PZC} was discussed, and which Corapcioglu and Huang [171] had rejected as an explanation for lead adsorption, these authors argued that the pH effect "is related to a surface charge which is very much dependent on the pH of the solution" and vice versa. This electrostatic interaction model was also endorsed by Periasamy and Namasivayam [236] who discussed it in terms of cation exchange on the oxygen functional groups of carbon.

In an ambitious paper [237], Reed defended his earlier proposal regarding lead removal mechanisms: "adsorption, surface precipitation, and pore precipitation." He insisted that "the dramatic increase in metal removal by the regenerated GAC columns was not caused by an increase in the number or type of adsorption sites but was due to the precipitation of Pb on the carbon surface or in the carbon pore liquid." He recognized that "[t]here was a large difference between the pretreated and virgin carbon suspension titration curves," but he attributed these differences "to the OH^- that was transferred during the filtration step, not from a change in surface characteristics." In subsequent studies by the same group [242,243], the authors were still not prepared to acknowledge the importance of the electrostatic interaction mechanism.

10. Uranium

Adsorption of uranium onto activated carbon is important from purification, environmental, and radioactive waste disposal points of view [244,245]. Most often uranium is present in the form of hexavalent UO_2^{2+} (as oxides and hydroxides). The speciation diagram for uranium has been presented recently by Qadeer and Saleem [246] and Park et al. [247]. Uranium hydrolysis begins at pH ca. 3, the principal product being the $UO_2(OH)^+$, as well as its dimer $(UO_2)_2(OH)_2^{2+}$. Pentavalent uranium disproportionates rapidly into U(VI) and U(IV) but persists as UO_2^+ at $2 < pH < 7$. Tetravalent uranium is present as U^{4+}, which is hydrolyzed more easily than UO_2^{2+}; hydrolysis starts at acid concentrations > 0.1 M, and with increasing pH several different mononuclear species are formed.

Only a few recent studies have been performed on the adsorption of uranium by activated carbons, and, as with most of the older studies (see, e.g., Ref. 248), none of them address specfically the role of carbon surface chemistry. Saleem et al. [244] reported a maximum uptake at pH = 4–5 on a commercial charcoal and attributed it to the maximum "availability of free uranium ions." When anions (e.g., nitrate, oxalate) were also present in solution, in most cases there was a decrease in uptake; the authors attributed this, rather vaguely, to "lower affinity of [these] complexes for adsorption." Acetate was the exception: the authors speculated that "the anionic complex of acetate ion with uranium is more strongly adsorbed . . . then the uranium ions themselves." In a follow-up study [246], this group presented the same results but attempted to analyze the pH effect in more detail. Their speciation diagram shows that "over the pH range 1–4, the predominant species is UO_2^{2+}," while beyond pH-5 the dominant aqueous species is $UO_2(OH)_2$. They concluded that "[a]s the pH of the aqueous solutions increases from 1 to 4, the adsorption of H_3O^+ ions decreases while that of the UO_2^{2+} ions increases." Above pH 4 the UO_2^{2+} ions start to hydrolyze, and the resulting species are presumably "weakly adsorbed" in comparison to UO_2^{2+}. Intriguingly, they fail to mention the directly relevant work of Abbasi and Streat [245] or the many relevant studies of the very issue that they do mention in the introduction in which it had been clarified how exactly the "pH can substantially affect the surface electric charge of the adsorbent." As shown in Fig. 12, Abbasi and Streat [245] found that "treatment with hot nitric acid oxidized the surface of activated carbon and significantly increased the adsorption capacity for uranium in near-neutral and slightly acidic nitrate solutions." Even though in none of these studies did the authors characterize the surface chemistry of their adsorbents, a straightforward explanation of their data is, of course, that adsorption of cations on a positively charged surface (pH $<$ pH_{PZC}) is not favorable, while it is indeed "likely that carboxylic groups with a pK value of 4–5 are largely responsible for the ion exchange or uranium from near-neutral aqueous solutions" [245].

Park et al. [247] did cite most of the studies discussed above in their attempt to elucidate the influence of (uncontrolled but recorded) pH on the adsorption of uranium ions. They also did not characterize the surface chemistry of their carbons, even though they noted that "surface-oxidized carbon has an adsorption capacity superior to that of untreated activated carbon." Instead, they invoked the monograph by Bansal et al. [35] to support the excessively generalized statement that the pH_{PZC} is "around 8 for activated carbon." To explain their results—

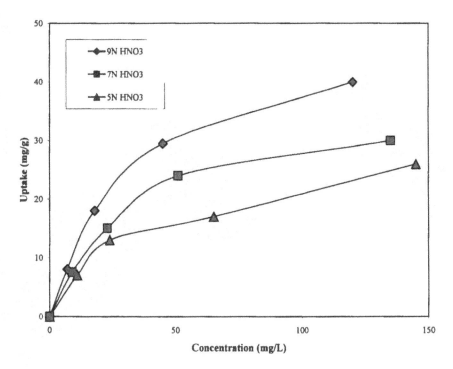

FIG. 12   Effect of surface chemistry on the adsorption isotherms of uranium. (Adapted from Ref. 245.)

large increases in uptake at pH > 3 (which are readily explained using electrostatic arguments and the likely scenario that the pH$_{PZC}$ of oxidized carbon is much closer to 3 than to 8)—the authors stated that ''(1) only cation forms of uranyl ions exist over wide pH ranges due to the high surface acidity of oxidized carbon, and (2) large amounts of surface oxides . . . lead to an increase of the adsorption capacity of uranium.''

11.   Gold

Gold forms a large number of complexes with various ligands; many of them are also strong oxidizing agents. The Au(III) species are generally more stable than the Au(I) species; a notable exception is the very stable [Au(CN)$_2$]$^-$ species involved in the cyanide (''carbon-in-pulp'') process of gold extraction [3,5,35,249–251].

Adsorption of gold on carbonaceous solids is technologically a very important topic [252–254], and a voluminous literature is thus available. An extensive review of the literature up to 1984 is available in Ref. 35. McDougall and Hancock

[252] and Huang [97] had reviewed some of the very early studies. Apart from the use of activated carbons (with reduction potentials of ~0.14 V) to precipitate gold from chloride [255,256] or bromide [257] solutions (with reduction potentials of 0.7–0.8 V), the central theme of many studies is the controversial mechanism of adsorption of $[Au(CN)_2]^-$. The difficulties in elucidating the mechanism(s) are a direct consequence of the uncertainties regarding the chemical surface properties of the activated carbons used. In particular, since the conditions of greatest practical interest (high pH) involve the interaction of anionic adsorbates with a mostly negatively charged adsorbent, elucidation of the character and number of basic sites on the carbon surface becomes a critical issue (see Section II).

Here we summarize the key developments with emphasis on the post-1985 period and the role of the carbon surface. Bansal et al. [35] have grouped the postulated mechanisms as follows: (1) reduction theory; (2) ion-pair adsorption theory; (3) aurocyanide anion adsorption theory; and (4) cluster compound formation theory.

Davidson [258] noted that gold recovery increased with decreasing pH in the presence of spectator Na^+ and Ca^{2+} ions and postulated that ''both the hydroxide and aurocyanide anions are competing for active adsorption sites on the charcoal surface;'' the author did not identify these sites, however. The presence of Ca^{2+} cations was more beneficial than that of Na^+ cations, and the author speculated that ''the adsorbed calcium aurocyanide complex is . . . more stable than the sodium aurocyanide complex.''

Based on scant experimental evidence [259,260], involving successful regeneration with NaOH and high uptakes for a charcoal activated in CO_2 at high temperature, Grabovskii and coworkers [259–261] speculated that gold adsorption is a surface reduction process,

$$-C_x + NaAu(CN)_2 + 2NaOH = -C_xO \cdots Na^+ + Au + 2NaCN + H_2O$$

where $-C_xO \cdots Na^+$ represents a cation adsorbed on the oxidized carbon surface.

In a much more through study [68,262–264], Pitt and coworkers examined the kinetics and thermodynamics of both gold and silver cyanide adsorption. They confirmed the decrease in cyanide uptake at high pH (see Fig. 13) and speculated that ''this must involve a change in the concentration of active adsorption sites,''

$$C_xO + H_2O = C_x^{2+} + 2OH^-$$

where ''in contact with water, the site is active enough to decompose water, produce OH^- ions, and provide a positively charged adsorption site.'' They then postulated that high pH conditions shift this equilibrium to the left ''with consequent elimination of active adsorption sites.'' This is an interesting proposal, in light of the discussion of carbon basicity in Section II, and especially because

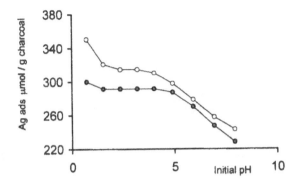

FIG. 13 Influence of pH on the adsorption of silver cyanide in the presence of NaCN at 40.81 mmol/L (O) and 16.32 mmol/L (●). (Adapted from Ref. 264.)

subsidiary slurry pH measurements on the commercial PCB carbon showed that the carbon used was indeed "basic" (pH$_{slurry}$ = 10.2; see Table 2), but it ignores the role of dissociated oxygen functional groups and electrostatic repulsion. Be that as it may, the authors' conclusions are worth quoting: "The overall surface charge of the charcoal particles is negative, but adsorption likely occurs on positively charged sites" [68]. These authors also compared the behavior of Au-vs. Ag-cyanide on the same adsorbent [264] and examined in more detail the effects of Na$^+$, Ca^{2+}, and CN$^-$ ions on Ag(CN)$_2^-$ uptake. Figure 14 reproduces the schematic representation of their adsorption model, which is primarily electrostatic and consistent with generally accepted views on the electrical double layer but also invokes the ionic solvation energy theory to account for both the higher uptake of the larger Au(CN)$_2^-$ ions and the beneficial effect of cations. The chemistry of the carbon surface was not discussed in any detail.

McDougall et al. [265] and McDougall and Hancock [252] provided a comprehensive review of earlier models of gold cyanide adsorption and, based on x-ray photoelectron spectroscopy and other results of their own, essentially endorsed the model of Davidson [258]. For example, they summarized the earlier work of Kuzminykh and Tyurin (see op. cit.), who apparently argued against electrostatic attraction of Au(CN)$_2^-$ to positive sites on the carbon surface; instead, these authors emphasized the importance of dispersion interactions with H[Au(CN)$_2$] in acidic solution and Na[Au(CN)$_2$] in basic solution. They noted that the uncertainty regarding the mechanism of adsorption is due to the "very little information [that] is available about the surface functional groups present on the carbon." They do not furnish such information, but they do argue to have provided evidence that "militates against the adsorption of Au(CN)$_2^-$ anions by simple electrostatic interactions in the electrical double layer, or with positively charged

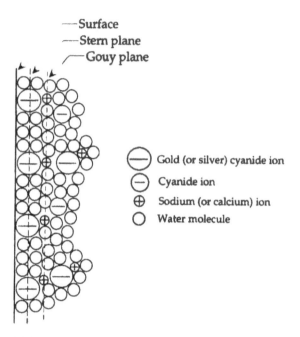

FIG. 14 Schematic representation of the adsorption model for silver cyanide on activated carbons. (From Ref. 264.)

sites on the carbon surface'' as well as against ''the idea that a simple ion-exchange process is involved.'' The authors concluded that the ''mechanism of adsorption of aurocyanide on carbon appears to involve adsorption of the aurocyanide as the less soluble $M^{n+}[Au(CN)_2^-]_n$ complex ($M = Na^+$, K^+, Ca^{2+}, H^+), which probably accounts for the initial adsorption stage, followed by a reduction step in which either a substoichiometric $AuCN(x)$ surface species or a cluster-type compound of gold is formed.'' Some years later, the same research group [266] presented another review of the (by then six different) mechanisms postulated for the adsorption of aurocyanide onto activated carbon. They compared the behavior of activated carbons with polymeric adsorbents and ion-exchange resins under a wide range of conditions (of pH, temperature, ion addition) and using techniques (of unclear usefulness) such as electron spin resonance spectroscopy and x-ray diffraction. They again reject the ion-exchange and electrical double layer theories of adsorption in favor of ''some kind of specific adsorption'' following the ''formation of an ion-pair species $[M^{n+}][Au(CN)_2^-]_n$ in the environment of the carbon.'' They also provide a more detailed description of this interaction: ''The condensed aromatic system [may] contribute to the extraction of aurocyanide by activated carbon" by "involving the donation of electrons from the condensed aromatic carbon structure to vacant orbitals in the gold atom."

Fleming and Nicol [267] reported that adsorption of "gold cyanide onto activated carbon is not an irreversible process," and thus confirmed similar findings for silver cyanide [263]. They also confirmed that the process, like adsorption of silver cyanide [263] is exothermic, in contrast to the behavior of some other inorganic ions [125,144,268,172,269] (see also Section III.B).

Investigation of the role of oxygen in the adsorption process has led to at least partial reconciliation of several of the proposed models of gold (and silver) cyanide adsorption. In the early study of Dixon et al. [68] it had been noted that "elimination of oxygen from the system also eliminates some of the adsorption sites." Van der Merwe and van Deventer [270] confirmed these findings, as did others [271,272], and addressed the issue in more detail by adsorbing gold and silver cyanide, in a closed vessel at pH = 8.5, on two carbons whose pH_{IEP} was measured to be less than 2. (The pH_{PZC} was not reported.) Table 12 summarizes some of the results of their experiments in which the "consumption [of oxygen from aerated water] increased drastically during the adsorption of anionic metal cyanides." (For a recent criticism of the experimental protocol used by van der Merwe and van Deventer, see Ref. 273.) In the absence of the cyanides, oxygen uptake was an order of magnitude lower; in the absence of oxygen, uptakes of both gold and silver were lower but still significant. Also, adsorption of the neutral $Hg(CN)_2$ was found to be unaffected by the presence of oxygen. Oxygen's participation in reduction of $Au(CN)_2^-$ to $AuCN$ was discarded: even after 3 weeks of exposure to $Au(CN)_2^-$, no reduction to $AuCN$ was detected at pH = 8.5. Partial reduction of gold cyanide on carbon B was confirmed by FTIR spec-

TABLE 12 Effects of pH and the Presence of O_2 on the Adsorption and Desorption of Gold and Silver Cyanide for Two Coconut-Shell-Based Activated Carbons

Carbon[a]	A	A	A	A	B
Adsorbate (M)	Au	Au	Au	Ag	Au
O_2 concentration (mg/L)	9.0	9.0	14.5	9.0	9.0
pH during adsorption	8.5	3.5	3.5	8.5	8.5
Adsorption time (days)	21	2	2	21	1
Uptake (mg M/g carbon)	36.6	22.6	22.6	33.6	28.4
Recovery by elution (%)[b]	100	85	85	38	80
$[MCN]/[M(CN)_2^-]$[c]	0.0	0.18	0.18	1.63	0.33

[a] Experimental carbons, both of which have an isoelectric point of ca. 2.0.
[b] $M(CN)_2$ eluted at room temperature at pH = 11 in the absence of free cyanide.
[c] Molar ratio of adsorbed species, determined from elution data.
Source: Ref. 270.

troscopy, as was the presence of $Au(CN)_2^-$ adsorbed species on both carbons. The authors thus postulated that the "gold (or silver) cyanide is adsorbed in two ways: (1) where oxygen is required for the oxidation of the surface functional groups used for adsorption, and (2) where the adsorption takes place without the use of oxygen." They thus endorsed the chromene oxidation mechanism of Garten and Weiss [41], but the results are also consistent with the Frumkin electrochemical model endorsed by Cho and Pitt [264]; in both cases electrostatic or ion-exchange adsorption is a necessary consequence.

In light of these results, Adams [274] and Adams and Fleming [275] revisited this issue by pointing out the differences in adsorption behavior under conditions of high and low ionic strength and were thus in a position to offer a "unified theory" [275]. Based on the negligible effect of oxygen in the presence of 0.1 M KCl, as well as other considerations, they concluded that under conditions of high ionic strength, i.e., "under normal [carbon-in-pulp process] conditions, aurocyanide is extracted in the form of an ion pair $M^{n+}[Au(CN)_2^-]_n$." (A similar mechanism was proposed for adsorption of $Au(CN)_4^-$ [276] and $Ag(CN)_2^-$ [277].) In contrast, "under conditions of low ionic strength, a portion of the gold is adsorbed by electrostatic interaction with ion-exchange sites formed through oxidation of the carbon surface by molecular oxygen." Several years later, Adams et al. [278] appear to maintain the same position by noting that "the most recent work . . . consistently points to a mechanism whereby the gold is adsorbed as a chemically unchanged $M^{n+}[Au(CN)_2^-]_n$ ion pair" (but they fail to mention the ion-pair mechanism in their conclusions). Among the references to this most recent work, they misquote the paper by Klauber [279], which actually supports a very different mechanism of gold cyanide adsorption. There is no mention of either ion-pair formation or the role of the cation M^{n+}; instead, the author suggests, based on XPS analysis, that "the linear $Au(CN)_2^-$ anion adsorbed intact on and parallel to the graphitic planes of the carbon, with the graphitic π-electrons taking part in a donor bond to the central gold atom, stabilized by charge transfer to the terminal nitrogen atoms."

Similar studies using XPS [265,279–283], as well as secondary ion mass spectrometry (SIMS) [284], Mössbauer spectroscopy [285,286] and FTIR spectroscopy [270] have been successful in shedding new light on the carbon/$Au(CN)_2^-$ interactions. Groenewold et al. [284] concluded that their "observations are consistent with the notion that gold is extant as the $Au(CN)_2^-$ complex on the carbon surface." Klauber [283] had reviewed the literature on x-ray photoelectron and Mössbauer spectroscopies and concluded that the "Mössbauer investigations have all been consistent with respect to the gold adsorbing primarily as $Au(CN)_2^-$ from alkaline solutions." The author's mechanistic conclusion based on XPS results, briefly mentioned above, is worth analyzing in more detail. It is based primarily on the measured N/Au stoichiometry of close to 2.0 for two different carbons over a wide range of adsorbate loadings (50–2000 µmol/g); furthermore, because the N 1s peak shapes remained unaltered upon adsorption, the author ruled out the surface-oxygen functional groups as specific adsorption sites. Thus, by elimination, "this leaves the graphitic planes of the activated carbons as the location onto which the $Au(CN)_2^-$ anions adsorb." Finally, this proposal was consistent with the "charge transfer of 0.3 to 0.5 of an electron to the central cationic gold atom" which "would logically come from the highly delocalized graphitic valence electrons." In a microcalorimetric study of heats of adsorption of gold chloride and cyanide complexes from aqueous solutions on both a graphitized carbon black and an activated carbon, Groszek et al. [287] reported results that "strongly support [these conclusions]" and concluded that "gold complexes are strongly adsorbed from their aqueous solutions on the basal planes of graphite." (For contradictory conclusions by the same research group, see Refs. 288 and 289, even though the study mentioned above [287] was cited in Ref. 289 as one in which it is suggested that "adsorption is more likely to take place at the edges of the graphitic crystallites.")

Ibrado and Fuerstenau [290] have tackled head-on the role of the carbon surface in the adsorption of gold cyanide by studying the effects of total oxygen content and phenolic and carboxylic acid group content, as well as the "aromaticity" of the adsorbent. They reported a sharp decrease in uptake at ca. 2 mmol O/g carbon and thus a detrimental effect of both phenolic and carboxyl groups. In contrast, there was a good positive correlation with carbon aromaticity (which ranged from 0.52 for a lignite to, surprisingly, 1.0 for both an activated carbon and a graphite) and a fair inverse correlation with the H/C ratio of the adsorbent. The authors interpreted this to be strong endorsement of the proposal by Klauber and coworkers (see above) "that adsorbed gold cyanide resides at, and interacts with, the graphitic plates of activated carbon." In a subsequent surface spectroscopy study [291] (which has been criticized by Vegter et al. [273]), the authors conclude that "the interaction between gold cyanide and activated carbon is a bond involving a partial donation of the delocalized π-electrons from activated carbon to gold" and that "the adsorbed species is most likely the unpaired dicyano complex" (rather than the ion pair). If it is assumed that graphene layers are also a source of carbon basicity, as argued in Section II, then the study of Papirer et al. [292] could be interpreted as further endorsement of this proposal. These authors leave open the question whether basic "groups, probably of the pyrone type, intervene directly in the mechanism of $Au(CN)_2^-$ fixation, acting as source of π electrons . . . or if the same groups influence the repartition of π electrons in the polyaromatic structures on the surface of active carbons." The analogy of this analysis to the discussion of adsorption of aromatics (see Sections IV.B and V) is remarkable, and it remains to be shown whether it is more than coincidental.

As late as 1997, however, Kongolo et al. [293] noted that "the mechanism of [the] adsorption has not yet been completely elucidated." They point out the need to gain "information on the state of the activated carbon in aqueous suspensions." They emphasize the importance of surface charge but neither give a reference for the critical statement that the "protonation of carboxyl and phenolic groups may form positive centers" nor appear to be aware of the fact (see Section II) that the zeta potential (measured by electrophoresis) can be a misleading representation of the charge inside the pores of a microporous activated carbon. Yet another reconciliation of the proposed mechanisms has been offered by Lagerge et al. [288]: "[f]or very low equilibrium concentrations, gold is adsorbed [irreversibly] as $Au(CN)_2^-$ anions," while for relatively concentrated solutions, gold complexes (e.g., $KAu(CN)_2$) are physically (and reversibly) adsorbed "on the less active parts of the activated carbon surface." In a follow-up publication [289], the authors reinforced these same conclusions in a study of a much wider range of carbons, ranging from graphite to activated carbons. In particular, and in contrast to most previous studies, they invoke the existence of "ionized polar groups or structural deficiencies located at the edges of some graphitic ring systems [which] could provide positive surface charges effective for the irreversible adsorption of aurocyanide anions through electrostatic interactions;" they do not identify such groups, however. They do not consider the possibility of electrostatic repulsion and thus conclude that "most of the BET surface area is not available for the adsorption of gold complexes."

In the recent extensive study of Jia et al. [294] the issue of the relative importance of porous structure and surface chemistry was revisited, based on the contention that the "detailed mechanism of the adsorption of aurocyanide species on activated carbon is not fully understood, in particular, the relative contributions of adsorption on the carbon basal plane lamellar and heteroatom functional groups." More than a dozen different activated carbons were used, and one of them was treated in nitric acid and subsequently in ammonia at 800°C; unfortunately, however, the only surface chemical characterization of these carbons was their elemental composition. Both the gold ($Au(CN)_2^-$) and the silver ($Ag(CN)_2^-$) adsorption capacities increased with increasing extent of carbon activation (in steam), and there was a much better correlation with the total pore volume than with micropore volume. The authors concluded that the "specific surface area, micropore and total pore volumes cannot act as reasonable evaluation parameters of gold and silver adsorption capacities for the carbons from different precursors" and that "pore structure and pore size distribution are more important factors in controlling gold adsorption capacity;" the latter conclusion was supported by the intriguing result that the "porosity > 2 nm is probably responsible for adsorption" although the "dimension of $Au(CN)_2^-$ ion is ~500 pm in spherical cross-section by 1 nm long." (The recent study by Adams [101] shows large differences in cyanide removal efficiencies for four carbons possessing comparable surface areas, but they are also largely rationalized in terms of physical surface properties.) The effect of chemical pretreatments was more dramatic: while the as-received carbon was most effective, oxidation suppressed the uptake of

$Au(CN)_2^-$ much more than oxidation followed by ammonia treatment. The role of nitrogen functional groups was dismissed, however, based on ancillary studies using carbons whose N content varied from 0 to 9.5%. Before reaching their own conclusions based on these results, the authors did provide a detailed discussion of the adsorption mechanisms proposed earlier; however, they failed to recognize the key point, and several of their conclusions (e.g., "little evidence for carbon surface functional groups playing an important role"; "major role played by surface functionalities probably is to improve the wetting of the surface") are thus not valid. In stating that "no evidence of improvement of gold adsorption from incorporated oxygen and nitrogen functional groups was observed," the authors fail to realize that at pH = 10 the carbon surface is negatively charged (due to the dominant effect of dissociated carboxyl groups) and their results are thus in agreement with the expected (and well documented) *detrimental* effect of oxidation. Therefore the authors' conclusion that "the graphene layers are the main sites for gold adsorption," while supported by some of the earlier studies discussed above, is not necessarily supported by the results of their own study.

Due to increased environmental concerns related to the cyanide species, other lixiviants such as the historically more popular halides [295] and thiourea are being considered as well. Arriagada and Garía [296] compared the retention of aurocyanide (at pH = 10) vs. thiourea–gold complexes (at pH = 2–3) on activated carbons whose surface physics and surface chemistry had been characterized. Even though the authors recognize the need to study "materials with the same textural properties (apparent surface area and pore size distribution) but with different types and concentrations of surface functional groups," their results and conclusions ("oxygenated surface functional groups reduce the capacity to adsorb the [anionic] gold–cyanide complex and enhance the capacity to adsorb the [cationic] gold–thiourea complex") lend strong support to the electrostatic adsorption mechanism. Earlier, Jeong and Sohn [297] and Soto and Machuca [298] had also studied the adsorption of cationic Ag- and $Au-(CS(NH_2)_2)_2)_2$ species on commercial activated carbons, but they did not address this key issue.

Based on so many almost converging studies, it is very likely now that substantial enhancements in gold removal can be achieved by judicious tailoring of surface chemistry starting with an activated carbon that possesses an adequate pore size distribution. Such a study, where the dominant effect of surface chemistry can be exploited, has yet to be published, however.

12. Arsenic

Arsenic is a commonly occurring toxic impurity in natural ecosystems [299]. It also pollutes the waters from the chemical industry's effluents as well as from mine tailings. The speciation diagrams of As(III) [299,300] and As(V) [299,300,46] indicate that the predominant trivalent species is H_3AsO_3 (except at pH > 9), while the predominant pentavalent species are $H_2AsO_4^-$ ($2 <$ pH $<$ 7) and $HAsO_4^2$ ($7 <$ pH < 12).

Gupta and Chen [299] were among the first to study the effect of pH and adsorbent nature on the uptake of both As(III) and As(V). They reported that both activated alumina (210 m^2/g) and bauxite adsorbed more As(V) at pH = 6–7 than an activated carbon (500–600 m^2/g) at pH = 3.2. In contrast to the inorganic adsorbents, for which the highest uptake was at pH < 7, for the activated carbon there was a maximum at pH = 3–5. The authors mention that "[t]he reported zero point of charge values for activated alumina and bauxite are slightly basic," and they attribute this to the fact that "negatively charged molecules were removed effectively onto the slightly positive or neutral charge surfaces." However, they did not have the foresight to propose similar arguments for the activated carbon and thus left essentially unexplained why "activated carbon adsorbs As(V) better in the acidic pH range" and why "it exhibits poor adsorption capacity in comparison to activated alumina and bauxite even at pH 4." They simply suggested that there is "probably little affinity between carbon surface and the nonionic, divalent and trivalent forms of As(V)." In a contemporary and similar study using five different activated carbons whose pH values ranged from 4.1 to 7.9, Kamegawa et al. [300] concluded that the behavior of As(III) and As(V) supports their idea that "heavy metal ions are adsorbed on activated carbon in the form of complex anion," analogous to their proposed model for adsorption of $HgCl_4^2$ in the presence of Cl^-:

This argument was based on the finding that the "coexistence of anions interfered greatly with adsorption of As(V)." It is not clear from their summary how their proposed surface chemistry (see figure above) is compatible with the measured range of pH (and thus pH_{PZC}) values of the carbons and with the effect of solution pH on the uptakes (a sharp maximum at pH = 5–6). Interestingly, the efficiency of As(III) removal by most carbons was lower than that of As(V); the authors speculated that those carbons "having a higher affinity to oxidize As(III) showed a higher adsorption ability for As."

In a much more insightful and detailed study, Huang and Fu [46] reexamined the same effects using a series of 15 well-characterized commercial adsorbents and found an order of magnitude difference in the uptakes at optimum pH (4.0).

They attributed this to "the fact that the concentration distribution of arsenic(V) species and the development of surface charge on the activated carbon are pH-dependent phenomena." The authors misidentified L-type and H-type carbons and made no comment about the proposal of Kamegawa et al. [300], but their Fig. 3 (op. cit.) goes a long way toward explaining the mechanism of anion adsorption on carbon surfaces: their "adsorption density curve is almost parallel to the zeta potential curve as a function of pH," which "suggests that a positive carbon surface better adsorbs anionic As(V) species." In a more recent study using a commercial activated carbon whose surface chemistry was not characterized, Rajakovic [301] concluded, rather vaguely, that the "pH values of the water are important [for adsorption of As(III) and As(V)] because of the change in the ionic forms of both arsenic species" and that "the optimal pH range is between 4 and 9 which is a consequence of the apparent affinity between the carbon surface and arsenic species H_3AsO_3 and $H_2AsO_4^-$." The author ignores both studies and instead echoes the rather confusing argument invoked by Gupta and Chen [299] that there is "an apparent affinity between the carbon surface and the negatively charged arsenic species with oxo-function groups." Without an attempt to reconcile some of these arguably inconsistent statements, she does mention that electrostatic interaction seems to be "an important mechanism for arsenic adsorption." In addition to summarizing the effect of pH, Fig. 15 illustrates the beneficial effect of cation impregnation of the activated carbon, especially for As(III), which the author attributes to a chemisorption process and/or $CuHAsO_3$ precipitation. A similar beneficial effect of Fe^{2+}-treatment of activated carbon on As(V)

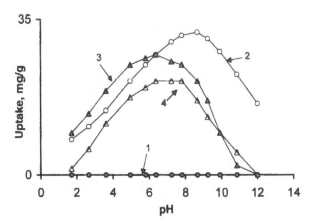

FIG. 15 Effect of pH on the uptake of arsenic by activated carbon impregnated with metallic copper: (1) As(III), no metal added; (2) As(III), impregnated with copper; (3) As(V), no metal added; (4) As(V), impregnated with copper. (Adapted from Ref. 301.)

uptake (e.g., a tenfold increase) had been reported by Huang and Vane [48], who attributed it to the development of positive charge on the carbon surface upon adsorption of Fe^{2+} ions and the consequent "formation of ferrous arsenate surface complexes" (and not the formation of precipitates).

In an ambitious paper on the mechanism of the adsorption of arsenic species on activated carbon, Lorenzen et al. [302] used three activated carbons of different origin (and ash content) to confirm many of the earlier findings: (1) arsenic was "most effectively removed from aqueous solutions at a pH of 6 using activated carbon pretreated with a Cu(II) solution;" (2) "carbon types with a high ash content are more suitable for arsenic adsorption" (in agreement with Ref. 303); (3) there was no correlation of uptake with the total surface area of the adsorbent (in agreement with Ref. 303). Rather anticlimactically, however, they ignore some of the forcefully argued mechanisms proposed in earlier studies. Instead, they merely state that adsorption of arsenic occurs "independently of the impregnated copper" and rule out the possibility of "ion pair formation," such as $[Na^+]_2[H_2AsO_4{}^{2-}]$. More recently, Raji and Anirudhan [269] also studied the virtues of a copper-impregnated carbon for the treatment of As(III)-rich water, but they ignored both of the above mentioned studies. Their explanation for the sharp increase in uptake at a pH of ca. 6–8 did invoke carbon surface chemistry, which was characterized by calculating pH_{PZC} using potentiometric titration. The authors argued that H_3AsO_3, which was predominant in the neutral and acid pH ranges, "was not removed by either adsorption or surface precipitation onto the positive (pH < pH_{PZC} = 8.2) or neutral surface of the adsorbent."

13. Phosphates

Organophosphoric esters are very effective and degradable pesticides [304], which makes them preferable to their chlorine-based counterparts. Their hydrolysis behavior is well known [305], the final product being the orthophosphate ion. But these compounds are very toxic to humans because of their inhibiting effect on acetylcolinesterase: as little as a few ppm can be lethal [306].

The important early paper of Jayson et al. [307], who studied the removal of triethyl-dimethyl- and orthophosphate ions using a commercial activated charcoal cloth, is discussed in detail in Section IV.B.1. Ferro-García et al. [308] studied the removal of orthophosphate using a range of well-characterized activated carbons at an initial pH of 7. All the carbons were basic, with slurry pH values above 7.0. The uptakes for both series of carbons (one derived from almond shells and the other from olive stones) were of the order of 1–5 mg/g and were found to correlate with the degree of activation of the carbons, i.e., with their surface area and porosity. The slightly lower uptakes on the olive-stone-derived carbons at comparable degrees of activation were attributed to their lower basicity, in agreement with the fact that adsorption of anions is favored at pH < pH_{PZC}. This explanation was also supported by the fact that the adsorption capacity of a commercial sample, whose slurry pH was the lowest of the lot, was ca. 50% of that of a sample that had a very similar porosity.

B. Dominant Factors

It is well known that the "adsorption phenomena at mineral/water interfaces are usually controlled by the electrical double layer" [57]. So the main issue here is to what extent the same principles apply to the mineral/water/carbon interfaces. The review by Huang [97] contains the following conclusion: "The adsorption capacity of inorganics by activated carbon is determined by the nature of the inorganics and the physical-chemical [sic] properties of the activated carbon. The important characteristics of activated carbon are specific surface area, pore structure, electrophoretic property and surface acidity." For the present review we analyzed the post-1975 literature in search of evidence that supports, contradicts or clarifies these significant but rather general conclusions. In particular we focused on the role of the chemistry of the carbon surface. Here we summarize these findings and then very briefly discuss the advances made in modeling the adsorption process.

In principle, the adsorptive capacity for inorganic compounds in aqueous solution is determined by both the physics (primarily the surface area) and the chemistry of the adsorbent. These are in turn defined by the choice of carbon precursor and its activation conditions. Among the process parameters, solution pH is the key factor: (1) surface charge density and the cationic or anionic nature of the surface depends on it (see Fig. 2); (2) it determines to a large extent the solution chemistry. Therefore the construction of appropriate speciation diagrams (see Appendix) is a necessary condition for understanding or predicting the uptakes. In constructing these diagrams, it should be kept in mind that the effective pH on a charged surface (pH_s) may differ substantially from that in the bulk of the solution (pH_b), according to the Boltzmann-type expression [204],

$$pH_s = pH_b \exp\left(\frac{-F\phi}{RT}\right)$$

where F and ϕ are the Faraday constant and the potential across the interface. One consequence of this fact is that surface precipitation (and enhanced apparent adsorption) of a metal hydroxide on carbons possessing high pH_{PZC} occurs at a lower bulk pH. An example of this phenomenon is discussed by Reed and Matsumoto [204].

Whether the knowledge of surface charge and its variations with pH is a sufficient condition for predicting uptakes requires closer scrutiny. In this regard, the conclusion by Corapcioglu and Huang [171] that "[e]lectrostatic interaction plays an insignificant role in adsorption reaction" certainly needs to be reexamined. These authors preferred the surface complex formation model and thus postulated the existence of "COH and CO-surface hydroxo groups" on L-type carbons vs. "COH$_2^+$ and COH as major hydroxo groups" on H-type carbons. They did not comment on the likelihood of protonation of these (or other) oxygen functional groups in aqueous solution (see Section II). Furthermore, they made no attempt to correlate (at least qualitatively) the calculated stability constants of surface complexes with the presence of such functional groups.

In light of the now well-recognized "asymmetric" chemistry of the carbon surface (see Section II)—and despite the fact that many clues have been abundant in the literature, both explicitly and implicitly—the key heretofore unresolved issue has been the following: Which surface sites are responsible for adsorption of inorganic species, and under which conditions? Whether the entire adsorbent surface is available for adsorption depends basically on three factors:

1. Is the surface electrostatically accessible, i.e., has electrostatic repulsion been eliminated?
2. Is the affinity of the uncharged surface for the adsorbate sufficient to attract the (most commonly) charged species in the absence of electrostatic attraction, i.e., by dispersion forces alone?
3. Even in the case when electrostatic interactions are unfavorable, can the enhanced dispersive interactions overcome the electrostatic repulsion?

To answer these questions we summarize here the information available on the uptake of cationic and anionic adsorbates. One obvious (yet often neglected) clue resides in the analysis of the fraction of the carbon surface covered by the various adsorbed species.

Table 13 from the study of Ferro-García et al. [187] provides such information, which turns out to be typical for cationic species (see, for example, Refs. 189,

TABLE 13 Summary of Physical and Chemical Surface Properties of Selected Activated Carbons and Their Metal Adsorption Capacities

Carbon[a]	pH	S_{CO_2} (m²/g)	$V_T{}^b$ (cm³/g)	$V_{micro}{}^b$ (%)	Fraction of surface occupied[c]			
					Zn(II)	Cd(II)	Cu(II)	Pb(II)
A	8.6	1103	0.707	38.2	0.041	0.007	0.039	0.027
H	8.2	1316	0.473	60.2	0.026	0.014	0.037	0.023
P	7.4	876	0.442	65.4	0.030	0.011	0.043	0.026

[a] The activated carbons A, H, P were prepared from almond shells, olive stones, and peach stones; their aqueous suspensions all had pH > 7.
[b] The surface area, total pore volume, and micropore volume (in pores <7.5 nm) were determined by CO$_2$ adsorption and Hg porosimetry.
[c] Calculated using Langmuir adsorption capacities obtained at 293 K and pH = 5.
Source: Ref. 187.

190, and 205). In the earlier work of Huang and Smith [44], the authors calculated for Cd(II) "a hydrated radius of approximately 25 Å which is larger than the reasonable size of hydrated cadmium ions." A much smaller fraction of the adsorbent surface is typically covered by inorganic adsorbates than by organic (especially aromatic) adsorbates (see Section IV). Therefore the access of usually charged inorganic pollutants to the abundant adsorption sites, especially in the case of microporous activated carbons, is hindered not so much by the size of the pores but by strong electrostatic repulsion. A logical inference is that, in contrast to the case of organic adsorbates (see Section IV), the graphene layers are not involved in the adsorption of inorganic cations. It is interesting to note also that Jayson et al. [222] assumed that, at the saturation adsorption capacity at pH = 5.5, mercury exists as a cationic species and thus calculated a surface coverage of 3.5% for the unhydrated cation; however, if a single sphere of hydration was assumed to surround it, then an unusually high coverage of ~90% was obtained.

Clearly, this is the direction in which further fundamental studies should be oriented. For example, it will be interesting to find out whether much higher surface coverages can be accomplished on a carbon whose maximum number of (cation-exchangeable) adsorption sites, e.g., ~3 mmol COO^-/g C, is not only created but made electrostatically accessible by adjusting the solution chemistry. Under these conditions, for example, the theoretical uptake of a divalent cation is 1.5 mmol/g, which translates into 450 m^2/g, which in turn is a large fraction of the total surface area. This is obtained by assuming a radius of 0.4 nm for a hydrated divalent cation, which is usual for heavy metals [309]. Indeed, in the study of Cr(III) adsorption by activated carbon MO (see Table 3), the surface covered by a monolayer of the adsorbed cations was 196 m^2/g on a sample whose N_2 and CO_2 surface areas were 164 and 537 m^2/g, respectively.

Alternatively, it will be important to verify whether the highest uptakes can be achieved under conditions of electrical neutrality of both the adsorbate species and the adsorbent surface, e.g., at some optimum pH. This has often been observed for organic adsorbates (see Section IV), because departure from these conditions led more often to electrostatic repulsion rather than electrostatic attraction. Not as many such reports are available in the literature on the adsorption of inorganics (see below), and their discussion does not follow this line of reasoning. It is worth noting, however, the findings of Bhattacharyya and Cheng [173] for adsorption of Cd-Ni chelates on a commercial activated carbon. The authors reported a set of intriguing results: when the metals were complexed with EDTA, a maximum in uptake was observed at intermediate pH; in contrast, when they were complexed with a ligand that yields a positively charged species over the entire pH range (4–12), a monotonic increase in uptake was observed.

The most consistent results seem to be the ones for mercury adsorption. For example, Huang and Blankenship [221] have reported pronounced uptake maxima vs. pH for Hg(II) removal by several activated carbons, but they preferred to discuss them in terms of adsorption/reduction mechanisms. Ma et al. [228] have confirmed the same trend, with a maximum occurring at pH ~7; their somewhat cryptic interpretation was that "the formation of HgOHCl could be involved." So have Namasivayam and Perisamy [212], but they offered no explanation.

Devi and Naidu [183] have reported maxima not only for the adsorption of Hg but also for Co, Cu, Zn, and Cd, after complexation with potassium ethyl xanthate. Both Pérez-Candela et al. [145] and Huang and coworkers [126,127] have reported a maximum uptake of Cr(VI) at pH ~ 3; the explanations offered were analogous to those of Huang and Blankenship [221]. An earlier study by Sharma and Forster [141] had shown, however, that such a maximum appears only when total chromium uptake (i.e., both Cr(VI) and Cr(III)) is plotted vs. pH, thus reinforcing the interpretation that it is related to the reduction of Cr(VI) to Cr(III) and a lower affinity of the adsorbent for Cr(III). Similar results were reported for the anionic As(V) species by Gupta and Chen [299] and Huang and Fu [46]; the rationalization offered by the latter authors was the presumable analogy with amphoteric oxide adsorbents "where a maximum adsorption occurs at a specific pH value then [sic] adsorption density decreases with further pH change." Rajakovic [301] confirmed these results, but she related them to "the change in the ionic forms" of the arsenic species. Saleem et al. [244] reported also that the uptake of uranium (from uranyl nitrate) exhibited a maximum at pH ~ 4; they explained the maximum by saying that "the availability of free uranium is maximum at pH 4," while beyond 4, precipitation starts, due to the formation of complexes in aqueous solution," and they speculated that "adsorption decreases since micropores are not formed." Finally, Wei and Cao [310] reported the same trend for the uptake of cyanide species and noted that "the chemical constitution both on carbon surface and in solution has great effect on the adsorption capacity;" they concluded that "electrostatic force is predominant in adsorption."

Regarding the uptake of anions, it is quite intriguing that in many cases they do not appear to be much higher, as one might expect if the positively charged graphene layers (see Fig. 2) were electrostatically accessible and "active" in adsorption. (Indeed, in a review of anion adsorption by Hingston [311], use of carbonaceous adsorbents is not even mentioned.) For example, in the study of Ferro-García et al. [308], the maximum uptakes of orthophosphates—under conditions that were favorable for electrostatic attraction (the initial pH of the solution was 7.0 and was lower than pH_{PZC})—amounted to equivalent surface coverages of much less than 100 m^2/g. Similarly, McDougall and Hancock [252] discuss the study of Clauss and Weiss (see op. cit.) in which it was "calculated that the surface coverage amounted to less than one gold atom per 5 nm^2 of available carbon surface." (Lagerge et al. [289] recently obtained very similar numbers for the aurocyanide species on a range of carbons whose surface areas varied from 1 to 1688 m^2/g.) The authors thus "excluded basal planes, carboxylic acid groups and basic oxides as so-called 'points of attachment' and proposed that the adsorption sites were probably quinone-type groups" or "special type of micropore[s]." Such proposals have not been verified, however. Indeed, in a recent comparison of uptakes of $[Cr(H_2O)_6]^{3+}$ cations and dichromate anions, Aggarwal et al. [154] did find that "the amount of Cr(VI) adsorbed is comparatively much larger than the amount of Cr(III) ions adsorbed under similar conditions." They speculated that this was due not only to the "smaller size of the Cr(VI) ion in water so that it can enter a larger proportion of the microcapillary pores" but also to the "existence of some positively charged sites where negatively charged Cr(VI) ions can be adsorbed." They did not identify these sites, however. Similarly, as discussed in Section III.A—and despite some controversy regarding the oxidation states of adsorbed gold species [281,282]—Klauber and coworkers [279,283], as well as Ibrado and Fuerstenau [290,291], have presented a compelling case for "a π-donor bond from the graphitic substrate to the gold atom of the $[Au(CN)_2^-]$ complex" [283].

Equally compelling, albeit indirect, evidence for a more favorable interaction between anions and a positively charged carbon surface had been provided by Carrasco-Marín et al. [312,22] in their study of catalytic activity of carbon-supported palladium: the adsorption of the anionic $PdCl_4^{2-}$ species on the positively charged surface (at pH = 0.1) was concluded to occur over a larger surface (thus leading to a smaller degree of sintering and higher catalytic activity) than the adsorption of $Pd(NH_3)_4^{2+}$ on the negatively charged surface (pH = 9.0). Similar results have been obtained recently in the analogous case of interaction between a carbon black and platinum complexes [313].

Following up on the important concept emphasized by Carrott et al. [202], that the equilibrium uptake is a function not only of the number of adsorption sites but also of the affinity for the adsorbate, the effect of temperature on (quasi?) equilibrium uptakes deserves special comment. (It also deserves a more detailed scrutiny, but such analysis is beyond the scope of our review.) In many cases and for several metals, it has been reported that the uptake increases with increasing temperature [44,171,138,124,125,201,172]. This effect, which had been reported as early as 1924 [314], is of course in conflict with elementary thermodynamic arguments [44]. Only a few attempts have been made to explain such trends. An obvious candidate is activated diffusion in the microporous adsorbents, but few studies have either addressed or dismissed it. Huang and Smith [44] speculated that the enthalpy change for the proton/carbon interaction is lower than that for the Cd^{2+}/interaction, which would explain larger temperature effects at low pH than at higher pH. Corapcioglu and Huang [171] argue in favor of enhanced carbon oxidation at elevated temperature (and low pH) whereby "an H-carbon . . . is progressively converted into an L-carbon thereby increasing the Cu(II) removal capacity." Huang and Smith [44] hypothesized that, at low pH, "a rise in temperature would reduce the tendency of carbon-H$^+$ formation and, as a result, decrease its competition with Cd^{2+} for the carbon sites;" at high pH (when the carbon surface is negatively charged), a similar temperature effect was attributed to the shift in the balance between the various species in solution toward the predominance of Cd^{2+} cationic species. Ferro-Garcia et al. [187] reported that while adsorption of zinc and cadmium appeared to be exothermic, that of copper appeared to be endothermic on two of the three carbons studied, thus confirming the findings of Corapcioglu and Huang [171]; they offered no explanation for the endothermicity of copper adsorption, however.

In clear contradiction to their own experimental results, Saleem and coworkers [113,244], Leyva-Ramos et al. [125] and Marzal et al. [201] use the van't Hoff equation to rationalize their results and derive a value for the endothermic heat of adsorption. Leyva-Ramos et al. [125] conclude that "adsorption was of a chemical type," but both they and the other authors fail to address the key issue: if the uptakes are indeed truly equilibrated (and not kinetically controlled), how can endothermic adsorption be a spontaneous process? Namasivayam and Yamuna [144] went through a similar exercise, but they did offer an explanation, albeit a very unconvincing one: "Positive values of [the entropy change] showed increased randomness at the solid/solution interface during the adsorption of Cr(VI) on [the biogas-residual-slurry-derived adsorbent]. The adsorbed water molecules, which are displaced by the adsorbate species, gain more translational entropy than is lost by the adsorbate molecules, thus allowing the prevalence of randomness in the system."

The state of the art in the modeling of these equilibrium adsorption trends has not advanced very much beyond the stage described by Huang and coworkers [97,171]. According to Huang [97], four variations of electric double layer theories [315,316,66] are available: (1) the surface complexation model; (2) the ion exchange model; (3) the solvated ions model of James and Healy; and (4) the Gouy–Chapman–Stern–Grahame model. However, the overwhelming majority of authors have preferred to ignore the electrostatic interactions and have simply adopted the Langmuir [299,44,206,167,98,187,317,136,188,189,113,301,244, 226,212,204,138,139,190,199,207,141,196,184,125,144,108] or Freundlich [216, 206,221,47,98,298,136,188,113,301,137,212,204,138,310,245,124,199,207,141, 184,144] formalisms. Of course, the parameters thus obtained, which are widely quoted, especially in the water treatment literature, are valid for a very limited set of operating conditions (e.g., constant pH). A typical and important example is the study of Reed and Matsumoto [204], who used both Langmuir and Freundlich isotherms to model cadmium adsorption, even though they are not capable of taking into account the key effect of the pH. They concluded that cadmium removal "is strongly related to the carbon's pH_{PZC}, acid–base characteristics, and surface charge–pH relationship" and that "surface area, an important adsorption parameter for organic adsorbates, does not appear to influence metal removal strongly." They did not make attempts to incorporate the former effects into their model. It has been difficult to strike the right compromise between the complexity of the relevant interactions and the needs of the water treatment industries. Thus none of the progress in the characterization of the surface chemistry of carbons (see Section II and references therein) has made its way into such models yet. An example of a comprehensive model with a reasonable number of adjustable parameters is presented in Section V.

It is fitting to close this discussion with a quote from a very recent communication by Huang's group [170]. After presenting selected results on the percent removal of Cd(II), Co(II), Cu(II), Pb(II), Ni(II) and Zn(II) ions as a function of pH for an H-type and an-L type carbon, the authors conclude that "[i]nformation on the surface acidity . . . is crucial to the satisfactory achievement of metal removal from aqueous solution. One should carefully consider the type of carbon (L-type or H-type carbon) to be used in the removal process as well as the chemical characteristics of the metal." The authors do not acknowledge explicitly the importance of electrostatic interactions, but they note that for the L-type carbon "the extent of adsorption is higher than for the H-type carbon in the entire range of pH studied" and that "L-type carbons . . . are more acidic and possesses [sic] negatively charged surface sites within a large range of pH, which favors the metal adsorption."

Finally, and at the risk of oversimplifying the phenomena reviewed in Section III, it is tempting to conclude that—in contrast to the case of organic (and especially aromatic) solutes (see Sections V and VI)—in most cases the uptake of inorganics on carbons is dominated by electrostatic (e.g., ion exchange) adsorbate–adsorbent interactions, even at $pH > pH_{PZC}$ (where the exact origin of the positive charge on the carbon surface has yet to be defined in precise quantitative terms). What remains to be established is whether there are situations (e.g., when process conditions do not allow the use of a chemically tailored carbon that has a favorable electrostatic interaction with the inorganic solute) where enhanced dispersion interactions (e.g., achieved by careful design of the porous structure) can overwhelm the inevitable electrostatic repulsion (whose strength depends on pH and electrolyte concentration). This is predicted, of course, by the classical DLVO theory [318–320] when the ionic strength of the solution increases (which decreases electrostatic repulsion), as shown in Fig. 16. Indeed, investigations along the lines of the important study of Ibrado and Fuerstenau [253] would be beneficial. This line of reasoning could be particularly fruitful when pursuing the optimization of removal of microorganisms [321–326], as well as pollutants containing chelated complexes or other bulky ligands.

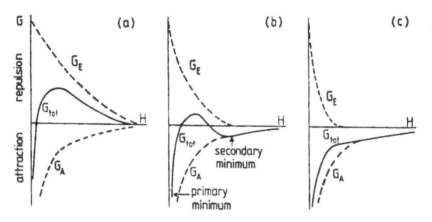

FIG. 16 Predictions of the DLVO theory regarding the balance between electrostatic repulsion (E) and dispersive attraction (A) of charged particles in solution: interaction energy (G) vs. distance between two particles of same charge at low (a), intermediate (b), and high (c) ionic strength.

IV. ADSORPTION OF ORGANIC SOLUTES

In spite of the typically larger uptakes than in the case of inorganic solutes [327], the removal of organic pollutants from aqueous streams is a challenge, especially as regards the specific role played by the carbon surface. Derylo-Marczewska and Jaroniec [28] have provided a very useful list of references containing experimental adsorption data for both single- and multisolute systems. In the review by Cookson [23] and the monographs by Perrich [26] and Faust and Aly [33], additional compilations of the voluminous experimental data, including the parameters in Langmuir or Freundlich isotherms, are available. Nevertheless, as in the case of adsorption of inorganics, these parameters should be used with caution, because the reliability of their extrapolation has not been demonstrated, and their general applicability has been questioned [328,329]. Furthermore, the well-documented surface chemistry effects are ignored when such an approach is taken.

In this review we concentrate on the studies that attempt to elucidate the importance of carbon surface properties in controlling the equilibrium uptakes of aromatic and aliphatic adsorbates. Rather than comparing model parameters, such as Langmuir or Freundlich constants, we examine the uptakes at comparable equilibrium concentrations and attempt to rationalize the differences observed under different conditions and on different adsorbents.

A. Phenomenological Aspects

Figure 1 shows that the available literature on this topic is almost overwhelming. The main focus of the necessarily selective analysis presented here is on the most important practical issue: the adsorption capacity of different carbons for different (mostly aromatic) adsorbates. Clues are sought to identify the key factors that can account for the widely varying uptakes of a given adsorbate on different adsorbents and of different adsorbates on the same adsorbent. An order-of-magnitude estimate of uptakes on a typical activated carbon (~ 1000 m^2/g) will prove to be quite revealing. If the uptake is limited to a monolayer, then it can be calculated as follows:

Monolayer adsorption capacity, g adsorbate/g adsorbent

$$= \left(\frac{1000 \text{ m}^2}{\text{g adsorbent}} \right) \left(\frac{1 \text{ molecule adsorbate}}{X \text{ m}^2} \right)$$

$$\left(\frac{1 \text{ mol adsorbate}}{6.022 \times 10^{23} \text{ molecules}} \right) \left(\frac{Y \text{ g adsorbate}}{1 \text{ mol adsorbate}} \right)$$

For a relatively small molecule such as phenol (~ 40 Å^2), the monolayer uptake amounts to ~ 0.4 g/g, and it remains in the same range even for much larger molecules (because of the obvious compensation between X and Y). For example, the uptake of humic acids (high-molecular-weight components of natural organic matter) can be as high as 0.1 g/g (at solute concentrations of <10 mg/L). It is well known [66] that "[m]ultilayer adsorption from solution is less common than it is from the gas phase, because of the stronger screening of interaction forces in condensed fluids." Lyklema [66] has recently discussed the results of Mattson et al. [24] regarding the "absence of well-established plateaus" in the uptake of phenol and nitrophenols on an activated carbon whose BET surface area was 1000 m^2/g. He assumed that the monolayer capacity (at the knee of the isotherm) was some 1.4 mmol/g and, using an adsorbed molecule cross-section of roughly 0.45 nm^2, concluded that the resulting coverage (380 m^2/g) is "lower than the BET (N$_2$) area but still so high as to suggest contribution [of pore filling] from pores." Similar comments and (re)calculations were made by Mattson et al. [330].

More specifically, we highlight the issues that should contain the answers to the following questions:

1. What is the maximum uptake of a given adsorbate on a high-surface-area carbon?
2. What fraction of the carbon surface is covered by the adsorbate at the maximum uptake?

3. If the entire adsorbent surface is not covered by the adsorbed species, is it because of molecular sieving effects (i.e., the porous structure of the adsorbent) or because of the incompatible surface chemistry (e.g., electrostatic repulsion between a negatively charged surface and anions in solution)?

A more detailed discussion of some of these issues is then presented in Section IV.B.

1. Phenol

In the past three decades this has been by far the most studied of all liquid-phase applications of carbon adsorbents, either in its own right [331–370,303,327,371–376,328,377–395] or in comparisons with other adsorbents [396,348,397–401,10,402–404] or, most commonly, as a baseline for investigations of other aromatics or substituted phenols [24,405–415,364,416–418,367,419–429,376, 430,403,431,328,329,379,432–436,386,437–445]. Interestingly, however, the monograph by Mattson and Mark [6] does not identify it as such. Clearly, increasingly stringent clean water legislation has had its effect: phenols were proclaimed "priority pollutants" and adopted as the main challenge. Mattson and Mark [6] emphasize one of the many complexities of phenol adsorption: the occurrence of a two-step isotherm, with the low-concentration first stage (<300 μmol/L) obeying the Langmuir equation (with uptakes of the order of 1 mmol/g, i.e., ~0.1 g/g) and the high-concentration second stage (up to ca. 100 mmol/L) conforming more to the BET equation (with uptakes of ~0.4 g/g).

Equilibrium considerations are of primary interest here, although several transport and kinetic [337,406,446,353,447,448,430,436,385,394,449] studies have also been of interest.

In the early study of Coughlin and Ezra [450], low surface coverages by phenol were reported (at unadjusted pH). The authors showed a similar trend to that reported for an anionic dye by Graham [451] (see Section IV.A.3): a decrease in the percent (BET) area occupied by phenol with increasing acidity of the adsorbent (as measured by Boehm titrations). Singh [334] examined the important issue of the orientation of adsorbed phenol molecules on the surface of two carbon blacks (the acidic Spheron-6, whose suspension pH was 3.8, and the basic Graphon, whose pH was 9.2), with a view toward using phenol monolayer uptake to estimate the surface area of adsorbents. The isotherms (at unadjusted pH) were of Type II. The molecular areas "calculated from the points of least slope on the isotherms" were 33 and 28 Å^2 for the two adsorbents. After comparing these values to the presumably much larger area (52.2 Å^2, according to the author) for the (intuitively more obvious) flat orientation, the author concluded that "phenol molecules, when adsorbed from aqueous solution by carbons, lie in an edge-on fashion on the surface so that the phenolic groups point toward the solution phase." Puri et al. [341] did not cite the study of Singh [334], but they expressed confidence that the phenol adsorption data "when analysed by modified B.E.T. equation permits calculation of specific surface area of each adsorbent." This confidence was based on remarkable agreement (see Fig. 17a) between monolayer (and reversible) uptakes of N_2 and phenol, which in turn was based on phenol's molecular area of 52.2 $\mathring{A}^2$ (and thus flat orientation on the surface). Farrier et al. [452] did cite the work of Singh [334] in a study of phenol adsorption on a macroreticular resin; when they used 28 $\mathring{A}^2$ as a surface area per phenol molecule, they calculated a surface coverage of 144%, but they made no comment about the reliability of the numbers used or the implication of such a high surface coverage. In a subsequent study of phenol removal from a benzene solution, Puri et al. [344] finally did cite the study of Singh [334] and addressed explicitly the issue of parallel vs. perpendicular orientation on the surface of as-received and chemically pretreated carbon blacks. Their results showed no consistent or significant trend in BET surface area changes upon sample outgassing and monotonic reduction in oxygen content; in contrast, there was a good and perhaps surprising correlation (see Section IV.B.1) between a decrease in phenol uptake (in mg/m^2) and the decrease in carbon's oxygen content. This led the authors to conclude that "as [surface] oxygen is gradually eliminated, the orientation of the molecules in the adsorbed phase tends to change gradually from perpendicular to parallel position." In a subsequent review paper [346], Puri sounds less compromising when discussing adsorption from aqueous solution: "The suggestion . . . of end-on orientation on oxygen-containing surfaces does not appear to be substantiated. It appears highly likely that attraction of phenol from aqueous solutions of moderate concentration by the carbon substrate lies in nonpolar forces operating over the entire phenol nucleus." Regarding adsorption from nonaqueous solution such as that of benzene, he reiterates the argument summarized above [344] and notes, quite convincingly, that the "perpendicular orientation [on oxygen-populated surfaces] is favored when there is little competition from the solvent for the surface."

Economy and Lin [340] investigated the phenol adsorption characteristics of high-surface-area activated carbon fibers; Fig. 17b shows that their correlation was not nearly as good, but the role of surface chemistry was not invoked. Additional evidence that the agreement shown in Fig. 17a is more often the exception [439] than the rule is contained in the study of Dondi et al. [342], who used a chromatographic method to determine low-concentration phenol adsorption isotherms on four different carbons, as well as in many other investigations (see, for example, Refs. 356, 382, 384, and 436).

Peel and Benedek [407] published a useful review of reported phenol adsorption isotherms on a popular commercial activated carbon and concluded that "the observed differences [as low as 100 vs. as high as 200 mg/g at 100 mg/L] are greater in magnitude than might reasonably be attributed to [variations in carbon properties and environmental conditions such as temperature, pH, and buffer

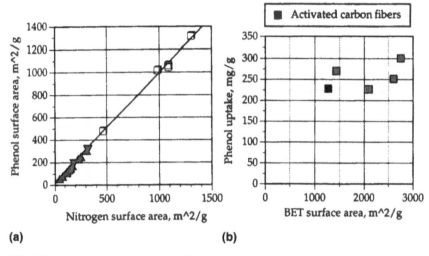

FIG. 17 Illustration of the relationship between phenol adsorption capacity and surface area of a series of carbon materials: (a) adapted from Ref. 341; (b) adapted from Ref. 340. □, activated carbons; ▲, carbon blacks; ▼, degassed carbon blacks.

strength].'' They set out to test the logical hypothesis that ''part if not all of the observed variations . . . were due to problems in determining attainment of the real equilibrium'' and confirmed that, ''[a]lthough about 60% of the final capacity was obtained after only 4 h, the remaining capacity was utilized very slowly.'' Both previous and subsequent research have shown (see Section IV.B.1 for more details) that, while the attainment of equilibrium can be a problem, the effects of pH and carbon surface chemistry variations can be dramatic as well. Thus, for example, Ogino et al. [351] confirmed the results of Coughlin and coworkers [450,331,332] by documenting a dramatic decrease (by a factor of 6 or so) in the Freundlich constant with increasing acidity of deliberately oxidized activated carbons. Mahajan et al. [345] also reported an order-of-magnitude decrease in phenol uptake upon oxidation of activated carbons. Similar results were reported by many other authors [350,408,356,357,361,363,365,371,394].

The reversibility of phenol adsorption [341], which is of great importance in adsorbent regeneration (or reactivation) [343,347–349,352,355,453,454] was first discussed in detail by Magne and Walker [360]. They reported that weakly adsorbed (physisorbed) phenol ''can become chemisorbed in the course of time or by increasing the temperature.'' On the basis of the finding that chemisorption (on a commercial activated carbon) was ''inhibited by the presence of oxygen surface complexes,'' the authors concluded that ''the sites responsible for phenol chemisorption are carbon sites of the active surface area,'' i.e., oxygen-free sites located primarily at the edges of the graphene layers, whereas physisorption occurs "on all the surface." Phenol chemisorption was also mentioned by Thakkar and Manes [413], in a study of adsorptive displacement, but they echoed the proposal of Mattson et al. [24] (see Section VI) and did not implicate the carbon active sites. Bhatia et al. [448] determined the effective diffusivity of phenol in water from the largely reversible desorption isotherms obtained with a commercial activated carbon. Xing et al. [402], among others, confirmed the reversible nature of phenol adsorption at neutral pH, while Yonge et al. [412] confirmed its irreversible nature under similar conditions.

Grant and King [417] revisited the issue of irreversibility and found evidence for "a chemical reaction that is slow compared to physisorption at 25°C." They note that a"popular hypothesis to rationalize irreversible adsorption of phenols on activated carbon is chemical reaction of sorbate molecules with oxygenated surface functional groups." In agreement with the conclusion of Magne and Walker [360] cited above, they argue that "surface functional groups are not primary sources of irreversible adsorption" and propose instead that irreversible adsorption is caused by oxidative coupling in the presence of molecular oxygen (see Sections IV.B.1 and V). Sheveleva et al. [420] studied the electrochemical regeneration of activated carbon fibers [340] and, as shown in Fig. 18, noted that much but not all of the phenol can be desorbed by increasing the pH and, especially, by applying a negative surface potential; they concluded that "the change in absorbability is affected by structural features and surface properties of the carbon fibers" but did not identify these properties.

Further insight into the role of the carbon surface in adsorbing phenol can be gained from a comparative study performed by Radeke et al. [10]. Their results

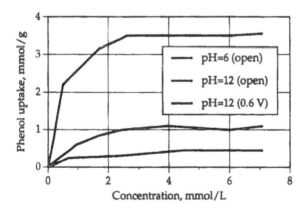

FIG. 18 Effect of pH on the adsorption isotherms of phenol on polarized carbon fibers. (Adapted from Ref. 420.)

are summarized in Table 14. Not only is there no correlation between phenol uptake and the total surface area of activated carbons, but the effectiveness of the hydrophobic zeolite DY2 and resin EP61 is also much lower than that of the activated carbons. This result was in contrast to the high suitability of dealuminated zeolites for the separation of hydrocarbons from moist air; the authors were thus forced to conclude that "it is not possible to extrapolate straightforward from the gas-phase adsorption of organics in the presence of water vapor to the separation of organics from liquid water." It should be noted, however, that as late as 1993, Yenkie and Natarajan [372] insisted that phenol adsorption can be used to "estimate the specific surface area" of activated carbons.

Additional insight has been provided by Király et al. [377] in a comparative study of adsorption of phenol, cyclohexanol and n-hexanol on a "graphitized" carbon black. From the break in the slope of the plot of adsorption enthalpy vs. surface coverage, they determined a molecular surface area of 326 m²/mmol (54.1 Å²/molecule) and concluded that "the homogeneous surface of [the adsorbent] is fully covered by hydrophobic C_6-ol molecules before formation of the second adsorption layer begins." Furthermore, from a comparison of adsorption enthalpies they concluded that phenol adsorption is the consequence of "only a weak specific interaction between the graphite surface and the benzene ring."

The study of Michot et al. [389] is an interesting attempt to determine where and how phenol adsorbs in activated carbons. These authors determined the uptake of phenol by a commercial activated carbon at pH $\sim$ 11.5 (i.e., under conditions of electrostatic repulsion) and concluded that "adsorption of phenol on activated carbon is mainly controlled by the porous structure of the adsorbent." They found a monotonic decrease in the adsorption of gas-phase argon with in-

TABLE 14 Comparison of Equilibrium Phenol Uptakes by Activated Carbons (AC), Zeolites, and Polymeric Resins

Adsorbent	Surface area (m²/g)	Hydrophobicity coefficient[a]	Phenol uptake[b] (mg/g)
AC (Filtrasorb 400)	880	0.69	~240
AC (AG3)	660	0.54	~140
Resin (Y77)	1250	0.55	~100
Resin (EP61)	350	1.0	~17
Zeolite Y (DY1)	690	0.08	<3.2
Zeolite Y (DY2)	500	0.54	~3.2

[a] Defined as $1 - \Delta H_{im,w}/\Delta H_{im,b}$, where $\Delta H_{im,w}$ and $\Delta H_{im,b}$ are the specific immersion heats of the adsorbents in water and benzene.
[b] At 293 K and 100 mg/L (unspecified pH).
Source: Ref. 10.

creasing preloading of the adsorbent by phenol. Intriguingly, this led them to postulate that adsorption of phenol occurs by a volume filling mechanism. In contrast, Teng and Hsieh [391,455] argued that the different uptakes observed on partially oxidized (activated) bituminous coal chars are reflections of fractional surface coverage (i.e., less than a monolayer, assuming 30.2 $\mathring{A}^2$ for the cross-sectional area of phenol) and obtained consistent results: at low carbon burnoff the uptake was close to 100% surface coverage, while at ca. 60% burnoff (when the surface is populated with the highest concentration of oxygen functional groups) it decreased to ca. 40%. They did not attempt to relate this to carbon surface chemistry (see Section VI) and, on the basis of a particle size dependence of the uptake, attributed it to the "increase in diffusion path with the burnoff level" [391]. Leyva-Ramos et al. [393] also noted a particle size dependence in aqueous solution, but not in cyclohexane, and speculated differently: "the surface area available for adsorption increased when the particle diameter was reduced." Very recently, Nevskaia et al. [394] have taken a similar approach to that of Teng and Hsieh [391] and have found the same trend: a decrease in surface coverage with increasing surface oxidation, in agreement not only with Leng and Pinto [386] (cited by the authors), but with many other studies as well (see above).

2. Substituted Benzenes and Phenols

Here we focus on the same questions posed in Section IV.A.1 and add the following: Which substituents on the benzene ring lead to suppressed uptakes and which ones lead to enhanced uptakes? This discussion introduces the issue of the importance of dispersive adsorbate–adsorbent interactions, which is then taken up in more detail in Sections IV.B.2 and V. In other studies [456–467,427,468, 429,469–477,438,478–480] these issues are not discussed, but other topics of both fundamental and practical interest are covered.

In an early and rarely cited study (13 citations through 1999), Wheeler and Levy [481] reported that "[l]arge alkyl groups ortho to the hydroxyl group greatly reduce the adsorption" of phenols (from cyclohexane solution) on a commercial activated carbon whose surface, they claimed, "contains neither strongly acidic nor basic groups." Based on the assumption that "phenols are adsorbed with the benzene ring parallel to the surface of the adsorbent," the authors concluded that the "fraction of the available surface covered" by phenol, p-tert-butylphenol, o-dimethylphenol, o-diisopropylphenol, and o-tert-butylphenol was 1/6, 1/8, 1/11, 1/17, and 1/62, respectively, and that "the increasing steric shielding of the hydroxyl group [suggests] that this group is largely responsible for the adsorption." The authors do not normalize their results with respect to the solubility of the adsorbate (see below) but they do discuss the various possible mechanisms of adsorption, including hydrogen bonding and "interaction between the π-electrons of the phenol rings and those of the 'aromatic structure' of the graphite surfaces" (see Sections V and VI). The relatively low uptakes were rationalized by suggesting that adsorption "takes place only at certain 'active centers' on the adsorbent," but these were not identified. In the pioneering study of Coughlin and coworkers [450,331,332], which surprisingly does not cite the work of Wheeler and Levy [481], the focus is on the varying uptakes of the same adsorbate on chemically modified and better characterized adsorbents (see Section VI). The authors make no comments about the relative uptakes of nitrobenzene and phenol on the same adsorbent; despite the huge aqueous solubility difference, e.g., 0.01 M vs. 0.47 M [24], the uptake of nitrobenzene exceeded that of phenol only by a factor of two or less. Similar results were obtained by Mattson et al. [24]. These authors address this issue by normalizing the adsorption isotherms using Traube's rule [482,3,483,408,484], i.e., by using "reduced" equilibrium concentrations (divided by adsorbate solubility); however, neither the normalized isotherms for p-nitrophenol at different temperatures nor those of nitrobenzene and phenols at the same temperature obeyed Traube's rule. Mattson and Mark [6] further discuss a 1946 study of Gupta and De in which phenol and aniline on a "purified, activated sugar carbon" conformed to Traube's rule— this was confirmed by Singh [334] for a "graphitized" carbon black—but the uptakes of benzoic and salicylic acids were much lower. The pH of the adsorption experiments was apparently not controlled; the authors thus surmise that the higher degree of dissociation of the latter two, and thus their enhanced solubility, can account for this discrepancy. Snoeyink et al. [333] reported a similar result, p-nitrophenol showing a higher uptake than phenol for one of the two carbons studied, and concluded that "somewhat different surface effects are involved for the two carbons in the lower regions of surface coverage." Puri [346] examined the behavior of the same two adsorbates on an activated carbon and a carbon black, but focused on their monolayer surface coverage; the molecular area of PNP turned out to be 55 Å^2, from which the author inferred that adsorbed PNP, like phenol, was oriented parallel to the surface. In a later study, Kamegawa and Yoshida [485] went as far as proposing the use of PNP to determine the surface area of the carbon portion of a carbon-coated silica gel. The applicability of the BET equation in this context was discussed also by González-Martín et al. [486]. The possibly dramatic effects of solution and surface chemistry on PNP uptakes are evident, however, from the studies of Gómez-Serrano and coworkers [487,488] and Hobday et al. [489].

Komori et al. [490] analyzed the adsorptivities of substituted benzoic acids on two commercial activated carbons. They concluded that "Traube's rule was applicable to the adsorption of the m-substituted benzoic acid except isophthalic acid." The trends for the o-substituted compounds were difficult to rationalize and were attributed to "ortho effects," suggesting the importance of both electron-withdrawing and electron-donating substituents in affecting adsorbate acidity. El-Dib and Badawy [491] reported Freundlich-type behavior of o-xylene, ethylbenzene, toluene, and benzene on a commercial activated carbon (in the concentration range 1–30 mg/L); even though the parameters obtained were in qualitative agreement with Traube's rule, the authors concluded, rather vaguely, that "adsorption of undissociated aromatics from dilute aqueous solutions is not clearly defined" and that "[i]nteractions at the carbon surface" [332] and the presence of alkyl groups may affect the adsorption process. On the basis of up-takes of a wide range of adsorbates, including phenol, nitrobenzene and o-chloro-phenol, Martin [492] discussed the activated carbon selection criteria for water treatment and concluded that, rather than the magnitude of surface area, "the distribution of surface area or pore volume with change in pore size is likely to be of much greater use to the carbon user."

Oda et al. [350] studied the adsorption of benzoic acid and phenol on several wood-based activated carbons and reported comparable uptakes in most cases, although the former is much less soluble than the latter [408]. They found that "the presence of acidic functional groups retards the adsorption of the adsorbates, but the effect is larger for phenol than benzoic acid;" from competitive adsorption experiments, they concluded that adsorbed benzoic acid cannot be displaced eas-ily by phenol, while substitution of the latter by the former proceeds easily. All these findings pointed in the direction of dominant surface chemical effects, but the authors only determined the total acidity of their carbons. Urano et al. [408] also discussed the "applicability of Traube's rule and influence of surface oxides of activated carbon on the adsorption potentials" using five different activated carbons and ten substituted benzenes; they found that "[a]ll the isotherms appar-ently do not superimpose [when normalized concentrations are used], and those for phenol and substituted phenols lie markedly above those for other com-pounds." They interpreted these findings for phenols (and to a lesser extent for aniline, but not for benzoic acid) in terms of specific interactions "with active groups on the carbon surface" (see Section V). The electrosorption study of Woodard et al. [493] also revealed complexities in the relative uptakes of substi-tuted benzenes and other adsorbates; the contributing factors mentioned are "ad-sorbate concentration and solubility, as well as electrostatic and dispersion inter-actions between the electrode and adsorbates."

Yonge et al. [412] analyzed the effect of adsorbate functional groups on their degree of irreversible adsorption on a commercial activated carbon (at pH = 7.0). Based on the findings summarized in Table 15, they proposed that "the more intense inductive effect [of methyl, ethyl, and methoxy groups] could result in a stronger sorbate/sorbent bond, and therefore, a higher degree of irreversibil-ity." The subsequent study by Caturla et al. [414] also emphasized the importance of the substituent groups. The results are summarized in Table 16. For the more microporous lower-burnoff carbons, molecular sieving effects suppressed the up-takes of substituted phenols. The interesting trends observed at higher degree of activation were attributed to the complex effects of surface chemistry and solution chemistry (see Section V).

TABLE 15 Equilibrium Adsorption and Desorption Results for Substituted Phenols on a Commercial Activated Carbon[a]

Adsorbate	Freundlich constants[b]		Irreversible uptake[c](%)
	K	n	
4-Isopropylphenol	1.204	0.124	87
Phenol	0.313	0.296	85
2-Ethylphenol	1.954	0.069	91
o-Cresol	2.035	0.066	94
o-Methoxyphenol	2.200	0.059	97

[a] Filtrasorb 400 (Calgon Carbon Corp.).
[b] Uptake = $KC^{1/n}$, with K in μM, at uncontrolled pH.
[c] At 500 μM, after rinsing with adsorbate-free buffer solution (pH = 7.0) until the resultant adsorbate concentration was <0.3 mg/L.
Source: Ref. 412.

Mostafa et al. [416] noted that the uptake of substituted phenols did not follow Traube's rule but instead did decrease in the order of their increasing pK_a values: o-chlorophenol > hydroquinone > resorcinol > phenol > catechol > pyrogallol. Even though the experiments were performed at (unspecified) pH where "the adsorbates were predominantly in an undissociated form," the authors noted that "the acidity of the adsorbate is an important factor in determining its adsorption capacity;" this was not surprising to them because, presumably, "the ability [of steam-activated and thus alkaline carbons] to remove phenols is expected to

TABLE 16 Uptakes of Substituted Phenols, Expressed as Normalized Surface Areas (with Respect to Phenol) as a Function of the Degree of Burn-Off (BO) for Activated Carbon Prepared from Olive Stones

Sample	Unbuffered solution			pH > 13		
	S_{PCP}/S_P	S_{DCP}/S_P	S_{DNP}/S_P	S_{PCP}/S_P	S_{DCP}/S_P	S_{DNP}/S_P
8% BO	0.95	0.78	0.66	0.95	0.75	0.67
19% BO	1.06	0.96	0.87	0.88	0.82	0.84
34% BO	1.05	1.11	1.10	0.94	0.96	0.96
52% BO	1.10	1.16	1.23	0.98	1.03	1.07

PCP = 4-chlorophenol; DCP = 2,4-dichlorophenol; DNP = 2,4-dinitrophenol; P = phenol.
Source: Ref. 414.

decrease with the decrease in their acidic strength.'' Kaneko et al. [363] compared the uptakes of benzene, nitrobenzene, benzoic acid, and phenol on untreated and H_2-treated activated carbon fibers and in both cases reported the order expected, based on their solubility: benzene > nitrobenzene > benzoic acid > phenol; at pH = 6, the uptake difference between nitrobenzene and phenol was of the order of 2–3 (see above).

In an ambitious study, Shirgaonkar et al. [422] analyzed the uptake of 24 phenols by a commercial activated carbon and attempted ''to correlate the experimental data (Freundlich parameters primarily) with molecular parameters, with a view to understanding the process of adsorption.'' However, they concluded rather vaguely that ''[t]o understand the phenomena of adsorption, an insight into the interaction of adsorbate with adsorbent and adsorbate with solvent are [sic] necessary.'' They did note that ''the nature of substituents also affects the adsorption,'' but their results did not reveal the nature of this dependence.

Sorial et al. [466] compared the uptakes of p-chlorophenol on five commercial carbons whose BET surface areas and other properties (e.g., iodine number, manganese content, total pore volume) varied considerably. The more than threefold differences in the Freundlich constants (under anoxic conditions) did not correlate with any one of these properties. Deng et al. [433] compared the uptakes of both p-chlorophenol and other substituted phenols on a range of activated carbons derived from elutrilithe by treatment with K_2CO_3. In contrast to earlier reports about the beneficial effect of $ZnCl_2$ treatment, they found higher uptakes for the K_2CO_3-treated carbons. Pretreatment in nitric acid and posttreatment in nitric acid and sodium hydroxide further increased the uptakes; this was attributed to beneficial changes in surface area and porosity as well as increased hydrophobicity of the adsorbents. For a given adsorbent, the uptake increased in the following order: phenol < 3-cresol < 2,3-dimethylphenol < 4-chlorophenol < 4-nitrophenol. In accounting for some of these trends, the authors commented that the presence of ''chloro- and nitro-groups on the phenol structure decreases the electron density of the aromatic ring, and as a result the interaction of the adsorbates and the carbon is enhanced, which naturally leads to an increase in adsorption capacity for these compounds'' and cited the study of Caturla et al. [414] in support of this intriguing statement (see Section IV.B.2 and V).

In the context of a study of foam flotation of powdered activated carbon (PAC), Zouboulis et al. [143] noted large and different pH effects when an anionic surfactant was used instead of a cationic one. For the cationic surfactant, best recovery (at low surfactant concentration) was achieved at the highest pH, in agreement with electrostatic arguments (see Section IV.B.1); for the anionic surfactant, an intermediate pH was the best. The authors also measured the zeta potential of the carbon in the presence and absence of the surfactants and concluded that the ''specific chemical nature and the dissociation of each surfactant, as well as the fact that the surface of the PAC particles was originally negatively charged and that in acidic pH the surface charge can be modified by the adsorption of protons, should be considered in order to evaluate the observed results.''

Nelson and Yang [494] studied the uptakes of a series of chlorophenols (with pK$_a$ values ranging from ca. 5 to 9) on a commercial activated carbon and found them to be consistent with a surface complexation model that introduces mutually independent equilibrium constants for acidic and basic surface groups. Dargaville et al. [431] compared the adsorption of phenol to that of multinuclear phenolic compounds (dimers to octamers) from ethanol solvent and reported that compounds having only ortho linkages were adsorbed to a greater extent than their para analogues. In addition to observing that there was a ''definite correlation between the maximum amount adsorbed and the pH of adsorbate species,'' the authors argued that these differences were due to poorer solvent properties for the ortho-linked compounds.

Leng and Pinto [386] analyzed the effect of carbon surface properties and the presence of dissolved O$_2$ on the adsorption of phenol, o-cresol, and benzoic acid at pH = 7.0 (see also Section VI.B.1). There was no correlation between the Freundlich K values and the BET surface areas under either oxic or anoxic conditions. The authors defined a polymerization factor (PF) to quantify the uptake enhancement under oxic conditions, which has been attributed to oxidative coupling [417,495], and reported its decrease with increasing surface oxygen content. Under both oxic and anoxic conditions there was also a decrease in the amount adsorbed with increasing surface oxygen content. The uptake order in both cases was o-cresol > phenol > benzoic acid, and the authors attributed this not to solubility or surface chemical effects on adsorption but to the propensity for surface polymerization. The authors concluded that adsorption on phenol(s) is a ''combination of physisorption and surface polymerization.'' In a subsequent publication by the same group [392], a similar study was extended to the role of KCl on the adsorption of phenol, toluene, and benzene at pH = 11.6 by carbons whose pH$_{PZC}$ values ranged between 2.8 and 9.2. There was a maximum in uptake as a function of KCl concentration for phenol, a monotonic decrease for toluene, and a minimum for benzene. The relative uptakes of these three adsorbates under identical conditions, in terms of Freundlich equilibrium constants, were toluene ≫ phenol > benzene; the difference between toluene and benzene was explained by invoking their octanol–water partition coefficients (K$_{ow}$ = 537 vs. 131.8), but why phenol uptake (with K$_{ow}$ of ca. 30) was intermediate was left unexplained. The authors offered explanations for these complex trends by invoking electrostatic interactions, ion exchange, competitive water adsorption, the ''salting out'' effect, as well as direct interaction between Cl$^-$ anions and the surface. In a study that also explored the impact of surface oxygen and dissolved O$_2$ on the uptake of aromatics, Tessmer et al. [496] argued that ''the presence of acidic surface functional groups hinders the ability of activated carbon to adsorb phenolic compounds under oxic conditions by reducing its effectiveness in promoting adsorption via oxidative coupling reactions."

Zaror [497] compared the uptakes of four phenols on a commercial activated carbon at a pH of 2 and reported both relatively low uptakes (<0.1 mmol/g at 0.3 mmol/L) and relatively small differences, in spite of the substantial differences in the nature of the substituents. Wang et al. [439] reported that the uptakes of *p*-nitrophenol, *p*-chlorophenol, and phenol were consistent with the notion that "materials of high molecular weight are adsorbed to a more considerable extent than those of low molecular weight for compounds of similar chemical constitution;" Traube would be happy to learn that a century later [498] his rule lives on and that, under certain (perhaps very limited) conditions, the complexities discussed elsewhere in this review can be ignored. A clear example of the inapplicability of Traube's rule, however, is the study of Mostafa et al. [416]; the authors did not attribute this to electrostatic effects or to changes in π- electron density (see below) but to "the difference in the ability of hydrogen bond formation of the different phenols."

As recently as 1998, Zhou et al. [445] analyzed the widely varying uptakes of substituted benzenes and phenols (nitrobenzene > benzaldehyde > nitro-4-phenol > 4-cresol > phenol > aniline) using models such as those of Langmuir, Freundlich, and preferably—when the solute concentration range was large—Redlich-Peterson and Jossens-Myers, in which the parameters are not related to the properties of the adsorbent; instead, they are related to the solubility and the Hammett constant of the solutes (see Section IV.B.2).

3. Dyes

The practical and fundamental interests in dye adsorption stem from three considerations: (1) their ubiquity in industrial waste waters [499–511,395]; (2) their use as surrogates for natural organic matter [512], and (3) their use for the determination of porous structure of adsorbents [451,513,514,346,515–520,512].

This last application is well known [451,520], but it is arguably overused. Because of the widely varying surface chemical properties of carbon adsorbents, great care must be taken if reliable estimates of surface area are to be obtained using dye adsorption. It is interesting that even in the pioneering work of Graham [451], it is clearly stated that "the surface area accessible to the first adsorbed monolayer [depends] on the interactions of the adsorbate molecules with specific substituents in the carbon surface." This is illustrated in Fig. 19. Despite the now well known uncertainties in the determination of surface areas of microporous carbons using N_2 adsorption at 77 K and the puzzlingly small decrease in carbon acidity upon heat treatment in N_2 at 900°C (from 1.22 to 1.04 meq/g), the correlation between the decreasing uptake of the anionic metanil yellow and increasing carbon acidity is quite good ($R^2 = 0.89$). The author does not mention that the pH of the solution and the pH_{PZC} of the adsorbent are the key issues, but he does

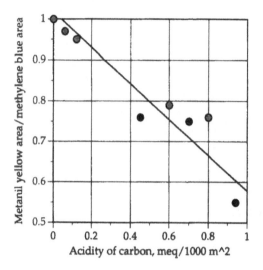

FIG. 19 Relationship between adsorption capacities of dyes and the surface chemistry of carbon. (Adapted from Ref. 451.)

state that "acidic groups in the carbon surface tend to reduce the capacity . . . for anionic adsorbates in general."

It is disturbing to note that not a single one of the subsequent publications in which dye adsorption is used as a probe of carbon surface physics (surface area, pore size distribution) cites this early study by Graham [451]. Even the main advocates of this approach [514] have pointed out, however, that there are "inconsistencies" in the use of dyes "for measuring the specific surface of powders," including the "effects of specific dye-substrate bonding" and "ageing effects on the solid surface." When faced with the pH effect reproduced in Fig. 20, Giles et al. [514] noted that "the pH of the test solutions is important," and advised "against tests being made from solutions which are outside the range of pH ca. 5–8;" they did not mention why this precise pH range is appropriate. Based on the discussion presented in Section II and further analysis in Section IV.B, it should be obvious that the nature and distribution of surface functional groups on the carbon surface will often dictate to what extent the surface is accessible to the dye solute species in aqueous solutions.

Mattson and Mark [6] had summarized the 1942 study of Sastri in which, for a number of different carbons (having adsorption capacities ranging from 800 to 50 mg/g), a trend of decreasing methylene blue uptake was observed as the pH decreased from 8 to 3. The authors commented that "the formation of positive surface sites [at the lower pH] certainly would not enhance cationic adsorption."

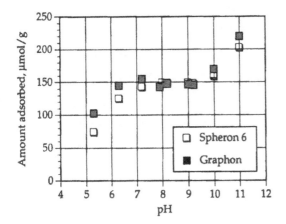

FIG. 20 Effect of pH on the adsorption of methylene blue by two carbon blacks. (Adapted from Ref. 514.)

Nandi and Walker [499] compared the uptakes of two acidic and one basic dye by coals, chars, and activated carbons and noted that "the area covered by a dye molecule would depend on the nature of the solid surface." They concluded that the removal of acidic groups by heat treatment had an effect on dye uptake for one activated carbon but had no effect for another. The authors did not report the pH of the adsorption measurements, nor did they characterize the surface chemistry of the adsorbents. McKay [500] did report the pH and even studied the effect of pH (in the range 5.2–8.5) on the rate of adsorption of Telon blue (an acidic dye), but the effect was neither clarified nor found to be significant. In contrast, Perineau et al. [501] noted major effects of pH on the adsorption of both an acidic and a basic dye and concluded that "pH values of about 2 are the best for the adsorption of acid dyes whereas less acidic values (pH > 5) are to be preferred for the removal of basic dyes."

In the virtually ignored study by Dai (with three nonself citations in five years) [521], the paper by Graham [451] is not cited either, but the key issue is both identified and clarified. Based on the results summarized in Fig. 21, the author concluded that "electrostatic interaction between cationic dyes and the surface of activated carbon has a great effect on adsorption capacity." Below the isoelectric point of the activated carbon (when the positive zeta potential was above 60 mV), "the capacity is significantly reduced due to electrostatic repulsion between cationic dyes and the carbon surface." In a follow-up study, while still failing to acknowledge earlier important contributions to the resolution of the key issues, Dai [522] reinforced and confirmed the electrostatic attraction vs. repulsion arguments. The author used anionic dyes (phenol red, carmine, and titan yellow) and

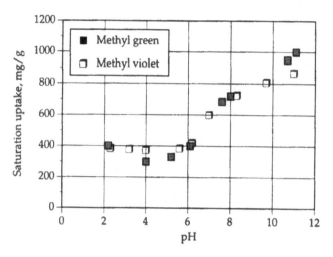

FIG. 21 Effect of pH on the adsorption of dyes by a commercial activated carbon. (Adapted from Ref. 521.)

obtained the expected results: for an activated carbon whose pH_{PZC} was 6.2, as pH increased from ca. 3 to 11 the uptake decreased monotonically. In implicit agreement with the (again uncited) early work of Radke and coworkers [523–525] (see Section IV.B.1), the author concluded that the "adsorption forces are the sum or difference of dispersion and electrostatic forces."

Juang and Swei [508] appear to be ignorant of both these studies and claim that "the effect of dye nature on its adsorption on activated carbon has not been fully clarified." In a study of uptakes of one anionic and one cationic dye by an apparently unusual activated carbon (having a BET surface area of 13 m^2/g and a pore volume of 0.04 cm^3/g), they arrived at even more puzzling conclusions. The higher uptake of the basic dye was attributed to "electrostatic interaction between the electron-acceptor groups on the [adsorbent] surface and the positive charges on the basic dye molecules." In support of this proposal, they cite the study by Mattson et al. [24] (see Section VI) in which, presumably, it has been shown that, upon activation, the existence of many conjugated, nonlocalized and highly active π bonds on the surface makes possible the formation of electron-accepting carbonyl and carboxyl groups. As discussed at length in Section VI, this is yet another misinterpretation of the adsorbent–adsorbate interactions. What Mattson et al. [24] proposed is in fact almost exactly the opposite: not an electrostatic interaction but the formation of a "donor-acceptor complex" in which the carbonyl oxygens on the adsorbent surface are the electron donors and the aromatic rings of the adsorbate are the electron acceptors.

Gupta et al. [507] cite only one study for "dye removal from wastewater" using adsorption on activated carbon and also ignore the key papers by Graham [451] and Dai [521]. More puzzling is the following finding of these authors: a decrease in % removal of malachite green (a basic dye) as pH increases from 2 to 10. Even more puzzling is their interpretation of this result, which is in conflict with the trends shown in Figs. 20 and 21: "Malachite green (pK_a = 10.3) becomes protonated in the acidic medium and deprotonation takes place at higher pH. Consequently, the positive charge density would be more on the dye molecule at lower pH; this accounts for the higher uptake on the negatively charged surface of carbonaceous adsorbent." The authors do not provide any data on the chemistry of the adsorbent surface but it is highly unlikely that the surface is negatively charged at pH = 2. In the recent studies of Al-Degs et al. [526] (see below) and Yasuda and coworkers [527–529], the crucial role of surface charge is recognized. In the latter study, the effectiveness of activated carbon fibers in the removal of a series of acid, direct, and basic dyes was consistent with the electrostatic arguments described above; the authors also report that a mesoporous carbon has a higher adsorption capacity for direct dyes than a microporous one, because of the large size of these molecules. This latter point is also discussed by Khalil and Girgis [530].

In an ambitious recent study, Krupa and Cannon [531,512] used dyes of varying molecular size (and charge) in an attempt to characterize the porous structure of a thermally regenerated activated carbon. They studied the effects of pH on the adsorption isotherms of two cationic dyes (methylene blue and crystal violet) and an anionic dye (Congo red). To address the concerns voiced by Giles et al. [514], when possible, adsorbates were "adsorbed in their neutral state to overcome the effects of specific bonding" [531]. However, the resultant shapes of the isotherms were quite unusual regardless of the pH used: there was no increase in uptake over a one-order-of-magnitude increase in adsorbate concentration for Congo red, while nonlinear upward-curving log–log plots were obtained for crystal violet in the same concentration range. (Methylene blue exhibited the expected Freundlich-type behavior.) The former effect was tentatively attributed [531] to Langmuirian behavior (implying that saturation had been reached at <10 mg/L), as well as to the "presence of charged functional groups on the carbon surface" and adsorption "governed by ion exchange;" no evidence was given, however, for the (unlikely?) presence of positively charged ion-exchangeable groups on the carbon surface. In discussing the complex pH effects (increase in uptake of Congo red at pH = 10 and especially at pH = 7 relative to the unbuffered solution; increase in the higher-concentration uptake of crystal violet at pH = 7, relative to unbuffered solution), Krupa [531] concluded, rather vaguely, that "significant chemistry differences" existed between the carbons used and that these "differences in adsorption behavior . . . may give an indication of differences in the net surface charge of each carbon." These investigators appear to have been unaware of both earlier [514,6,501] and contemporary [521] studies of this issue.

Very recently, Al-Degs et al. [532] compared the effectiveness of five commercial activated carbons in adsorbing Remasol Reactive Black B, a large anionic dye pollutant. The adsorption capacity decreased in the following order: F-400 (pH_{PZC}, ca. 7.3; iodine number, 1050 mg/g) $>$ C207 (7.20; 1000) $>$ EA207 (7.30; 900) $\gg$ Centaur (5.82; 800) = Chilean lignite (pH_{PZC} = 5.90) and did not correlate with either surface area, iodine number, bulk density, or ash content. The observed differences were attributed to the "surface charge developed on the carbon particles at the equilibrium pH." In addition to confirming the electrostatic arguments discussed above—namely that the more effective adsorbents had a net positive charge during adsorption, while the opposite was true for the ineffective adsorbents—the authors rescued the study of Graham [451] from undeserved oblivion.

4. Natural Organic Matter (NOM)

In the treatment of drinking water, the conventional role of activated carbon is to remove taste- and odor-causing contaminants [533–535,436,536–538]. Nevertheless, the existence of fundamental relationships between the physics and/or chemistry of the activated carbon surface and its effectiveness in removing species such as 2-methylisoborneol or geosmin appears to be both uncertain [538] and insufficiently documented. On the other hand, the need to remove natural organic matter is more recent, although it has long been known to impart undesirable color to drinking water; it is due to the carcinogenic nature of halogenated organic compounds formed during chlorination of water in the presence of humic substances, as well as to the interference of NOM with the mandatory removal of synthetic organic matter (SOM). Much research has thus been conducted in the past two decades [539–551,495,552–566,536,567,568] in attempts to identify the desirable properties of activated carbons for maximum removal of these complex organic species. A more comprehensive review of this problem, in which the challenge is to understand the crucial roles of *both* pore size distribution and surface chemistry, is in preparation [569].

Natural organic matter arises from the decay of plant and animal residues and from other biological activities of microorganisms in the environment, and it is ubiquitous in natural waters at concentrations up to 40 mg/L [562]. Humic and fulvic acid are its dominant components, and they are typically described [570,559] as polymeric structures containing aromatic rings with acidic functional groups attached to them, i.e., not unlike some activated carbons themselves. It is therefore intriguing that the early literature on the subject has shown that "the effectiveness of carbon adsorption processes in removing humic substances and other types of organic matter from natural waters is limited" [542]. In this brief review, we analyze the literature in search of confirmation or re-evaluation of this statement. In Section IV.B.1 we analyze in more detail the key role of electrostatic interactions in NOM adsorption.

Youssefi and Faust [541] noted a low humic acid affinity of a commercial activated carbon, both in the rapid column breakthrough and by analyzing the adsorption isotherms. Weber et al. [540] confirmed the low uptakes on a different commercial carbon (in the range of 3–40 mg of humic acid per gram of adsorbent at 1–5 mg/L) but noted that water constituents such as Ca^{2+}, Mg^{2+}, and OCl^- ions enhance the uptake by altering the adsorption affinity. Since the majority of humic substances are very large molecules—e.g., in the molecular size range 4.7–33 Å and molecular weight range 500–10,000 [564]—Lee et al. [542] investigated the role of carbon's pore size distribution in the uptake of humic and fulvic acids. The results that confirm the importance of adsorbents' physical surface properties are summarized in Table 17. While the correlation coefficients for the dependence of the adsorption constant on the total surface area of seven commercial activated carbons were very low (r^2 varied between 0.24 and 0.56), they improved as the size of the maximum pore radius increased and were quite good

TABLE 17 Goodness of Fit in Correlations Between Adsorption Constants and Pore Volumes of Activated Carbons in Pores Smaller than a Certain Radius

R_{max}[a] (Å)	Correlation coefficient (r^2)			
	Humic acid	Fulvic acid	Low MW fulvic acid[b]	High MW fulvic acid[c]
20	0.38	−0.54	0.11	−0.64
40	0.86	−0.13	0.53	−0.23
60	0.93	0.42	0.89	0.36
80	0.95	0.67	0.95	0.62
100	0.95	0.73	0.96	0.69
200	0.96	0.84	0.94	0.80
300	0.97	0.86	0.94	0.83
500	0.97	0.89	0.93	0.85
1000	0.97	0.91	0.93	0.87
V_{total}	0.97	0.90	0.88	0.92

[a] Maximum pore radius determined using a combination of N_2 adsorption and Hg intrusion.
[b] Molecular weight less than 1000.
[c] Molecular weight greater than 50,000.
Source: Ref. 542.

for the total pore volume. Two wood-based carbons possessing the largest pore volumes were excluded from this analysis because of "anomalous behavior," presumably related to the (unspecified) role of surface chemistry.

Randtke and Jepsen [544] confirmed the important effect of salts (cations) on the uptake of fulvic acids; the presence of Ca^{2+} was particularly beneficial, especially at higher pH (see Section IV.B.1). Weber et al. [545] examined the same issues, reported similar findings (e.g., higher uptake at lower pH, higher uptake for larger-pore-size carbons) and concluded that "solute heterogeneity, pH, carbon dosage and type, types and concentrations of cations in solution, and adsorbent particle size" all have a significant influence on both the rates and the capacities of adsorption of humic substances. Ogino et al. [546] used the same methodology as Lee et al. [542] to further evaluate whether pore sizes (of both as-received and chemically treated activated carbon fibers and carbon blacks) are a limiting factor in the adsorption of humic substances. Their results, reproduced in Fig. 22, are a confirmation that "pore-size distribution is an important factor in removing humic substances effectively." (The deviations shown in the figure were attributed, albeit somewhat vaguely, to "the chemical properties of ash and surface oxygen groups on the adsorbents.") Subsequent work by Weber and coworkers [562–564], using a commercial adsorbent whose narrow micropores (<10 Å) accounted for as much as 86% of the total surface area, further confirmed these findings; the authors concluded that "adsorption of humic substances was governed, in large part, by molecular size distribution in relation to adsorbent pore sizes" [564] and that "differences observed in [the] uptake of different [NOM] fractions can be attributed principally to physical size effects" [562].

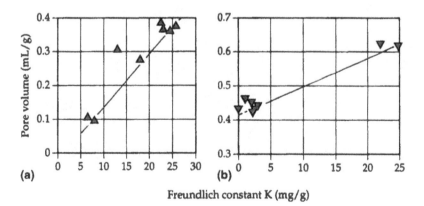

FIG. 22 Effect of porosity on the removal of humic substances by a series of carbon adsorbents: (a) humic substances fractionated at pH = 2.8; (b) humic substances fractionated at pH = 6.8. (Adapted from Ref. 546.)

Newcombe et al. [536] also concluded that the "adsorption of four NOM ultrafil-tration fractions on to activated carbon was consistent with the pore volume distri-butions of the carbons and the hydrodynamic diameters of the fractions."

The study of Lafrance and Mazet [549] not only provided further confirmation of the beneficial effect of cations (Na^+) but also clarified its mechanism by ana-lyzing the effect of sodium salt concentration on the zeta potential of a commer-cial powdered activated carbon (see Section IV.B.1). The reported uptakes of a commercial sodium humate remained <100 mg/g (at 10 mg/L). In the study of Annesini et al. [556], even at humic acid concentrations of 20 mg/L the uptake did not exceed 10 mg/g. These uptakes should be contrasted with those reported recently by Newcombe and Drikas [566] and illustrated in Fig. 23: at low pH, when electrostatic repulsive forces (see Section IV.B.1) are eliminated, the uptake at comparable concentrations can be as high as 200–300 mg/g.

The beneficial effect of dissolved O_2 was reported by Vidic and Suidan [495]. By analogy with the oxidative coupling (surface polymerization) of phenols, it was attributed to "conglomeration of NOM under oxic conditions." The results of Warta et al. [560] and Cerminara et al. [557] confirmed these findings, but the role of the adsorbent remained unclear. (This beneficial effect for NOM re-moval must be weighed of course against the consequent reduction in the adsorp-tion capacity for other water contaminants [536,567].) Karanfil et al. [561] ad-dressed the same issue in more detail and linked it to the well-established effect of molecular and pore sizes: in the case of "oxygen-sensitive humics," the "greater accessibility of smaller organic macromolecules to active surface sites [which

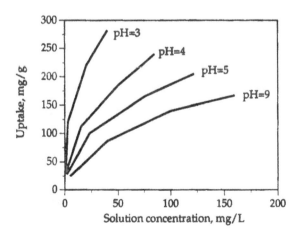

FIG. 23 Effect of pH on the adsorption isotherms for natural organic matter on a com-mercial activated carbon. (Adapted from Ref. 566.)

promote surface polymerization] significantly increases the impacts of [dissolved oxygen] on their sorption behavior."

In a recent review of technological alternatives for NOM removal, Jacangelo et al. [558] presented field data that illustrate wide variability in adsorption capacities of activated carbon in a single location (e.g., exhaustion periods between 41 and 182 days). They concluded that "these results are evidence of the site-specific nature of [dissolved organic carbon] removal by [activated carbon]" and that "the concerns regarding reliability of treatment practices to meet the new [regulations] have a sound basis." Clearly, much fundamental work remains to be done to understand fully the complex nature of these adsorbent/adsorbate interactions and thus be able to optimize both the physical and the chemical accessibility of the carbon surface to natural organic matter.

5. Other Organics

Studies of adsorption of nonaromatic organic compounds are not as abundant but they are very diverse (and are cited here only to illustrate and emphasize this diversity). They go back to the interwar period (see, for example, Refs. 571–573, when activated charcoal was being developed for chemical and biological defense and its properties were intensely studied [2,3]). More recently, the removal of trihalomethanes [574–579], amines [580,581], acetic acid [582], the pesticide thiram [583], chlorinated organic compounds [584,552,585,586,403, 587–591], alcohols [572,592,593–597], carboxylic and fatty acids [592,483, 598,594,599], N-acetylcysteine [600], amino acids [601], benzo-15-crown-5 ether [602], quaternary ammonium compounds [603], organophosphates and organophosphonates [604] has been studied. Adsorption of surfactants has also been of continued interest [605–614,143,615–617].

B. Importance of Carbon Surface Chemistry

The brief review of the vast literature on the phenomenological aspects of adsorption of aromatic solutes has highlighted studies that provide clues, either explicitly or implicitly, to the optimization of carbon surface chemistry for removal of specific pollutants from aqueous streams. Here we make an attempt to synthesize the available information. In Section V we then offer suggestions regarding a comprehensive model of adsorption of organic (and inorganic) solutes.

Based on extensive experimental evidence regarding the importance of pH and surface chemistry, it is obvious that the same arguments discussed in Section III.B in the context of adsorption of inorganic solutes are applicable also—and indeed are required—for understanding the adsorption of organic solutes, many of which are weak electrolytes. What does need careful consideration is the answer to the following two questions:

1. Which specific features of carbon surface chemistry need to be known, in
    both qualitative and quantitative terms, in order to be able to optimize an
    activated carbon for a specific pollution control task?
2. What are the *relative* roles of electrostatic and dispersive interactions in the
    various adsorbate/adsorbent systems of interest?

In the preface to their landmark monograph, Mattson and Mark [6] wrote:
"Carbon researchers have, for the most part, considered the surface chemistry
of activated carbon to be in such a state of disarray that they want to avoid lengthy
discussions of surface phenomena." More than two decades later, with vast improvements in our knowledge of the role of surface chemistry in carbon gasification [37,618] and in the use of carbons as catalyst supports [22], it is argued here
that this is no longer true.

There are probably no more contentious issues in the use of activated carbons
than the role of surface oxygen functional groups in the aqueous phase adsorption
in general and adsorption of weak aromatic electrolytes in particular. In 1980,
Suffet [619] made the following (under)statement and posed the key question:
"There is some confusion in the literature regarding the major sites of adsorption
[on granular activated carbon]. Is an oxidized surface adsorbing material on the
oxidized functional groups, or is it adsorbing on the nonpolar carbon surface?"

Even though the components of the total interaction potential between such
complex adsorbents as solid carbons and a wide range of adsorbates can be
grouped in many different ways [315,316], it is convenient and meaningful to
consider only the London dispersion (induced dipole) forces and the electrostatic
(double-layer) forces [620,621,76,77].

1. Role of Electrostatic Interactions

Here it would be perhaps more convenient to review the mainstream adsorption
literature separately from the relevant subset of the vast electrochemistry literature. Furthermore, some of the key concepts and experimental findings of interest
here were first reported in the colloidal (carbon black) stability literature. It is
unfortunate that not enough cross-fertilization has yet occurred between these
three disciplines, as evidenced by the preponderance of narrowly focused literature citations in the respective publications. One of the objectives of this review
is to highlight the connecting points between these disciplines. We thus follow
the key developments more or less chronologically, jumping back and forth from
one discipline to another.

Two phenomenological aspects, and their explanation, are of primary interest:
(1) the demonstration that carbons are amphoteric solids, albeit quite complex
ones (see Section II), and (2) the analysis of the effects of pH on uptakes of weak
electrolytes. The studies in which the latter effect is merely noted are reviewed

first. (It should be noted also that there are too many published studies in which the pH at which the adsorption tests were carried out is not even mentioned, let alone controlled.) The much more consequential studies in which both effects are explored and/or clarified are then discussed in some detail.

Mattson et al. [24] noted a decrease in p-nitrophenol uptake with increasing pH from 2 to 7, but attributed it mainly to increasing solubility rather than electrostatic repulsion. Vaccaro [335] noted that the pH effects "originate from a variety of mechanisms, many of which remain inadequately understood" but chose not to address the relevant issues. Zogorski et al. [446] noted a decrease in the rate of adsorption of substituted phenols as pH increased. Peel and Benedek [407] noted that the "influence of pH is complex and potentially could have . . . a significant effect on the results of studies conducted in the vicinity of the pK_a." Fuerstenau and Pradip [409] reported a small but consistent decrease and a small but consistent increase for two coals, with pH_{PZC} values of 3.6 and 4.6, respectively, in the uptake of o-cresol ($pK_a = 10.2$); they dismissed these trends by stating that "no effect of pH on adsorption was observed" and that "[t]his is not surprising since o-cresol is a neutral molecule and presumably adsorption is occurring through hydrophobic interactions of the benzene ring with the coal surface." Puri [354] discussed the decrease in the pH values of the aqueous suspensions of oxidized activated carbons and its effect on the uptake of ammonia but did not consider this effect in the adsorption of aromatic species such as aniline. McGuire et al. [358] reported that as the pH was increased from 1.9 to 7.7 the uptake of phenol also increased slightly. Sheveleva et al. [420] noted that "the increase in the pH of the medium favours the ionization of weak organic acids" which in turn "should suppress the adsorption." Grant and King [417] reported a large suppression of reversible phenol uptake as the pH increased from 1.8 to 12.1, but they attributed this primarily to a solubility effect ("phenolate anion has more affinity for the aqueous phase than does phenol"); electrostatic repulsion was not associated with the charge on the adsorbent surface but rather to "anions in the surface layer." Vidic et al. [423] reported similar results for 2-chlorophenol uptakes under anoxic conditions but invoked the same unconvincing explanation offered by Rosene and Manes [622], "in terms of physical adsorption of the undissociated acid," which does not recognize the importance of carbon surface chemistry (see below). Nakhla et al. [427] also studied the effect of pH on activated carbon adsorption of phenolics in both oxic and anoxic systems. Surprisingly, they state that "[l]ittle is known about the influence of parameters such as pH . . . on the retention capacities" of activated carbons (under conditions when surface-promoted chemical reactions such as oxidative coupling are possible). Their findings confirmed those of Grant and King [417] and Vidic et al. [423], and the explanations offered are similar, eluding any reference to the role of carbon surface chemistry, especially under anoxic conditions. Analysis of pH effects on the kinetics of adsorption of phenols by the same research group [430] also ignores the role of surface chemistry. Similar findings and conclusions are reported by Cooney and Xi [426], with some notable exceptions. These authors do state that "ionized forms of species adsorb less effectively to activated carbon than do their undissociated forms," but do not give an explanation for this fact; and they discuss the role of electron-donating substituent groups in accelerating oxidative coupling, an important issue which we address in Section V. Ha et al. [623] studied the adsorption equilibrium of the protein bovine serum albumin (BSA) on surface-modified carbon fibers. They report that "the saturation amount adsorbed is significantly affected by surface characteristics as well as pH." Maximum uptakes were typically observed at intermediate pHs, "near the isoelectric point of BSA." Curiously, however, the authors affirm that carbon fibers usually possess a net negative charge over the pH range of their experiments [3–9]. They do conclude that the "surface charge attributed [to] the surface functional groups" or the "dissociability of the surface functional groups" is one of the "controlling factors for BSA adsorption," but they leave the reader somewhat confused about how exactly such control is exerted. Longchamp et al. [624] studied the pH effects on the adsorption behavior of plasmin (Pln, an enzyme) on the surface of nujol/graphite paste. They also concluded that "maximal efficiency in the adsorption of Pln was attained at pH 7.7," which is close to the isoelectric point of the "major fraction of Pln," and that "the local isoelectric potential distribution of the supporting surface is also of concern" but did not elaborate on this latter point. Dargaville et al. [431] state that the "pH of the adsorbent/ adsorbate system could be critical in determining the amount adsorbed" and that "the affinity for adsorption increases (hence larger amounts adsorbed) as the acidity of the adsorbate increases," but they neither provide an explanation for these effects nor do they direct the reader to the appropriate references where such explanations are offered.

Moreno-Castilla et al. [69] have shown that the adsorption of substituted phenols on activated carbons depends on solution pH. Thus at acidic pH the amount adsorbed remained practically constant or increased slightly with increasing pH. When the pH increased further, there was a decrease in the amount adsorbed; the pH at which this decrease took place depended on the difference between the external and internal surface charge density as measured by electrophoretic and titration measurements, respectively. A sharp turn toward a more substantive discussion of *coupled* pH and surface chemistry effects thus occurred in the mid-1990s, and these publications are analyzed below. The seeds for such a discussion were planted much earlier, however, and we analyze first how and why it took decades for them to flourish.

As early as 1932, Miller [625] stated that "[i]t is clearly possible to have charcoal either positively or negatively charged in oxygen, depending upon the conditions of the previous heat treatment." A few years later, in a paper whose dubious distinction is that it has never been cited, in almost 40 years (according to the Science Citation Index, 1961–1999), Frampton and Gortner [55] studied charcoals using electrokinetic techniques. They did not report the variations in the point of zero charge as a function of charcoal pretreatment (e.g., H_2 at 1000°C, N_2 at 900°C, HNO_3, etc.); instead they measured the electrophoretic mobilities in water and found them to be all negative (of the order of 3–4 μm/s/V/cm) and insensitive to treatment conditions. They concluded therefore that (1) "electrokinetic data do not indicate a difference between 'basic' and 'acidic' charcoal" and (2) "apparently the adsorption of weak electrolytes from aqueous solution on charcoal is molecular rather than ionic." In the first edition of *Active Carbon*, Hassler [3] cites this work to affirm that the "electrokinetic behavior of carbon does not appear to be altered by the adsorption of weak organic acids or bases;" in the subsequent edition [5], however, the same statement is made without any references. Also intriguing is the omission from this second edition of a graph showing the influence of pH on the electrophoretic mobility and the amphoteric character of activated carbons. In fact, the relevant changes between the two editions appear in hindsight to have hindered our understanding of electrostatic interactions (see pp. 89–92 in the former and pp. 231–233 and 360 in the latter), because for a long time henceforth (see below), when surface charge has been invoked, if at all, carbonaceous adsorbents have been assumed to possess only a negative charge.

As late as 1968, Snoeyink and Weber [626] affirmed, based on electrophoretic mobility measurements, that "active carbon has a negative surface potential" and speculated on the possible origins of positive surface sites for reaction with acids (hydronium ions). In a comprehensive treatment of solid acids and bases, however, Tanabe [627] lists "charcoal heat-treated at 300°C" as an acid and "charcoal heat-treated at 900°C or activated with N_2O, NH_3, or $ZnCl_2-NH_4Cl-CO_2$" as a base. The role of acidic surface groups in promoting adsorption of cations was emphasized in the early important study of Tomita and Tamai [628], but it wasn't until the mid-1990s that the origin of the positive charge was discussed in the context of the adsorption of organic compounds from aqueous solutions. Indeed as late as 1981, Hingston [311] has reviewed anion adsorption and has not considered the positively charged carbon surfaces to be important.

Of course, surface charge development has been well documented in the carbon black literature [36], where particle coagulation is the key issue. Thus, for example, Kratohvil and Matijevic [56] measured electrophoretic mobilities and concluded (without alluding to the work of Frampton and Gortner) that "[c]arbon particles carry a positive charge at lower pH values." They also studied the pH dependence of the adsorption of a nonionic and an anionic surfactant as well as of electrolytes with counterions of varying charges.

Murata and Matsuda [629] compared the electrophoretically determined zeta potentials and pH values of a series of carbon blacks (see Fig. 24) and correlated them with the concentration of acidic surface functional groups. Similarly, Tobias

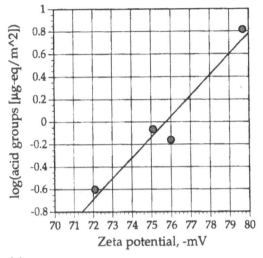

(a)

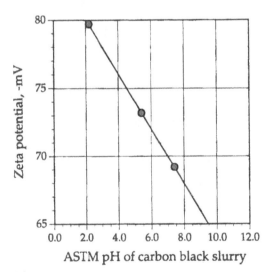

(b)

FIG. 24 Relationship between the surface chemistry and the electrochemistry of commercial and chemically pretreated carbon blacks: (a) correlation between the concentration of acidic functional groups (determined by titrations) and zeta potential (determined by electrophoresis); (b) correlation between zeta potential and the slurry pH. (Adapted from Ref. 629.)

and Soffer [630] found a correlation between the immersion potential (equivalent to the potential of zero charge given vs. a reference electrode) and the degree of surface oxidation of a furnace carbon black. In another key contribution from the carbon black literature, Lau et al. [70] have demonstrated the importance of positively charged carbon surfaces as well as the often subtle role of the ubiquitous surface oxygen in the electrokinetic behavior of carbons. Their results are reproduced in Fig. 25.

Since the 1980s a lively discussion on the role of electrostatic interactions in adsorption (or vice versa) has been developing in the electrochemistry literature [631] in conjunction with the key role of surface-modified carbon electrodes in such diverse applications as electroanalysis [632,633], electrocatalysis [634–636], and in-vivo voltammetry [637]. Indeed the field of "environmental electrochemistry" is now emerging [638], and carbon materials have much to offer in it. The importance of surface chemistry in electroadsorption had been anticipated

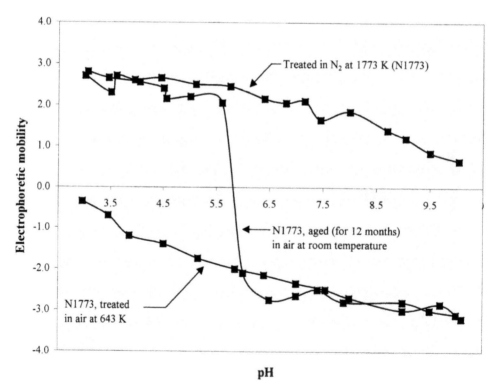

FIG. 25 Electrophoretic mobility (μm/s per V/cm) vs. pH for a series of chemically modified and weathered carbon blacks. (Adapted from Ref. 70.)

by Panzer and Elving [639]: "[A]n appreciation of the constitution of graphite electrodes in terms of surface compounds and of the possible interactions of these compounds with solution species or their reaction products, when the electrodes are used in electrolytic processes, is essential for their optimum utilization." An early study by Randin and Yeager [640] did not highlight the need to determine the point (or potential) of zero charge of the carbon but did emphasize "the pH dependence of the oxidation and ionization states of these functional groups." Similarly, as illustrated in Fig. 26, Fabish and Hair [641] emphasized the inverse exponential relationship between the average work function and the slurry pH of carbon blacks.

The key papers in this regard are those of Oren and coworkers [642–646], in which the importance of the pH_{PZC} is emphasized. They were discussed in Section III.B. Representative results of Cohen et al. [645] are summarized in Table 18. Of special interest here is the coupling of adsorption, especially of polar aromatic compounds, with the electrochemical reactions on (electro)chemically modified carbon surfaces. This issue has since been considered by many investigators [647–652,493,653–669]. However, the factors governing the propensity for and the extent of adsorption have not been discussed in much detail, except to make the (often vaguely stated) point that the solution pH and the nature of the carbon surface are important. The studies of Vasquez et al. [650] and Woodard et al. [493] have gone farthest in this direction; they are discussed in more detail in Section VI. The investigations summarized below serve to illustrate both the

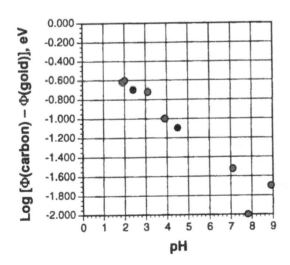

FIG. 26 Relationship between the work function (Φ) and the pH of a series of carbon blacks. (Adapted from Ref. 641.)

TABLE 18 Saturation Alumina Uptakes by Chemically Different Fibrous Electrodes

Potential (mV vs. SCE)	Estimated point of zero charge (mV)	Amount adsorbed (g/g)		No. of monolayers	
		550	−200	550	−200
Carbon felt	150	0.58	0.56	8.8	8.4
Oxidized carbon felt[a]	350	0.44	0.63	6.6	9.5
Graphite felt	−250	0.09	0.21	1.3	3.1

[a] For 1.5 h in H_2SO_4/HNO_3 at 100°C.
Source: Adapted from Ref. 645.

challenges and the opportunities for solving important scientific and technological problems by combining the efforts of carbon surface scientists, adsorption experts, and electrochemists.

Deakin and coworkers [647,651] analyzed the effect of pH on the redox behavior of the $Fe(CN)_6^{3-/4-}$ couple and the mechanisms of oxidation of catechols and ascorbic acid at carbon electrodes. They concluded that "some [electrode] property other than microscopic area changes or electrostatics plays a role in the observed effects," and suggested that "the hydrophobic nature of unactivated surfaces is a major contribution to the observed results."

Michael and Justice [654] studied the oxidation of catechols at carbon fiber disk electrodes. Based on the fact that at neutral pH the amine group in dopamine is cationic and the finding that dopamine was specifically adsorbed on the carbon surface, they suggested an "electrostatic scheme of adsorption" in which there is "a more specific chemical interaction of the amine side chain with the electrode surface." They finally noted that the "[v]erifiction of this idea would require knowledge of the point of zero charge for the particular carbon surface being used." The specific adsorption concept at carbon electrodes was endorsed by Saraceno and Ewing [657] in a similar study of catechol adsorption.

Bodalbhai and Brajter-Toth [655] undertook a closer investigation of the effect of electrode treatment on small adsorbing biological molecules, such as 2,6-diamino-8-purinol (DAPOL), in order to "shed light on the relative importance of . . . adsorption" in their electrochemical behavior. On the basis of effects such as the pH dependence of adsorption rate, they concluded that "the chemistry of the surface may be important."

Eisinger and Keller [670] discuss electrosorption, or potential-swing adsorption, not as an ancillary effect in the performance of electrodes but as a process in its own right, e.g., for the removal of dilute organics from salt-containing aqueous solutions. They highlight the virtue of high-surface-area and high-electrical-conductivity carbon materials for such applications. Their results for ethylenediamine (EDA) removal as a function of potential are summarized in Fig. 27. They attribute the enhancement of adsorption at negative potentials to an electrostatic effect, "even though the species most readily adsorbed from solution is the undissociated EDA." No evidence is provided for the latter statement, but they go on to postulate that a "dipole effect will induce a partial positive charge in the nitrogen atoms" and "the charged nitrogen atoms will then be more strongly attracted by a negatively-charged [sic] carbon surface." In contrast, in an explicit effort to "define better the characteristics of an active carbon surface," Allred and McCreery [665] concluded, in the case of the aromatic catechols on fractured glassy carbon electrode surfaces, that the adsorption "is not charge specific and probably involves the catechol ring rather than the side chain." These results, together with those of McCreery et al. [671], are discussed in Section IV.B.2. Similar contrasting results for the adsorption of aromatic vs. aliphatic amines were reported recently by Surmann and Peter [672]. A possible reconciliation of these effects, in terms of the relative importance of electrostatic vs. dispersion interactions, is discussed in Section V. In an attempt to "counter the problem of protein adsorption during voltammetric measurements," Downard and Roddick [667] used differential pulse voltammetry to examine the effect of electrochemical pretreatment on the adsorption of proteins on glassy carbon electrodes. The results for bovine serum albumin (BSA) as the model protein are summarized

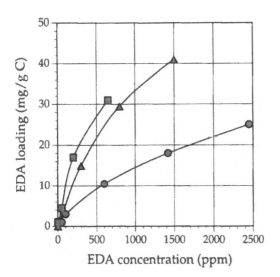

FIG. 27 Adsorption isotherms for ethylenediamine on a commercial activated carbon (Nuchar WV-M, Westvaco): ●, no electric potential applied; ▲, −0.5 V; ■, −1.0 V. (Adapted from Ref. 670.)

in Table 19. In contrast to acid pretreatments, "pretreatments in base . . . are effective in reducing albumin adsorption." The authors correlated this behavior to the finding that "[p]retreatments in acid and base give significantly different carbon surfaces" and concluded—by invoking the review of McCreery [673]—that the electrochemical treatment "which leads to the formation of a graphite-oxide surface increases BSA adsorption," without suggesting a mechanism how this might occur. These authors further argued that treatment in base (1 M NaOH) "increases the density of surface oxides (compared to that of a polished electrode) without the formation of [electrochemical graphite oxide]," but they did not provide evidence for this puzzling statement.

The jury seems to be still out on both the precise role of carbon surface chemistry in electrode processes and its influence on adsorption phenomena on electrodes, and a more comprehensive scrutiny of this issue is long overdue.

The key missing point in these discussions among electrochemists was available since the early 1980s in the important work of Müller and coworkers [523–525] regarding pH effects on the adsorption of weak organic electrolytes. Some of their key results applied to p-nitrophenol [674] are summarized in Fig. 28, while a variation of their model for single-solute systems is discussed in Section V. It is seen that "certain combinations of solute pK, solute concentration, and surface charge" are predicted to result in "an adsorption maximum with hydrogen ion concentration" [523], this being a frequently reported experimental finding. In particular such "adsorption maximum is found in a pH range where the

TABLE 19 Effects of Acid-, Base-, and Electrochemical Pretreatment of Glassy Carbon Electrodes on the Adsorption of Bovine Serum Albumin

Pretreatment	Ratio of anodic peak current in presence of BSA to anodic peak current in absence of BSA
None	0.80
One activation cycle in acid[a]	0.74
Three activation cycles in acid	0.57
Acid pretreatment for 90 s at 1.8 V	0.21
Base pretreatment at 1.0 V, 5 min[b]	0.69
Base pretreatment at 1.4 V, 5 min	0.73

[a] 30 s at 1.8 V followed by 15 s at −0.2 V in 0.1 M H_2SO_4.
[b] Potential scan (in 1 M NaOH) at 100 mV/s from 0 V to upper limit shown.
Source: Ref. 667.

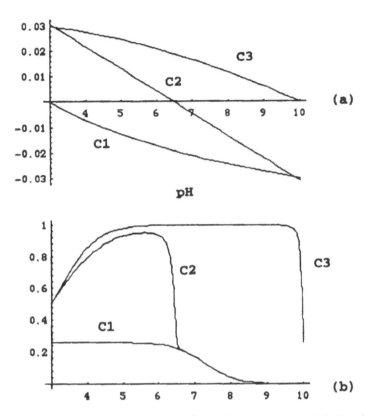

FIG. 28 (a) Surface charge (in C/m²) vs. pH for three (hypothetical) activated carbons: C1, "acidic" carbon; C2, typical as-received carbon; C3, "basic" carbon. (b) Effect of pH on the *p*-nitrophenol surface coverage of carbons having different surface chemistry. (From Ref. 674; see also Fig. 32 and Section V.)

solute exists mostly as a counterion and where the effect of pH on ionization of the solute is stronger than that on the decrease of the magnitude of the surface charge" [523]. Clearly, the emphasis here is on electrostatic interactions. The authors acknowledge that "[a]romatic solutes adsorb onto activated carbon by interaction of the aromatic ring with the carbon surface" and that "changes in the electron density of the aromatic ring affect the affinity to adsorb" [524], although their reference for the latter statement should be the work of Coughlin and coworkers [450,331,332] and not the cited study of Mattson et al. [24], as discussed in Section VI; nevertheless, the adjustable parameters in their model are not subjected to a scrutiny that would reveal the importance of this nonelectrostatic contribution to the overall adsorption potential. The dominant effect of electrostatic interactions was illustrated by showing (1) a dramatic decrease in the uptake of benzoic acid (pK_a = 4.2) as the pH was raised from 4.5 to 11.0 for a carbon whose pH_{PZC} was 6.7; (2) a gradual increase in the uptake of aniline (pK_a = 4.2) by the same carbon as the pH was raised from 2.0 to 10.8; and (3) a gradual decrease in the uptake of p-nitrophenol (pK_a = 7.3) as the pH was raised from 2.05 to 10.6 for a carbon whose pH_{PZC} was 5.8.

In the mainstream literature on the adsorption of organics, on the other hand, this appreciation of the importance of electrostatic interactions, and its further development, was a very slow process. It had been hampered by a lack of knowledge of the key aspects of the chemistry of the carbon surface and by efforts to find a direct relationship between specific features of this chemistry and carbon's adsorptive properties. In the pioneering work of Snoeyink and Weber [675], it was concluded that "[t]here is strong indication of the need for continuing research on the definition of the nature of surface functional groups, and on the particular conditions of preparation which will produce an active carbon best suited for a specific application." Snoeyink et al. [333] were somewhat more specific a few years later, after establishing that the uptakes of both phenol and p-nitrophenol are suppressed as pH increases; but they still concluded that the "precise reasons for the capacity differences [of a coal-based carbon vs. a coconut-shell carbon] with respect to the two solutes are not known." Thirty years later, a much better understanding of surface chemistry has been achieved, but, as argued below, it was not necessary to wait for its development in order to be able to optimize the adsorption of organics on activated carbons.

Chapter 7 of the landmark review by Mattson and Mark [6] had emphasized the pH dependence of the uptake of acetic acid, n-butylamine, substituted benzoic acids, anionic benzenesulfonates, and methylene blue. In particular, the suppression of methylene blue uptake at pH below the isoelectric point of the carbon was explained by stating that "the formation of positive surface sites in that region certainly would not enhance cationic adsorption." The important work of Getzen and Ward [676,677] is also cited, in which the concept of the isoelectric point (or point of zero charge) is not mentioned—presumably because their work is based in turn on the historically significant concept of hydrolytic adsorption—but which clearly emphasizes the importance of electrostatic adsorbate/adsorbent interactions: "There is a distribution of residual forces on the surface of the charcoal (which may be varied by changes in pH) such that they may exert an attraction for both electropositive and electronegative ions as well as for the uncharged molecules . . . They are, in the case of ions, the coulombic forces and, in the case of uncharged molecules, van der Waals and similar forces."

Even though Getzen and Ward state that the "existence of such forces is recognized today and their description is generally understood" [676], the subsequent three decades of published work in this area have shown too many examples to the contrary.

Wang et al. [607] studied the adsorption of dissolved organics from industrial effluents onto a commercial activated carbon. As illustrated in Table 20, they place emphasis on the pK_a, pK_b or isoelectric point of the adsorbate and state that "the pH effect upon the effectiveness of carbon adsorption mainly depends upon the nature of the adsorbed substance." Based on their own work and analysis of the literature, they postulate that maximum adsorption of organic acids and bases occurs around their respective pK_a or pK_b value, even though they acknowledge, at least "as the ionic organic compounds become more complex," that "electrostatic adsorption forces between the adsorbent and the ionic adsorbate appear to govern."

The historically popular concept of hydrolytic adsorption, which in fact obscures the key role of carbon surface chemistry, has often been used to account for the pH effects. Thus, for example, Rosene and Manes [622] used a Polanyi-based model of competitive adsorption between benzoic acid and sodium benzoate. They criticize the approach taken by Ward and Getzen [677], especially with regard to anion interaction with a positively charged surface, and essentially ignore the amphoteric character of the activated carbon.

As a forerunner to the landmark publications of Müller and coworkers [523–525], Jossens et al. [678] presented an ideal-adsorbed-solution (IAS) model for

TABLE 20 Adsorption Results for a Range of Organic Compounds on a Commercial Activated Carbon[a]

Organic compound	pK_a(or pK_b)	pH_{IEP}[b]	pH range	pH_{max}[e]
Benzoic acid	4.20	—	3;7;11	3
Phenoxyacetic acid	3.03	—	3;7;11	3
2,4-Dichlorophenylacetic acid	3.92	—	3;7;11	3
Phenol[c]	9.89	—	2;5.6;7.5;10.6	7.5
Nicotine[c]	10.9	—	1–12	8.2–12
Urea[c]	13.9	—	7–11.8	11.8
Albumin[d]	—	4–5	2–11	4–5
Gelatin[d]	—	4–5	2–11	4–5
Sulphanilamide[d]	—	~6	1.3;5.9;12.2	5.9

[a] Aqua Nuchar A (Westvaco).
[b] Isoelectric point of the adsorbate.
[c] Organic base.
[d] Ampholyte.
[e] pH at which maximum uptake was observed.
Source: Ref. 607.

adsorption of aromatic acids from dilute aqueous solutions. However, they detected "systematic deviations between calculated and observed results," which they attributed to "systematic effects of acidity on adsorption" and thus realized the importance of performing adsorption experiments under controlled pH conditions. In a similar approach to the prediction of phenol and p-nitrophenol uptakes, Myers and Zolandz [679] did control the pH but they did not account explicitly for the nature of the carbon surface. Some of their results are reproduced in Table 21. Based on the significant discrepancies between experimental and predicted values and especially the failure of the IAS theory to predict the uptake maxima at intermediate pH, they concluded "that electrostatic interactions between the surface and sorbate ions are very important at high pH values, but that specific physical interactions become more important with decreasing pH."

The studies by Martin and coworkers [680–683,492,684] are also instructive. They first inappropriately cite the work of Wang et al. [607] for the statement that "[i]t is now generally agreed that carbon surface properties can change with pH" [682]; as mentioned above, these authors place emphasis on changes in solute charge with pH and incorrectly assert that activated carbon "is generally accepted as being negatively charged." Martin and Al-Bahrani [682] echo this last misconception in their study of the adsorption of 2-methylpyridine ($pK_b = 8.03$) and o-cresol ($pK_a = 10.2$) on a commercial activated carbon; it has since been well documented [171] that many activated carbons, including some from the same company (Calgon Carbon Corporation)—see Table 2—have very high

TABLE 21 Experimental and Predicted[a] Uptakes of Phenol and p-Nitrophenol by a Commercial Activated Carbon[b] as a Function of pH

pH	PNP conc. (mmol/L)	Phenol conc. (mmol/L)	PNP uptake (mmol/g)		Phenol uptake (mmol/g)	
			Exp.	Predicted	Exp.	Predicted
2.11	0.0011	0.188	0.243	0.240	0.409	0.399
2.13	0.0029	0.054	0.476	0.485	0.110	0.092
2.16	0.0055	0.214	0.445	0.489	0.148	0.236
9.27	0.0035	0.145	0.465	0.0424	0.830	0.484
9.28	0.0008	0.134	0.244	0.0158	0.933	0.524
9.32	0.0023	0.012	0.477	0.0854	0.530	0.015
11.98	0.0187	0.213	0.0663	0.0361	0.149	0.316
12.01	0.0405	0.213	0.0941	0.0711	0.155	0.271
12.02	0.0367	0.060	0.132	0.120	0.0525	0.104

[a] Using the ideal adsorbed solution (IAS) theory.
[b] Darco (ICI Americas).
Source: Ref. 679.

pH_{PZC} values. The uptake of both adsorbates decreased as the pH was lowered from 6.8 to 5.0 (for 2-methylpyridine) or 3.5 (for o-cresol), while it remained essentially constant as the pH was increased to 9.5. The latter finding was rationalized by saying that "both ionised and unionised molecules of both compounds were adsorbed on activated carbon." The rationale for the former finding was more convoluted, however; a much more straightforward explanation is offered in Section V. A subsequent paper by Martin and Iwugo [684] is more enlightening. It has not been cited much (5 non-self-citations in 15 years), and it fails to refer the reader to the study of Müller et al. [523]. It does not mention the influence of pH on surface charge; but it does contain the following key statement: "[I]onization of adsorbates, changes in electrical attractive or repulsive forces between adsorbate and adsorbent and chemical reactions could all affect the adsorption characteristics of organics."

A contemporary study by Jayson et al. [307] does cite the paper by Müller et al. [523] and thus combines electrophoretic measurements on an activated carbon cloth (ACC, $pH_{IEP} = 2.7$) with the analysis of pH effects on uptakes of triethyl and dimethyl ($pK_a = 2.12$) phosphates. Table 22 summarizes the Langmuir monolayer capacities obtained in an effort to unravel "the controlling mechanisms governing adsorption" of these pesticides. While the low values of adsorption heats (<5 kJ/mol) suggested that "the adsorption process is purely physical

TABLE 22 Effect of pH on the Uptake of Organic Phosphates by an Activated Carbon Cloth[a]

Adsorbate	pH	Temperature	Monolayer capacity[b], $\mu mol/g$
Triethyl phosphate (TEP)	2.7	293	2420
		313	2420
	6–7	293	2450
		313	2370
	10.5	293	2420
		313	2380
Dimethyl phosphate (DMP)	2.7	293	344
		313	315
	6–7	293	107
		313	103
	10.5	293	91
		313	82

[a] ACC, Chemical Defence Establishment (1300 m²/g, pH=6.0).
[b] Langmuir constant.
Source: Ref. 307.

in nature,'' the pH dependence of the ionizable DMP uptake was concluded to ''arise because of a combination of [its] pK value and the charge on the charcoal cloth surface in aqueous media.'' More specifically, the maximum uptake of DMP at the isoelectric point of the adsorbent was attributed to the facts that ''coulombic repulsion between the (normally) negatively charged ACC surface and DMP anions in solution no longer occurs'' and ''[b]oth molecular and ionic species of DMP are, therefore, capable of adsorbing.'' A small decrease in the uptake at pH = 2, where electrostatic attraction between the positively charged surface and a few remaining anions could have been expected, was rationalized by postulating the interfering preferential adsorption of the more abundant chloride ions.

Because they span the period of time before and after the publication of the paper by Müller et al. [523], it is interesting to follow the development of arguments of Abe and coworkers [685–691] about the ''adsorptive mechanism on activated carbon in the liquid phase'' [686,687]. In their early studies the pH of the aqueous solutions used was not specified. Apparently following to-this-date popular trends in the adsorption in soils [692,397,7,8] and in the use of linear free energy relationships [405,680,693,694,79,695–700,12,701,702]—where the general applicability of octanol–water partition coefficients and other structure-property relationships is being scrutinized—emphasis is (mis?)placed on adsorbate properties such as molecular weight, molar attraction constant, and a ''modified organicity/inorganicity value,'' and it is concluded that ''the adsorption of the hydrocarbon portion of the [mostly aliphatic monofunctional compounds] takes place by means of the precipitation of the hydrocarbon portion of the solute from the aqueous solution and that the adsorbability depends mainly on the volume of the hydrocarbon'' [687]. In a later study of the adsorption of amino acids [689], the authors acknowledge that ''no special effort [was] made to control [the] pH,'' but they insist that the ''adsorption isotherms of amino acids can be predicted from the molecular refraction of the parachor alone.'' They do cite the study of Müller et al. [523] to state that ''the solution pH influences the adsorption of ionized solutes'' and concede that their ''prediction methods may not be applicable for adsorptions at pH values extremely removed from the isoelectric point'' (presumably of the adsorbates).

Similar analysis can be made of the series of contemporary papers by Urano and coworkers [703,408,594,704], in none of which has the work of Müller et al. [523] been mentioned. In an early paper on the adsorption of strong acids and bases [703], the authors provide a very useful determination of the pH_{PZC} of a commercial activated carbon. Subsequently, uptakes of a large number of both aliphatic and aromatic adsorbates by five commercial carbons were reported, but among the properties of the adsorbents there was no mention of their pH_{PZC} nor of the pH of the solutions [594]. Instead, and surprisingly, emphasis was placed on the adsorbent's physical surface properties, and the authors concluded that they can "predict the adsorption capacities of organic compounds in aqueous solutions for any activated carbon whose pore size distribution is known." Also surprising is the introduction of new arguments regarding the mechanisms of adsorption in the later papers [704,599], even though many of the same adsorbates are reconsidered. In particular, the authors [704,599] emphasize the "π-electron combination on the carbon surface" (see Sections IV.B.2 and VI). They acknowledge the influence of "several electrophilic groups ... on the surface of activated carbon" [599] and that "the influence of the pH on the adsorptions of organic compounds in water has not been made clear" [704] but then attribute the decrease in uptakes of weak anionic electrolytes with ionization in basic solutions only to the fact that "the apparent solubilities of the ionized forms increase much more than those of the non-ionized forms" [704]. They emphasize the *sui generis* behavior of phenolic compounds and conclude, rather unconvincingly, that "the deviation of phenol [from predictions based on Freundlich isotherm parameters for the nonionized and ionized species] may be due to the combination of the hydroxide ion with the electrophilic groups on the surface of activated carbon."

An ambitious theoretical analysis of the "selective adsorption of organic homologues onto activated carbon from dilute aqueous solutions" [705] is worth mentioning at this point. Altshuler and Belfort [705] conclude that "dispersion forces dominate during activated-carbon adsorption" (see Section IV.B.2), especially for branched alkyl alcohols, but they also mention the potential importance of a "second type of surface interaction which occurs at the more reactive edges of the microcrystallites [and] can be characterized by positive physical (and maybe even chemical) attractive interactions due to hydrogen bonding and electrostatic forces." Obviously, the electrostatic interaction (either repulsion or attraction) between the positively charged basal plane surfaces (see Section II) and organic adsorbates has been an elusive concept to grasp (see Section V). Indeed, in a theoretical discussion in print [706] that focused on treatment of water by granular activated carbon—despite the optimism expressed by the discussion moderator—there was much talk about concepts and approaches such as the octanol–water partition coefficients, ideal adsorbed solution theory, and Dubinin–Radushkevich theory, but the only consideration of surface chemistry was disguised in the concept of "heterogeneity," while the (by then clarified) importance of electrostatic interactions [523] remained buried in the exponentially growing literature on the subject.

Both practical and fundamental implications of the "influence of pH on the removal of organics by granular activated carbon" were discussed by Semmens et al. [707]. Their results on the purification of the coagulated Mississippi River water are reproduced in Table 23. Operation at low pH "improved the performance of the GAC columns and resulted in lower organics concentrations in the

TABLE 23 Effect of pH on the Column Breakthrough Time for
Organic Compounds Adsorbed on a Commercial Activated Carbon[a]

Adsorption pH	Breakthrough time (in h) determined from TOC[b] analysis	Breakthrough time (in h) determined from analysis of UV-absorbing organics
8.7	47	46
7.0	106	96
5.0	225	225

[a] Nuchar WV-G (Westvaco).
[b] TOC = total organic carbon.
Source: Ref. 707.

effluent." These "difference[s] may stem from the increased solubility of the
ionized species or may be explained in terms of the mechanism of adsorption."
The authors went on to identify the key controlling factors: "As the pH is increased, the carbon surface will become more negative, and weakly acidic groups,
such as phenolic groups, on the organics may dissociate, rendering the organics
more negative. This will tend to bring about a reduction in adsorption capacity
as the pH is increased." This led them to the practically important conclusion
that the "combined effects of pH on coagulation and on adsorption highlight the
need for effective pH control."

By this time, a brief review from an authoritative industrial source [708] had
identified the reason for the fact that "quite a number of phenomena [including
adsorption of polar substances and electrochemical properties] cannot be explained at all with the usual carbon characterizations" and had postulated that
"the presence and the configuration of oxygen-containing functional groups are
responsible for the differences hitherto not explained." The author briefly discussed the zeta potential but concluded, rather anticlimactically, that oxidized
carbons adsorb a polar compound such as methanol in preference to toluene and
in proportion to the amount of surface oxygen present. In a more comprehensive
treatise on activated carbons, Bansal et al. [35] do not do justice to this issue:
their 17-page section on the removal of organics from water (in the active carbon
applications chapter) cites only 13 papers from the pre-1983 period, none of
which illustrate the important role of pH in the adsorption of weak organic electrolytes. The authors point out "the importance of the surface chemistry of activated carbons for [the] removal [of organics]," but how exactly and which features of the surface chemistry are important was not settled. Even in the more

fundamental chapter on the characterization of active carbons, electrostatic interactions are mentioned only in passing, as they relate to heat of immersion measurements. In this respect, Kinoshita's monograph [36] is more enlightening: it contains a detailed discussion of the zeta potential and surface charge, but not in the context of liquid-phase adsorption phenomena.

After analyzing the recovery of acetic acid from aqueous solutions on both activated carbons and polymeric adsorbents, Kuo et al. [582] concluded that the "effect of pH of the aqueous phase upon capacity can be interpreted through a two-solute competitive adsorption model, involving ionized and un-ionized species." These authors cite the first study of Müller et al. [523]—though not the follow-up papers [524,525]—but the context in which it is cited obscures the key point (pH influence on the surface charge): "A model allowing for surface heterogeneity, such as that of Müller et al. [523], may indeed fit the data better but will contain an additional adjustable parameter." The authors discuss the nature of carbon surface chemistry and even make the very interesting statement (see Section V) that "the effects of particular groups upon adsorption from solution depend on the relative donor–acceptor capabilities of all species involved." But then they dismiss the importance of carbon surface chemistry by stating that "acetic acid concentrations were high enough so that surface charge should have had only a minor effect." Instead they place emphasis on the heteroatom content $(O + S + N)$ as "a crude measure of the amount of surface functionality" and show a rough correlation between its increase and the increase in selective acetic acid uptake.

In another study that underestimates the importance of carbon surface chemistry, Helmy et al. [709] set out to provide a "pH/pK$_a$ relationship [that] permits the individual isotherms to be obtained for the charged and neutral sorbate species." They studied the uptakes of quinoline (pK$_a$ = 4.9) and 8-hydroxyquinoline (pK$_a$ = 5.0) on a commercial charcoal: the former reached a plateau as pH increased from 2 to 7, while the latter exhibited a maximum at pH = 6 and decreased thereafter. The authors' theory led them to conclude that "the surface of charcoal prefers the neutral over the charged molecule," which was confirmed by noting that their respective isotherms were of type I and type III. Intriguingly, the reasons for this preference were discussed only in terms of the repulsion between charged adsorbate species, while the electrostatic adsorbate–adsorbent interactions were ignored. Not surprisingly, the work of Müller and coworkers [523–525] is not cited.

In yet another such study, which does cite the early Müller et al. paper [523], albeit inconsequentially, Cooney and Wijaya [710] present compelling evidence for the crucial importance of electrostatic adsorbate–adsorbent interactions. Figure 29 summarizes their key results: the uptake of aromatic carboxylic acids monotonically decreased with increasing pH, while the uptake of aniline did ex-

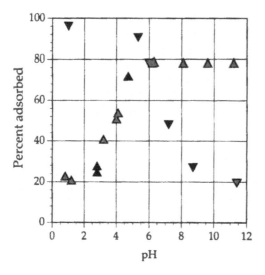

FIG. 29 Effect of pH on the uptake of benzoic acid (▼) and aniline (▲) by a commercial activated carbon. (Adapted from Ref. 710.)

actly the opposite. The authors do not mention the source of the activated carbon used. They correctly emphasize that "[o]ne must also consider the effect of pH . . . on the nature of the carbon surface," but then dismiss the issue by echoing the misconception that "the surface carries a strong negative charge, as do most activated carbons." So they end up invoking exactly the same argument as Helmy et al. [709]: "electrical repulsion of adjacently adsorbed molecules." They fail to point out, or clarify, the obvious contradiction presented by the aniline uptake results: they do say that "at low pH, where [it has a positive charge], the adsorbabilities are low," but they neither discuss nor explain this. Obviously, at low pHs, which are probably below the pH_{PZC} of the carbon, the electrostatic repulsion between the anilinium cations and the positively charged surface is responsible for the suppressed uptakes (see Section V).

A report on the biomedical application of activated carbon adsorption [600] is also revealing. The authors analyzed the uptakes of an aromatic compound, acetaminophen (active ingredient in Tylenol, $pK_a = 9.5$), and an aliphatic one, N-acetylcysteine (which provides a protective effect against acetaminophen overdose; $pK_a = 3.3$), under both gastric (pH = 1.2) and intestinal (pH = 7.0) conditions. Their results are reproduced in Table 24.

Acetaminophen   N-acetyl-L-cysteine

Apparently unaware of the mainstream discussions in the literature regarding the role of electrostatic interactions (the vast majority of the references being from medical and pharmacological journals), the authors discuss the Langmuirian behavior of both adsorbates and conclude that "the acetaminophen molecule will maximize pi-electron and van der Waals interaction by adsorbing on the surface in a flat configuration" while the "N-acetylcysteine surface configuration seems to change depending on pH conditions and presence of acetaminophen which may block surface sites." The absence of electrostatic interactions in this pH range for the highly basic aromatic compound is not surprising, of course, while the authors do not explain why they "would predict . . . that N-acetylcysteine should be adsorbed more at the gastric than the intestinal pH."

TABLE 24 Effect of pH on the Uptake of Two Biomedically Relevant Solutes by a Commercial Activated Carbon[a]

Adsorbate	pH	Percent adsorbed[b]
Acetaminophen	1.2	97
($pK_a > 7.5$)	7.0	97
N-acetylcysteine	1.2	77
($pK_a = 3.3$)	7.0	54

[a] Nuchar SA (Westvaco).
[b] At 37°C using 25 mL of solution (0.02 M) and 0.5 g of adsorbent.
Source: Ref. 600.

Kamegawa and Yoshida [612] studied the pH dependence of the adsorption of several surface-active agents by chemically modified carbons. The uptake of nonionic polyoxyethylene nonylphenyl ether (PO) was not affected by pH, but it decreased with increasing outgassing temperature of the carbon. The uptakes of both the cationic octadecyltrimethylammonium chloride (OT) and the anionic sodium dodecylbenzenesulfonate (DB) exhibited a shallow minimum at intermediate pHs. The authors concluded that "the van der Waals force was dominant in the adsorption of PO," while electrostatic attraction and repulsion, respectively, contributed to the behavior of OT and DB.

Leyva-Ramos [613] also studied the effect of pH on the adsorption of anionic DB but reported a contradictory result: "The adsorption capacity is considerably affected by pH and there exists an optimal pH, near pH = 7, where the maximum adsorption takes place." The author also reported an uptake increase with increasing temperature, stating that "temperature promotes the adsorption of the detergent" but leaving some doubt regarding the attainment of adsorption equilibrium in the water-filled pores (despite the 100 h equilibration times). The pH effect was attributed, rather vaguely, to "the hydrophobicity of the surfactant, the solubility of the surfactant, and the ions of the buffer solution." Even though the author acknowledges that "the relative charge on the carbon particles depends on pH," the explanations for the conclusion that the "surfactant is bound chemically onto the surface and the uptake is greatly dependent upon pH and temperature" remained ambiguous.

The relevant studies by Mazet and coworkers [611,549,614,601,711] published in a span of six years, are both illustrative and instructive. In the first of these papers, electrostatic interactions—between a powdered activated carbon (PAC) and calcium cations—and zeta potential measurements are discussed, but the emphasis is on the beneficial effect of Ca^{2+} in neutralizing the negative charge on the carbon surface. The last paper focuses much more on the chemical nature of the carbon surface and its impact on adsorption of a wide range of organics. Both positive and negative zeta potentials are reported, with pH_{IEP} values ranging from below 3 to above 6 [711]. Particularly instructive are the results for phenol and benzoic acid—see Table 25—both showing monotonically decreasing uptakes with increasing pH in agreement with the reversal of zeta potential from positive to negative in the same pH interval. The authors attribute these results to the effects of solubility, adsorbate dissociation (ionization), surface charge, and electrostatic repulsion and conclude "that the electric charge of PAC is an important parameter for the adsorption of organic solutes." Table 25 also illustrates the neglected point that, on a given adsorbent, the uptake of phenol is comparable to that of benzoic acid, although the latter is much less soluble (see Section V); the authors mention the "existence of a polarity due to π electrons (in phenol for example)," but they interpret this also as an electrostatic rather than a dispersion effect (see Section V).

TABLE 25 Effect of pH on the Uptakes of Phenol and Benzoic Acid by Chemically Different Activated Carbons

					Adsorption capacity, mmol/g	
Adsorbent[a]	Acidic groups (meq/g)	Basic groups (meq/g)	pH	Zeta potential (mV)	Phenol	Benzoic acid
Heat-treated	0.12	0.377	3.0	14	0.99	1.16
(800°C, 4 h)			5.6	3	1.08	1.06
			7.2	−4	0.93	0.82
As-received	1.05	0.26	3.0	−4	1.02	1.0
AC			5.6	−20	0.78	0.71
			7.2	−28	0.73	0.65
Oxidized	1.92	0.10	5.6	−26	0.43	0.57
(in HNO₃)			7.2	−32	0.37	0.50

[a] AC=Filtrasorb 400 (Chemviron).
Source: Adapted from Ref. 711.

The mechanism of adsorption and the effectiveness of a mesoporous mineral-carbon material was studied as a function of pH by Korczak and Kurbiel [365]. Their important results are summarized in Table 26. The authors concluded that the "interaction between sorbent electrostatic charge and dissociated functional groups was ascertained. An increase in the ion exchange capacity of the acidic functional groups caused a distinct decrease of sorption of phenylacetic and caproic acids which were in the form of anions." Their explanation for the increasing uptake of cationic methylene blue with increasing pH was more convoluted, however, because of the failure to recognize the importance of surface charge. Instead of invoking increasing electrostatic repulsion as pH decreases, the authors argue in favor of "concomitant partial dissociation of the basic groups on the sorbent," but they do not identify these groups. Similarly, Mollah and Robinson [474] reported a decrease in the uptake of pentachlorophenolate anions as the pH was raised from 6 to 11 and noted that this "may have been due to changes in the chemical nature of the [activated carbon] surface at different values of pH," but they failed to be more specific than that.

The important role of electrostatic interactions is clearly seen in the study of Moreno-Castilla et al. [69] which dealt with the adsorption of substituted phenols (whose pK_a ranged from 7.13 for *p*-nitrophenol to 10.17 for *p*-cresol) on a series of well characterized activated carbons, all of which exhibited the expected decrease in uptake at high pH values. Table 27 summarizes the key results. Sample CP-10, with the highest surface area and the most developed porosity, had the highest adsorption capacity for all phenols. However, the pH at which the adsorp-

TABLE 26 Effect of pH on the Uptake of Organic Compounds by a Mineral-Carbon Sorbent[a]

Adsorbate	pH	Adsorbate form	Adsorbent's ion exchange capacity[b] (meq/g)		Uptake (mmol/g)
			Acidic	Basic	
Benzene	2.01	neutral	0.000	0.065	0.0705
	4.30	neutral	0.025	0.000	0.0698
	7.75	neutral	0.475	0.000	0.0678
Phenylacetic acid	2.23	neutral	0.000	0.045	0.0542
(pK_a = 4.31)	4.19	neutral/anion	0.019	0.000	0.0540
	5.40	anion/neutral	0.125	0.000	0.0506
	7.20	anion	0.380	0.000	0.0327
Caproic acid	2.14	neutral	0.000	0.050	0.0437
(pK_a = 4.86)	4.16	neutral/anion	0.017	0.000	0.0420
	5.46	anion/neutral	0.013	0.000	0.0401
	7.70	anion	0.470	0.000	0.0212
Methyl violet	1.82	cation	0.000	0.075	0.0608
	3.95	cation	0.000	0.000	0.0675
	5.78	cation	0.155	0.000	0.7070
	8.10	cation/neutral	0.495	0.000	0.8140

[a] Prepared from spent Fuller's earth (58 m²/g, 0.293 cm³/g, 13.3% microporosity; 0.235 meq/g carboxyl groups, 0.45 meq/g lactone groups, 0.62 meq/g phenolic groups).
[b] Determined using Boehm's titration method (see Ref. 38).
Source: Ref. 365.

tion capacity decrease occurred was found to be dependent on the difference between the external and internal surface charge density, as measured by electrophoresis and titrations (see Section II). Thus, in carbon CP-5 with a pH_{IEP} lower than that of CP-10, the decrease begins at a pH about two units lower than in the case of CP-10; this is a consequence of the fact (see Table 27) that a larger number of negatively charged adsorption sites (located predominantly at the external surface) are generated at lower pH in carbon CP-5. Even though the adsorption model of Radke and coworkers [523–525] is not invoked here, the conclusions that "the variation in the surface charge density of the activated carbons affects their adsorption capacity" and that "the amount of compound adsorbed depends also on the surface charge of the carbon" are indeed a clear demonstration of its validity.

The more recent studies of King and coworkers [712,384] have addressed both the effects of pH and the surface properties on the uptakes of lactic and succinic

TABLE 27 Adsorption of Phenol and Substituted Phenols on Two Coal-Derived Activated Carbons

Sample	S_{N2} (m²/g)	V_2 (cm³/g)	V_3 (cm³/g)	V_{H2O} (cm³/g)	pH_{IEP}	pH_{PZC}	Monolayer uptake (mg/g)			
							phenol	p-cresol	MAP	PNP
CP-5	905	0.10	0.12	0.50	3.4	10.1	196.7	233.0	174.0	256.0
CP-10	1114	0.13	0.14	0.65	5.0	10.4	218.0	289.0	218.4	286.4

Note 1: Textural properties of the adsorbents, S_{N2}, V_2, and V_3 were determined by the BET method, with V_2 and V_3 corresponding to pores within 3.7–50 nm and >50 nm, respectively.
Note 2: Adsorption experiments were performed at uncontrolled pH; reported uptakes are Langmuir constants (MAP = *m*-aminophenol, PNP = *p*-nitrophenol).
Source: Ref. 69.

acid as well as the adsorption and regeneration of phenol. In many cases, uptakes decreased as pH increased and all oxidized carbons "exhibited a significant reduction in uptake at the lower solute concentrations" [384]. Table 28 summarizes the relevant results. The observed uptakes were interpreted as a complex net effect of several potentially conflicting trends: (1) higher extent of oxidative coupling at high pH; (2) lower extent of oxidative coupling on oxidized surfaces; (3) suppressed uptake at high pH "because of an increase in the negative charge on the adsorbent surface and an increased degree of phenol ionization; (4) a decrease in available surface area upon surface oxidation due to preferential adsorption of water; (5) a decrease in the adsorbate–adsorbent interaction energy for oxidized surfaces, presumably as a consequence of suppressed π–π overlap, as originally argued by Coughlin and coworkers [450,331,332] (see Sections IV.B.2 and VI) and not, as the authors state, "proposed by Mahajan et al. [345]." Thus, for example, even though the data shown in Table 28 are not unambiguous, the authors invoke the dominant effect of oxidative coupling for the observation that the "[u]ptake at pH 9 was either higher (F400, RO, CG6) or remained essentially the same (WVB) as that at pH 2" [384]. The authors characterized the surface chemistry of their carbons and cited the work of Müller et al. [524] in this context, but they did not determine the surface charge variations with pH.

The recent studies by Newcombe and coworkers [553,713,554,555,566,536] of the complex fundamental issue of adsorption of natural organic matter (NOM)—see Section IV.A.4—are crucial contributions to our understanding of electrostatic interactions. These authors performed measurements of both the isoelectric point (pH_{IEP}), using electrophoresis, and point of zero charge (pH_{PZC}), using titration, and investigated their relationship with the adsorption and regeneration behavior of several commercial activated carbons. Table 29 summarizes the surface charge characteristics of a virgin and a spent commercial activated carbon. The significance of the differences between pH_{PZC} and pH_{IEP} was discussed in Section II. Of interest here is the decrease in pH_{PZC} with increasing adsorption of humic and fulvic acids, as well as the increasing negative charge of the adsorbent. This result led the authors to conclude [553] that "the surface charge properties of the GAC are largely determined by the adsorbed material, rather than the carbon surface itself." When applying these concepts to chemical regeneration of GAC from an operating water treatment plant, the authors [713] noted that in prior studies "no attempt was made to investigate the surface chemistry of the interactions taking place at the carbon–water interface. As a result the findings have often been poorly understood, incorrect conclusions have been drawn and the possibility of chemical regeneration has not been fully explored." To remedy this situation, the authors [713,554] analyzed the surface chemistry of virgin, spent, and regenerated activated carbon and speculated that the effectiveness of the selected regeneration protocol was due to the creation of "an environment of lower pH within the pore structure, enhancing adsorption of or-

TABLE 28 Freundlich Parameters for Phenol Uptakes by Commercial and Chemically Modified Activated Carbons

Adsorbent	Treatment	pH	K_F[a]	n[a]
F400	As-received	2	51.1	0.198
	As-received	9	47.1	0.223
	Acid-washed[b]	9	75.9	0.168
	AW, H1[c]	9	60.7	0.183
	Ox 2/50[d]	9	56.1	0.207
	Ox 9/70[e]	9	28.6	0.235
	Ox 9/70, H6[f]	9	57.7	0.189
WVB	As-received	2	12.7	0.345
	As-received	9	11.5	0.353
	Acid-washed[g]	9	14.2	0.338
	AW, H1[h]	9	40.7	0.207
	Ox 2/50[i]	9	38.4	0.212
RO	As-received	2	57.4	0.185
	As-received	7	84.7	0.158
	As-received	9	68.7	0.179
CG6	As-received	2	42.6	0.211
	As-received	9	41.3	0.231

[a] Uptakes expressed in $\mu g/m^2$ and equilibrium concentrations in ppm.
[b] Uptakes of HCl, NaOH, Na_2CO_3, $NaHCO_3$ (meq/m^2) = 0.34, 0.14, 0.0, 0.0.
[c] Uptakes of HCl, NaOH, Na_2CO_3, $NaHCO_3$ (meq/m^2) = 0.37, 0.0, 0.0, 0.0.
[d] Uptakes of HCl, NaOH, Na_2CO_3, $NaHCO_3$ (meq/m^2) = 0.29, 0.60, 0.19, 0.05.
[e] Uptakes of HCl, NaOH, Na_2CO_3, $NaHCO_3$ (meq/m^2) = 0.11, 1.71, 0.96, 0.61.
[f] Uptakes of HCl, NaOH, Na_2CO_3, $NaHCO_3$ (meq/m^2) = 0.24, 0.55, 0.22, 0.05.
[g] Uptakes of HCl, NaOH, Na_2CO_3, $NaHCO_3$ (meq/m^2) = 0.18, 0.57, 0.24, 0.12.
[h] Uptakes of HCl, NaOH, Na_2CO_3, $NaHCO_3$ (meq/m^2) ~ 0.26, 0.31, 0.09, 0.0.
[i] Uptakes of HCl, NaOH, Na_2CO_3, $NaHCO_3$ (meq/m^2) = 0.17, 0.76, 0.40, 0.19.
Source: Ref. 384.

TABLE 29 Surface Charge Variations with pH in an Activated
Carbon Upon Adsorption of Increasing Amount of Humic Substances

Sample	pH$_{PZC}$	pH$_{IEP}$	Surface charge, C/m²	
			pH = 7.0	pH = 8.0
Virgin GAC	7.3	3.0	+0.0003	−0.0006
Spent GAC (3.5 mg DOC/g)	6.4	2.7	−0.0019	−0.0046
Spent GAC (10.0 mg DOC/g)	4.5	2.4	−0.0053	−0.0084

Source: Ref. 554.

ganics'' by virtue of ''a large decrease in the electrostatic repulsion between the
surface and the adsorbing species.''

In a more incisive study of the same issue, Newcombe [555] set out to determine ''how surface concentration and pH affect the degree of ionization, the
conformation, and the change in the free energy of dissociation of the adsorbed
material, and how the effects may be explained in terms of the properties of both
the humic material and the carbon surface.'' She reports that ''[t]reatment of
[spent carbon] with NaOH facilitates the removal of the organic matter.'' In order
to understand this finding, surface charge measurements were carried out after
both adsorption and NaOH-treatment experiments, and estimates of the degree
of ionization of the adsorbed material (vs. the adsorbate in solution) were made.
The results are summarized in Table 30. The effective charge of the adsorbed

TABLE 30 Effect of pH on the Surface Charge of an Activated Carbon Before and
After Adsorption of Humic Substances

pH	Charge of adsorbed species (eq/g × 10⁻⁵)	Charge of virgin carbon (eq/g × 10⁻⁵)	Adsorbed species effective charge	Calculated effective surface charge (eq/g × 10⁻⁵)
4.5	−10.2	16.7	0.0	16.7
5.0	−12.0	14.1	0.0	14.1
5.5	−12.7	12.0	−0.7	11.3
6.0	−13.4	10.3	−3.1	7.2
6.5	−13.7	7.0	−6.7	0.3
7.0	−13.7	3.7	−10.0	−6.3
7.5	−13.8	0.0	−13.8	−13.8
8.0	−14.0	−2.9	−14.0	−16.9
8.5	−14.0	−4.7	−14.0	−18.7

Source: Ref. 555.

species was calculated by taking into account the maximum possible "formation of ion pairs between the negative groups on the humic material and the positive groups on the carbon surface" at pH < pH_{PZC} of the virgin carbon. For example, at pH = 4.5 (<pH_{PZC}) all the carboxyl groups on the adsorbed species are "neutralized" (and thus made unavailable for titration) by the positive charge on the adsorbent surface, and the effective charge is zero, while at the other extreme (pH = 8.5 > pH_{PZC}) it is assumed that there are either no positive groups on the surface or that they are unavailable (see, however, Section II), and thus the effective charge of the adsorbed species remains high due to the presence of dissociated carboxyl groups. As seen from Table 30, the adsorbent and effective adsorbate charges were assumed to be additive. Good agreement between calculated and experimental surface charge vs. pH was obtained when the adsorbate uptake was low. The poor agreement at higher uptakes was attributed to a "reorientation of adsorbed molecules [which] led to a greater electrostatic interaction between carboxyl groups and consequent lowering of degree of ionization upon adsorption." Most recently, the same research group tackled the key issue of the relative importance of electrostatic and nonelectrostatic effects [566] in the adsorption of NOM by studying the effectiveness of electrostatic screening (using an added salt) in the case of two chemically different commercial activated carbons. Their rationale was based on two scenarios:

1.   If adsorbate–adsorbent electrostatic attraction is dominant and the surface concentration is low, an increase in ionic strength should suppress adsorption (screening reduced regime).

2.   If dispersion (nonelectrostatic) forces govern adsorption or if electrostatic interaction is repulsive, or if the surface concentration is high (and thus lateral interactions become important), an increase in ionic strength should enhance adsorption (screening enhanced regime). These experimental conditions conform to the predictions of the DLVO theory, as shown in Fig. 16.

The properties of the two microporous carbons used (both with 0.30 cm^3/g in micropores <0.8 nm) are summarized in Tables 31 and 32. At low pH, where electrostatic interactions were minimized because the adsorbate species were predominantly in molecular form, the uptake was controlled by the surface physics (pore volume in sizes between 0.8 and 50 nm). As the pH increased, the screening enhancement (using NaCl) became more dramatic, especially for the more acidic carbon W. At the same time, the uptake decreased because of increasing electrostatic repulsion. As expected based on the development of carbon surface charge, the decrease was more dramatic in this pH range for the more acidic carbon W; this is illustrated in Fig. 30. Perhaps surprisingly, however, the authors attribute these uptake decreases with increasing pH primarily to increased repulsive NOM_{ads}–NOM_{ads} and NOM_{ads}–NOM_{soln} interactions. To dismiss the intuitive expectation that such interactions should become *less* important than the carbon–

TABLE 31 Effects of Physical and Chemical Surface Properties of Commercial Activated Carbons on the Adsorption of Natural Organic Matter

| Carbon | Surface area (m^2/g) | Pore volume (cm^3/g) | | pH_{PZC} | Carbon surface charge[a] $(mmol/g)$ | Adsorbent/adsorbate complex charge[a] $(mmol/g)$ | Adsorbed NOM charge[a] $(mmol/g)$ |
		Micropores $(0.8 < w < 2$ nm$)$	Mesopores $(2 < w < 50$ nm$)$				
W	1197	0.30	0.49	4.5	−0.22	−0.88	−0.013
C	1018	0.17	0.07	7.5	+0.05	−0.27	−0.008

[a] at pH = 7.
Source: Adapted from Ref. 566.

TABLE 32 Effect of pH on the Surface Charge of an
Activated Carbon in the Presence and Absence of Adsorbed
Natural Organic Matter

pH	C_{trans}[a] (mg/g)	Adsorbent surface charge (mmol/g)	Adsorbed NOM charge at C_{trans} (mmol/g)
4	32	+0.24	−0.26
7	14.5	+0.05	−0.23
9	12.8	−0.09	−0.22

[a] Surface concentration of adsorbate at which transition from screening-enhanced to screening-reduced regime occurs.
Source: Ref. 566.

NOM interactions as the adsorbate surface concentration decreases with increasing pH, the authors argued that the dramatic decrease noted above is due to the fact that the "Carbon W–NOM$_{ads}$ complex displays a more negative surface charge than the Carbon C–NOM$_{ads}$ complex and the NOM$_{ads}$(Carbon C) displays a decreased degree of ionisation relative to NOM$_{ads}$(Carbon W)." Apparently,

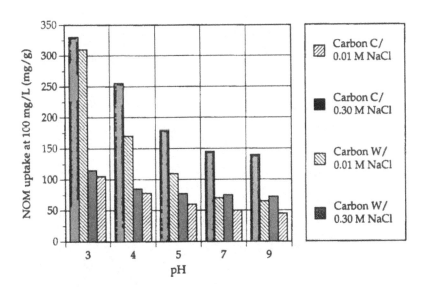

FIG. 30 Effect of pH on the uptake of natural organic matter by two commercial activated carbons at two different ionic strengths. (Adapted from Ref. 566.)

NOM_{ads}–NOM_{ads} and NOM_{ads}–NOM_{soln} interactions are suppressed only at much lower concentrations. Under these conditions (typically <10 mg/L, but dependent on pH and the carbon nature), as the solution concentration decreased the authors observed a transition from screening-enhanced to screening-reduced regime, especially for carbon C. The transition occurred at the point (see Table 32) at which "the negative charge from the adsorbed NOM balanced the positively charged sites on the activated carbon surface," in agreement with the authors' previous conclusion that "at low NOM surface coverage the carboxyl groups [on the adsorbate] form ion pairs with positive surface sites and directly compensate the surface charge" [566]. The authors argued that carbon W has no positive surface sites; this conclusion is supported by their titration data, but it may require additional confirmation in light of the proposals [67] that a major contribution to positive charge on carbon surfaces is from the graphene basal plane (see Section II) and not only from basic oxygen functional groups.

Most prior studies of humic acid adsorption, even those published after the landmark paper by Müller et al. [523], provide less evidence of attempts to achieve a fundamental understanding of the process. They have neither emphasized the importance of the chemical nature of the carbon surface nor attempted to clarify the exact role of electrostatic (vs. dispersive) interactions. For example, Weber et al. [540] concluded that "the adsorption mechanism involves formation of ion-humate-carbon complexes" but did not provide any relevant characterization of the carbon surface. Youssefi and Faust [541] noted "a slight effect of pH on humic acid adsorption" and speculated that it "may be due to competition of hydronium ions and humic acid for adsorption sites at the lower pH," but they did not attempt to identify such sites. McCreary and Snoeyink [539] reported that "[s]oil fulvic acid adsorption increased with decreasing solution pH" but did not relate this finding to any chemical property of the activated carbon adsorbent. Lee et al. [542] emphasized the pore size distribution as "an important parameter relative to the carbon's capacity for humic substances." They did note that a "decrease of pH rendered humic substances more readily adsorbed;" they went on to quote a study on montmorillonite adsorption to infer that "[a]s pH increases, more functional groups are ionized to yield a larger negative charge on the fulvic acid" and concluded that "[s]ince pH affects both adsorbate and adsorbent characteristics, the [correlations based on pore volume considerations] are valid only for a pH of 6." Randtke and Jepsen [544] evaluated the uptake of fulvic acids by a commercial activated carbon as a function of salt addition and pH. They noted that "[f]ew studies have made a serious attempt to elucidate the mechanisms responsible for enhanced or hindered adsorption in the presence of salts" and addressed in particular the issue of adsorbent–salt interactions by stating (albeit in an oversimplifying way) that "both the fulvic acid and the surface of the activated carbon are negatively charged and able to complex metal ions." Their experimental results showed the commonly observed uptake decrease with increasing pH (especially at pH ~ 3–4), but the authors made no attempt to relate this to carbon surface chemistry. Weber et al. [545] studied some of the same issues using three different carbons, reported similar findings, and invoked the same arguments to explain them ("humic acid solutions [are] mixtures of compounds with weakly acidic functional groups" and "at lower pH, more of these groups are in an uncharged, hence more adsorbable, state").

The work of Summers and Roberts [547,548] is a notable departure from the tone of these prior studies. In their study of adsorption of humic substances on five activated carbons, they emphasize the importance of the point of zero charge of carbons and cite the work of Müller et al. [523] as one of the two examples of "[a]dsorption models which incorporate electrostatic principles to account for the effects of pH and ionic strength on the activated carbon adsorption of dissociating organics solutes." The adsorption parameters at pH = 7, along with the physical and chemical properties of the carbons, are summarized in Table 33. The authors concluded that the behavior of the humic acid, "after normalization to accessible surface area [based on a size exclusion model], seems to be governed by the surface electrostatic characteristics." The authors emphasize the "strong correlation between the K_F values and the HCl consumption and the pH_{PZC} values for the five adsorbents examined" and note that at neutral pH the humic substances are negatively charged," as they are thought to have pK_a in the range of pH 3 to 6. Thus, for example, "F300 AC has the highest HCl consumption and [pH_{PZC}], indicating that it is the most positively charged adsorbent; as anticipated, it has the highest adsorption capacity of the negatively charged [humic acid]." The authors also conclude that "the results of experiments at different ionic strengths qualitatively adhere to electrostatic principles for both positively and

TABLE 33 Uptakes of Humic Acid on a Range of Commercial Activated Carbons (AC)

AC Adsorbent	SA_{total} (m²/g)	SA_{avail}[a] (m²/g)	pH_{PZC}[b]	HCl uptake (μeq/m²)	Freundlich parameters[c]		
					K_F	$K_{F'}$	$1/n$
F300	820	27.5	10.2	0.49	7.96	0.876	0.334
F400	890	21.5	10.0	0.35	6.74	0.792	0.302
CET	1460	290	9.6	0.15	26.7	0.633	0.340
CFT	1265	124	9.3	0.28	15.0	0.654	0.350
WV-B	1343	212	5.0	0.06	12.5	0.317	0.314

[a] Determined from considerations of adsorbate molecular size.
[b] See also Table 2.
[c] K_F is in "g DOC/kg AC/(g DOC/kg AC)$^{1/n}$" [sic]; $K_{F'}$ is in analogous mg/m² units.
Source: Ref. 547.

negatively charged adsorbents." A subsequent report by Lafrance and Mazet [549], which explored the reasons for uptake enhancement in the presence of sodium salts, reinforced the importance of electrostatic interactions and anticipated some of the conclusions of Newcombe and coworkers: the zeta potential was found to increase in the presence of adsorbed humic substances, and the role of Na$^+$ ions is to "reduce the repulsive interactions between the intramolecular functional groups of humate polyanions and between humate molecules and the surface sites of the PAC." Finally, in a recent study, Karanfil et al. [567] analyzed the effect of pH on the adsorption of dissolved organic matter, in both the presence and absence of O$_2$, and arrived at some of the same conclusions about the importance of electrostatic interactions. The relevant results are reproduced in Table 34. The higher uptake of the lower-molecular-weight fulvic acids (PMA and LaFA) at low pH was attributed to "size exclusion effects which prevent larger molecules from accessing smaller GAC pores." As pH increases, "repulsive forces between negatively charged GAC surfaces and DOM molecules start to control the extent of DOM sorption." The authors attributed the lower uptake of PMA, compared to that of LaFA, to the higher content of carboxyl groups and lower content of phenolic groups (implying perhaps that the latter may not be dissociated at pH ~ 10–11). The much lower uptake of PMA, compared to that of the higher-molecular-weight PHA, was taken as further support for the "the interpretation that repulsive forces between PMA molecules and GAC surfaces significantly affect sorption;" no comment was made about the possible role of the higher aromaticity of PHA.

Obviously, adsorption of complex mixtures of solutes such as NOM involves both electrostatic and dispersive interactions (see Section V), as well as molecular sieving effects, and it will be the ultimate challenge for both manufacturers of

TABLE 34   Freundlich Isotherm Parameters[a] for Adsorption of Dissolved Natural Organic Matter (DOM) in the Absence of Oxygen

DOM	pH ~ 4		pH = 7		pH ~ 10–11	
	K_F	n	K_F	n	K_F	n
PMA[b]	8.53	0.418	2.70	0.377	0.12	0.909
LaFA[c]	7.82	0.408	3.64	0.291	1.00	0.531
PHA[d]	3.03	0.358	1.66	0.406	0.97	0.514

[a] Using normalized solute concentration, with K_F in mg of dissolved organic carbon per gram of activated carbon (F-400, Calgon Carbon Corp., pH$_{PZC}$ ~ 9).
[b] Polymaleic acid (23% aromatic; 3.2 meq/g phenolic and 13.0 meq/g carboxyl acidity).
[c] Laurentian fulvic acid (28% aromatic, 9.3 meq/g phenolic and 11.7 meq/g carboxyl acidity).
[d] Peat humic acid (47% aromatic, 3.5 meq/g phenolic and 7.0 meq/g carboxyl acidity).
Source: Ref. 567.

carbon adsorbents and experts in the chemistry and physics of carbon surfaces. The encouraging development is that, by the mid-1990s, the role of electrostatic interactions in the adsorption of organics and how these are affected by the variability in carbon surface chemistry have become mainstream issues.

As mentioned in Section IV.A, Nelson and Yang [494] have proposed a surface complexation model "to describe the effect of pH on adsorption equilibria of chlorophenols on activated carbon." To account for the well known suppression of uptakes as pH increases, they introduce the ionization constants of *separate* acidic (A) (e.g., carboxyl) and basic (B) (e.g., pyrone) functional groups on the carbon surface:

$$-AH = -A^- + H_s^+ \qquad K^{intr}_a$$
$$-BH^+ = -B + H_s^+ \qquad K^{intr}_b$$

but they select values of the equilibrium constants in what appears to be an arbitrary fashion. Their model is essentially identical to that of Müller et al. [523–525], but they fail to take into account the specific surface chemistry of the commercial activated carbon used. Instead, they used an optimization procedure to back-calculate the concentration of AH and BH$^+$ sites on the surface, which turned out to be 0.886 and 1.853 mmol/g, respectively; no comment was made whether these are realistic values (see Section V). A recent study by Vidic et al. [478] (see below) reports total acidic and basic groups on the same adsorbent (as measured by titration) to be 71 and 122 µeq/g, or at least one order of magnitude less than the values used by Nelson and Yang [494].

In yet another study reporting suppressed uptakes of phenol in basic solutions, Halhouli et al. [374] invoke a rather vague explanation in terms of "repulsive forces prevailing at higher pH values between the adsorbed ions." It is indeed rather amazing that, after so many studies of this very issue, the authors of this peer-reviewed publication can get away with the conclusion that the "experimental results are explained in terms of the degree of ionization of the adsorbate which is, in turn, controlled by the pH of the solution," thus ignoring completely the key issue of surface charge development.

Liu and Pinto [437] have analyzed the effect of pH on the uptakes of phenol and aniline in an effort to assess the validity of the ideal adsorbed solution theory. The "large decrease in the amount adsorbed from pH 6.3 to pH 11.35" was attributed to both greater solubility of dissociated phenol at pH > pK$_a$ and "increased repulsion between dissociated form of the adsorbate and the carbon surface," while the "decrease in capacity from pH 6.3 to 3.07 is explained . . . as occurring due to increased proton adsorption on carbonyl oxygen sites, which suppresses phenol adsorption on these sites." The references provided for these interesting statements [333,35,417] are rather inappropriate, however; the cited studies are hardly representative, let alone the first ones, in which such arguments have been proven to be valid.

After several years of intense investigations of oxidative coupling in the presence and absence of molecular oxygen, Vidic et al. [478] tackled the issue of the impact of carbon surface chemistry, both "to determine if metal oxides on the surface of granular activated carbon are responsible for the increased adsorptive capacity for phenolic compounds under oxic conditions" and "to ascertain the influence of oxygen-containing surface functional groups on the ability of activated carbon to catalyze polymerization of phenolic compounds." No effect of the former was found, but the latter was confirmed to be the key factor. While they found "no relationship between surface functional group content as measured by the wet chemistry methods and irreversible adsorption," they concluded, somewhat contradictorily, that the "increased basic group content results in the highest oxic capacity and the greatest irreversible adsorption for [the carbon outgassed in argon at 900°C]." Some of the data on which this conclusion is based are summarized in Table 35. The effect of surface chemistry changes was smaller under anoxic conditions, which is surprising because the highly basic carbons (pH ≈ pH_{PZC} > 9) would be expected to attract phenolate ions—under both oxic and anoxic conditions—much more effectively than the carbons B2 and B2-9000x (and yet the trends under oxic and anoxic conditions seem to be different).

TABLE 35 Effects of Carbon Surface Chemistry and Dissolved Oxygen on the Uptake of 2-Chlorophenol

Carbon[a]	Acidic groups[b]	Basic groups[b]	S.A. (m^2/g)	Isotherm type	Total uptake[c]	Adsorbate recovery[d] (%)	Irreversible uptake[e]
B2	71	122	818	Oxic	246.0	6–38	0.284
(pH = 7.2)				Anoxic	128.2	82–100	
B2-500	41	159	735	Oxic	270.4	20–43	0.345
(pH = 9.8)				Anoxic	115.8	80–92	
B2-900	0	188	716	Oxic	297.1	13–31	0.424
(pH = 11.0)				Anoxic	115.0	94–100	
B2-1200	—	213	471	Oxic	111.2	54–60	0.199
(pH = 11.1)				Anoxic	88.3	82–99	
B2-9000x	134	392	—	Oxic	240.8	29–40	0.325
(pH = 6.2)				Anoxic	117.7	87–99	

[a] Surface treatment of commercial carbon (F400, Calgon Carbon Corp.) effected by heating in Ar to 500, 900 or 1200°C or by a 2-week soak in O_2-saturated water.
[b] Determined by titrations, in µeq/g.
[c] Expressed as Freundlich constant (mg/g)(L/mg)$^{1/n}$, determined at pH = 7.0.
[d] Percentage of adsorbate loaded onto the carbon surface that was recovered by Soxhlet extraction.
[e] Amount of adsorbate (mg/m^2) that could not be extracted (computed using oxic uptakes).
Source: Adapted from Ref. 478.

Instead of exploring the validity of such an argument, however, the authors attribute the "slightly higher (irreversible) anoxic capacity" of carbons B2-500 and B2-900 to a reduction of pore blockage. It seems appropriate here to recall the relevance of the four additional factors listed by King and coworkers [712,384] and thus conclude that a good understanding of the adsorption of phenols on carbons remains almost as elusive [418] as ever.

2. Role of Dispersion (van der Waals) Interactions

Nonspecific dispersion interactions between the adsorbent and the adsorbate are the dominant driving force for the adsorption of gases and vapors. Therefore, in the absence of molecular sieving effects, the entire carbon surface is "active" in the removal of gaseous or vapor-phase pollutants. For example, typical saturation uptakes of benzene or toluene on activated carbons are 0.005 mol/g from the gas phase and 0.002 mol/g from the aqueous phase at ca. 50 mg/L. It was seen in Section III that adsorption of inorganic solutes is thought to take place, for the most part, on specific sites on the carbon surface, by processes such as complex formation or ion exchange. Typical uptakes of metallic species are much lower (see Table 13).

Based on such comparisons and taking into account typical uptakes of aromatic solutes (of the order of 0.005 mol/g under most favorable conditions), it is tempting to postulate that the dispersion interactions—which allow a much more effective utilization of the carbon surface—may be the dominant driving force for the adsorption of organic solutes. In particular, aromatic solutes have a "natural" affinity for the graphene layers on the carbon surface (see Fig. 3) because of the possibility of π–π overlap. Indeed, π–π interactions and π–cation complexation are currently very powerful concepts and popular research topics [75–77,714,88,715–717,89,90,718,719]. (Thus, for example, the landmark paper by Hunter and Sanders [74] currently has close to 800 citations in the Science Citation Index.)

Curiously, however, the exact nature of the dispersive interactions between carbon surfaces and aromatic adsorbates has not been discussed in any detail. There has been much progress in the theoretical aspects, as noted above, but these have yet to be linked to the key experimental observations. And some of the key experimental observations are related to the effects of substituents on the aromatic rings, certainly those on the adsorbates but also those on the adsorbents. In Sections IV.A.1 and IV.A.2 we presented a brief overview of these effects. Here we attempt to provide a synthesis of these findings. In Section V we will present a model that takes them into account.

Chapter 7 of our landmark reference [6] discusses various aspects of the adsorption of weak electrolytes and nonelectrolytes from aqueous solution. In particular an attempt is made to elucidate the mechanism of adsorption of undissociated aromatic compounds from dilute aqueous solutions. As discussed in detail in Section VI, the authors concluded that the aromatic ring of the adsorbate interacts with the surface of the carbon "through the π-electron system of the ring," and proposed, perhaps surprisingly, not a π-π interaction but "a donor-acceptor complex with surface carbonyl oxygen groups, with adsorption continuing after these sites are exhausted by complexation with the rings of the basal planes." In the contemporary study of Coughlin and coworkers [450,331,332], the intuitively more appealing π-π interactions were invoked. More recently, McDermott and McCreery [720] used scanning tunneling microscopy in an attempt to unravel the adsorption mechanism on highly oriented pyrolytic graphite. They considered "[n]onspecific adsorption interactions [which] include dispersion interactions such as London and van der Waals forces, interactions between permanent dipoles, and hydrophobic effects." On the basis of the "correlation [between] the large extent [of] STM-observable electronic perturbation and the anomalously high adsorption," they concluded that "electronic perturbation near a step edge leads to partial charges in the HOPG, which attract the quinones" and that "these electrostatic interactions are more important than dispersion interactions or hydrophobic effects."

It is obvious from these preliminary considerations that the terminology used in the various disciplines in which adsorption is of interest may lead to some confusion. So here we briefly review first the intermolecular forces that we have lumped, for convenience, into dispersion interactions. At the most fundamental level, of course, all intermolecular interactions are electrostatic, but they are of three basic types [721,315]: long-range attractive and repulsive (coulombic) charge–charge interactions, which we have discussed in Section IV.B.1, and short-range attractive (van der Waals) interactions such as those among permanent and induced dipoles, which we shall consider here. Even though the term "dispersive interactions" is commonly reserved for the so-called London forces between induced dipoles—because they can be expressed "in terms of oscillator strengths that appear in the index of refraction for light" [721,722]—in liquid-phase adsorption, whose theory lags considerably behind that of gas-phase adsorption, such a distinction is thought to be premature.

A convenient point of departure is that of the increasingly popular quantitative structure activity relationships (QSAR) mentioned above [696,699,11], which derive adsorbate–adsorbent interaction indices from, for example, water solubility data, molecular connectivities [697], n-octanol-water partition coefficients, reversed-phase liquid chromatography capacity factors [723], or linear solvation energy relationships (LSER).

Among these the LSERs are thought to provide the greatest insight into the mechanism of adsorption [7]. In particular, the definition of acids and bases as hydrogen-bond donors and hydrogen-bond acceptors, respectively, has led Kamlet, Taft and coworkers [724,725] to develop the so-called solvatochromic parameters, whose coefficients reveal the relative importance of acid–base, dipole–

dipole, dipole–induced-dipole and induced-dipole–induced-dipole interactions. Thus, for example, Abraham et al. [726] showed, as expected, that "interactions between the gaseous solutes (that include alcohols and amines) and the four [commercial activated carbon] adsorbents involve just general dispersion forces." In contrast, Kamlet et al. [695] analyzed the factors that influence the adsorption of organic compounds from aqueous solution and reported the intriguing results summarized in Table 36. In discussing these results, the authors noted that "when the aromatic solutes were included [in their correlation], the quality of the correlation dropped dramatically." They "had considered that the differences might be attributable to specific interactions of the aromatic solutes with the π-electrons of the activated carbons, but this would have led to higher calculated [α] values, whereas in most instances the experimental α-values are lower than calculated." It turns out, however, that the columns in this table were mixed up by the authors; the corrected experimental and predicted values, prepared using both the original prediction [695] and one of its modified versions [727], are presented in Table 37. Now the LSER analysis suggests exactly the opposite trend and is largely consistent with the well-documented affinity of carbon surfaces for aromatic compounds, which is expected to be due to π–π interactions.

An early analysis of the adsorption of substituted phenols by soils [397] is both relevant and intriguing. The author concluded, for example, that "adsorption was enhanced by electron-donating substituents, indicating that the phenolic —OH group formed H-bonds by acting as a proton acceptor," but this conclusion

TABLE 36 Use of Linear Solvation Energy Relationship to Predict the Uptake[a] of Selected Aromatics on a Commercial Activated Carbon

Adsorbate	Experimental	Predicted[b]
Benzene	0.82	1.34
Toluene	1.30	0.51
Ethylbenzene	1.71	0.69
Pyridine	−0.91	−0.25
Benzaldehyde	0.60	1.25
Nitrobenzene	0.91	1.43
Acetophenone	0.71	1.66

[a] The values shown (α) represent the logarithm of adsorbability (or the partition coefficient of adsorbate between adsorbent and solvent at infinite dilution).
[b] $\log \alpha = -1.93 + 3.06\ V/100 + 0.56\ \pi^* - 3.20\beta$, where V is a measure of solute molar volume and π^* and β are the solvatochromic parameters that scale dipolarity/polarizabilities and hydrogen bond acceptor basicities of the adsorbates.
Source: Ref. 695.

TABLE 37 Use of Linear Solvation Energy Relationship to Predict the Uptake[a] of Selected Aromatics on a Commercial Activated Carbon

Adsorbate	Experimental	Predicted	
		Taft et al. (1985)[b]	Abraham et al. (Ref. 727)
Benzene	1.34	1.07	0.10
Toluene	0.51	1.58	0.54
Ethylbenzene	0.69	2.01	0.93
Pyridine	−0.25	−0.71	−1.06
Benzaldehyde	1.25	0.59	0.05
Nitrobenzene	1.43	1.13	0.52
Acetophenone	1.66	0.92	0.17

[a] The values shown represent the logarithm of adsorbability (or the partition coefficient of adsorbate between adsorbent and solvent at infinite dilution).
[b] *J. Pharm. Sci.* 74:807–814, 1985.
Source: Recalculated from Ref. 695.

is based on Freundlich parameters that were not normalized with respect to adsorbate solubility. In the important study of Caturla et al. [414] (see Section IV.A.2), the key question has been identified: Do the changes in surface chemistry have a greater effect on dispersive interactions or on electrostatic interactions? The authors hypothesize that "the interaction will be weaker as the basicity of the carbon increases because of the excess negative charge on the surface" and that "consequently the amount of phenol adsorbed is lower than if no surface complexes were present." Furthermore, they state that "[i]n the case of PCP and DCP, chlorine decreases the electron density of the aromatic ring and as a result the interaction of the system with the carbon will increase with increasing basicity of the carbon surface."

As mentioned in Section IV.A.2, Leng and Pinto [386] specifically addressed the effect of surface properties on the oxic and anoxic adsorption behavior of phenol, benzoic acid, and *o*-cresol. Commercial carbons were oxidized in air at 350°C, which is known [37] to introduce both CO- and CO_2-yielding surface groups; nevertheless, from FTIR spectra, they concluded that the main differences are due to relative quantities of surface carboxylic groups. Presumably because the experiments were performed at pH = 7.0, i.e., below the pK_a of phenol, their explanation for the decrease in uptake with increasing surface oxygen was not the electrostatic repulsion but "increased water cluster formation" as well as "increased removal of π electrons from the basal planes [450,674], which results in weaker dispersion interactions with phenol."

The main point to be made on the basis of this brief (and selective) survey of the key issues is the following: at least two competing mechanisms of noncoulombic adsorbate–adsorbent interactions are still competing for the favors of the investigators, and the attempts to study them in isolation from almost equally ubiquitous electrostatic interactions may be part of the problem (and not of the solution). These issues are taken up in turn in Sections V and VI.

V. TOWARD A COMPREHENSIVE MODEL

The theoretical background to the modeling of adsorption of organics from aqueous solutions has been provided by Weber and DiGiano [21] in a clear and concise manner. It is important to note that, when the adsorption isotherms for a range of adsorbates on a given carbon adsorbent are normalized with respect to the "escaping tendency" of the solute from solution, significant uptake differences remain, and these need to be explained. The isotherms normalized with respect to the aqueous solubility reflect the true affinities of the adsorbates for the adsorbent [66]. In this section we summarize the reasons for these differences. We also review very briefly the modeling literature that addresses them.

McGuire and Suffet [728] proposed the "calculated net adsorption energy" concept which is based on the solubility parameter of the adsorbate. They justified their approach by noting that the "interactions involved in the adsorption of nonpolar and polar compounds onto a nonpolar [activated carbon] surface are, for the most part, governed entirely by the dispersion forces." Their results are summarized in Fig. 31. Even on a log–log plot, the r^2 correlation coefficient is only 0.7. The authors cautioned against extrapolating such a correlation "to predict the adsorption capacities of other neutral organic compounds." Clearly, incorporation of model parameters that quantify the chemistry of the carbon surface is necessary.

Weber and Van Vliet [20] briefly described the Michigan Adsorption Design and Applications Model (MADAM), which includes both equilibrium and kinetic considerations, while Manes [729] advocated the use of the Polanyi adsorption potential theory, even to account for pH effects; neither of these approaches includes a description of the role of carbon surface chemistry.

In a recent paper entitled "Models and Predictability of the Micropollutant Removal by Adsorption on Activated Carbon," Haist-Gulde et al. [30,31] briefly reviewed the models available to describe adsorption equilibria in single-component and multicomponent systems. They emphasize empirical models such as that of Freundlich and the ideal adsorbed solution (IAS) theory and do not even mention the work of Müller and coworkers [523–525]. Very recently, Crittenden et al. [702] have proposed an ambitious correlation that allows one to "estimate the adsorption equilibrium capacity of various adsorbents and organic compounds using a combination of Polanyi potential theory and linear solvation energy relationships." When all organic compounds—halogenated aliphatics, aromatics and halogenated aromatics, polyfunctional organic compounds and sulfonated aro-

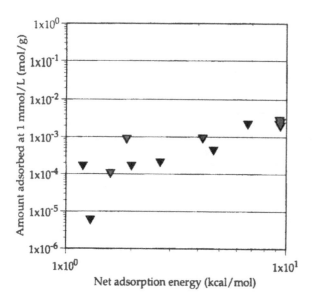

FIG. 31 Relationship between adsorption capacity for a range of organic compounds on activated carbon and the net adsorption energy. (Adapted from Ref. 728.)

matics—were correlated together (for a single adsorbent), "a good correlation could not be developed," but when the above mentioned classes were used, the authors found that the correlation "showed significant improvement over one that considers only van der Waals forces." Interestingly, however, the correlations were much worse for sulfonated aromatic compounds than for the other groups. The authors argued that, because of the anionic character of sulfonated aromatics and "[s]ince the anions do not adsorb to any significant degree unless their counterions are present, the sulfonates will take the protons (H^+) from the solution [at pH = 7] and then will adsorb on the surface." Their suppressed uptake was speculated to be due "to the existence of water in the adsorbed state, which causes the density of the solute in the adsorbed state to be smaller than is expected."

It is clear that there are quite a few possible theoretical approaches to the formulation of a comprehensive model for the adsorption processes discussed above. The most fundamental one would be based on the perturbation molecular orbital (PMO) theory of chemical reactivity [730,731] in which the wave functions of the products are approximated using the wave functions of the reactants. A key issue in the use of Klopman's PMO theory is the relative importance of the two terms in the expression for the total energy change of the system, ΔE_{pert}, which is taken to be a good index of reactivity, ΔE_{react} [732,733]:

$$\Delta E_{react} \sim \Delta E_{ch} + \Delta E_{orb}$$

The question is whether the extent of adsorption is determined by charge control (ΔE_{ch}, i.e., coulombic or electrostatic interactions) or by orbital control (ΔE_{orb}, i.e., attractive interactions between the filled orbitals of the adsorbate and the empty orbitals of the adsorbent).

In a recent important study, Tamon and coworkers [733–735] implicitly took the approach that orbital control is dominant. They analyzed the adsorption and desorption characteristics of a series of aromatic compounds at uncontrolled pH and used the semiempirical MINDO/3 method to determine the HOMO energy levels (E_H) for these adsorbates and the LUMO levels (E_L) for several adsorbents (including chemically modified activated carbons). The range of calculated $|E_H - E_L|$ values for adsorbates ranging from p-nitrobenzoic acid to aniline on a model activated carbon was 6.7–9.0 eV, as shown in Table 38. Low values (<7.3 eV) coincided with those systems for which irreversible adsorption was found: those adsorbates that possess electron-donating groups (e.g., aniline), i.e., with the highest HOMO levels (or lowest ionization potentials), could not be

TABLE 38 Results of Semiempirical (MINDO/3) Quantum Chemical Calculations of Energies of the Highest Occupied Molecular Orbitals (HOMO) and Lowest Unoccupied Molecular Orbitals (LUMO) for a Series of Aromatic Adsorbates and Model Carbon Clusters

	HOMO (eV)	LUMO (eV)
Phenol	−8.65	1.15
Nitrobenzene	−10.02	−0.36
Benzaldehyde	−9.61	0.22
p-Nitrophenol	−9.59	−0.21
Aniline	−8.10	1.44
p-Cresol	−8.62	1.09
b-Naphthol	−8.05	0.41
p-Nitrobenzoic acid	−10.37	−0.90
p-Nitroaniline	−9.09	−0.21
p-Chlorophenol	−8.61	0.44
m-Cresol	−8.69	1.02
AC	−6.19	−1.35
AC—OH	−5.85	−1.09
AC—O	−7.92	−1.25
AC—COOH	−6.61	−1.92
AC—OC2H5	−5.85	−1.11

Source: Ref. 735.

desorbed easily (using distilled water at 308 K). In contrast, those aromatic species that possess electron-withdrawing groups (e.g., nitrobenzene) were more weakly (reversibly) adsorbed. The polyaromatic cluster models used to calculate the energy levels of the adsorbents were a nonsubstituted phenanthroperylene (AC), AC substituted with two phenolic groups (AC—OH), with two carbonyl groups (AC—O), and with two carboxyl groups (AC—COOH). The LUMO energies for AC—OH and AC—COOH, with their respective electron-donating and electron-withdrawing groups, are seen to be higher and lower, respectively, in comparison to the nonsubstituted AC. Thus, by calculation of $|E_H - E_L|$ values for these cluster models and all the phenolic adsorbates, it is deduced that the values corresponding to AC—OH are in all cases higher than those for AC, and those corresponding to AC—COOH are all lower than those for AC. This in turn implies that adsorption of these phenolic compounds on AC—OH should be weaker than on AC, while adsorption on AC—COOH should be stronger than on AC. The former prediction was confirmed experimentally by Tamon et al. [735]: somewhat surprisingly, treatment of a commercial activated carbon in 16N boiling nitric acid produced adsorbents with a large majority of the weaker phenolic, rather than the stronger carboxyl, groups. Even though the oxidized carbons had lower uptakes (probably because of electrostatic repulsion), "the desorption characteristics were improved and the hysteresis [attributed to irreversible adsorption] disappeared." It would be important to confirm these experimental findings, especially because the relative effects of electron withdrawal from the adsorbate (predicted to result in weaker adsorption) and the adsorbent (predicted to result in stronger adsorption) are difficult to quantify using molecular models. Be that as it may, this interesting point was not addressed by Tamon and coworkers, and it could indicate the formation of donor–acceptor complexes in which the benzene ring of the adsorbate would be the donor and the graphene layers (and not the carbonyl groups) would be the acceptor.

The arguments presented by Tamon and coworkers are in agreement with the conclusion of Radovic and coworkers [736,674,737] that, when dispersion forces are dominant, "the electron-withdrawing effects of nitrogen and carboxyl functional groups suppress the interaction of the basal planes with the adsorbate's aromatic rings." Based on similar arguments, Tamon et al. [733] proposed a two-state model to explain the appearance of irreversible adsorption of electron-donating compounds, according to which the barrier for going from the precursor (reversible) adsorption state to the irreversible state is higher for the electron-withdrawing aromatic compounds than for the electron-donating compounds. They do not provide a justification for this interesting assumption; instead, they invoke the Hammond postulate, according to which "the structure of the transition state will resemble the product more closely than the reactant for endothermic processes whereas the opposite is true for exothermic processes" [733]. How exactly this is related to the exothermic adsorption process is not clarified. A more straightforward alternative theory is offered below.

Functionalization of either the adsorbate or the adsorbent that increases the π electron density leads to either enhanced or stronger adsorption when the adsorption process is governed by π–π (dispersion) interactions. The converse is also supported by available experimental evidence: functionalization (of the carbon adsorbent or the aromatic adsorbate) that decreases the π electron density leads to suppressed or weaker adsorption.

As abundantly documented in Section IV.B.1, some adsorption systems involving aromatic adsorbates are very much influenced by electrostatic interactions. Clearly, a model that takes into account *both* electrostatic and dispersion interactions is needed. Such a model has been presented by Müller and coworkers [523–525]. Radovic and coworkers [674] used this model to illustrate the possibly dramatic effects of modifications of carbon surface chemistry on equilibrium uptakes of *p*-nitrophenol; they have also extended it to evaluate the *relative* importance of electrostatic and dispersive interactions [738]. This approach is summarized next.

The wide range of surface chemical properties of activated carbon adsorbents, including the important lack of "symmetry" between the low-pH and high-pH regions, was documented in Fig. 4 and is illustrated schematically in Fig. 32 (see also Fig. 28). The "acidic" (C1) and "basic" (C3) carbons whose pH_{PZC} values are 3 and 10, respectively, can be easily designed to develop an electric charge of the order of ~ 0.03 C/m^2, which corresponds typically to a surface potential of ~ 150 mV. At the other extreme, high-temperature treatment and stabilization in H$_2$ of an "acidic" carbon [82] or low-temperature treatment in the presence of hydrogen atoms [83,84] produces a "basic" carbon whose pH_{PZC} remains high and constant over extended periods of exposure to room-temperature air.

The key equation describing competitive Langmuirian adsorption on such amphoteric carbons, in terms of fractional surface coverage θ of molecules (M) and ions (M$^+$ or M$^-$) of a partially dissociated organic solute on a homogeneous surface, is

$$\theta = \theta_M + \theta_{M^\pm} = \frac{K[M] + K'[M^\pm]}{1 + K[M] + K'[M^\pm]}$$

In the above isotherm equation, K is the equilibrium constant for the molecules (due to London dispersion interactions), while K' is the equilibrium constant for the ions and has the form

$$K' = K K_q = K_o \left[\exp\left(\frac{U}{RT}\right) \right] \left[\exp\left(\frac{z_\pm F\phi_s}{RT}\right) \right]$$

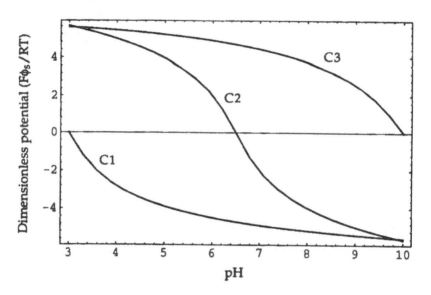

FIG. 32 Surface potential vs. pH for three (hypothetical) activated carbons: C1, "acidic" carbon ($pH_{IEP} = pH_{PZC} = 8.0$); C2, typical as-received (amphoteric) carbon ($pH_{IEP} = pH_{PZC} = 6.5$); C3, "basic" carbon ($pH_{IEP} = pH_{PZC} = 10.0$).

In the above expression, consisting of two driving forces for adsorption, $z_{\pm}$ is the valence of the charged species, F is the Faraday constant, ϕ_s is the interfacial electric potential due to the development of surface charge, and U is the potential arising from the dispersive adsorbent–adsorbate interactions. Implicit in this expression is the realization that the coionic solute species (e.g., an anilinium cation interacting with the surface at $pH < pH_{PZC}$) can adsorb only if the (attractive) adsorption potential U is larger than the repelling electrostatic potential $z_{\pm}F\phi_s$. Explicit exclusion of coionic species from the surface is included in the model for a heterogeneous adsorbent surface, which conforms to the Freundlich isotherm:

$$\theta = \frac{C_\alpha}{C_1}\left(\frac{C_1}{C_1 + K}\right)^n$$

with

$$C_\alpha = [M] + \frac{\alpha[M^{\pm}]}{K_q} \quad \text{and} \quad C_1 = [M] + \frac{[M^{\pm}]}{K_q}$$

Here again it is assumed that the electrostatic potential is of the same order of magnitude or lower than the dispersion potential, $(z_{\pm}F\phi_s)/RT \le U/RT$.

The remarkable potential flexibility of activated carbon adsorbents is readily apparent from inspection of Figs. 32–34. As is intuitively obvious from simple electrostatic arguments, an "acidic" carbon is most appropriate for adsorbing the weakly basic aniline (pK_a = 4.6), especially at low pH, when a large fraction of the adsorbent surface is unavailable for adsorption of coionic species. These figures show the results of a parametric sensitivity study of the effects of modified dispersion potentials, as a consequence of removal (Fig. 33) or incorporation (Fig. 34) of electron-withdrawing groups, as well as incorporation of electron-donating groups (Fig. 34) at the edges of graphene layers of an activated carbon.

Conversion of an "acidic" carbon (C1) to a "basic" carbon (C3—E) increases the point of zero charge from 3 to 10, and this is detrimental for the adsorption of anilinium cations at low pH; but it also enhances the π electron density in the graphene layers and thus increases the dispersive potential, say,

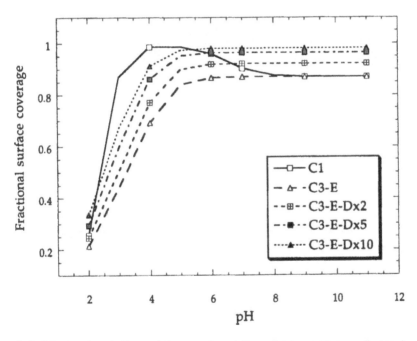

FIG. 33 Combined effects of electrostatic and dispersive interactions on changes in aniline uptake upon thermal treatment (e.g., in H_2 or NH_3) of an "acidic" carbon and its conversion to a "basic" carbon: C1, original "acidic" carbon; C3-E, "basic" carbon (only electrostatic interaction adjusted); C3-E-Dx2, "basic" carbon (dispersive attraction potential enhanced by a factor of 2); C3-E-Dx5, "basic" carbon (dispersive attraction potential enhanced by a factor of 5); C3-E-Dx10, "basic" carbon (dispersive attraction potential enhanced by a factor of 10).

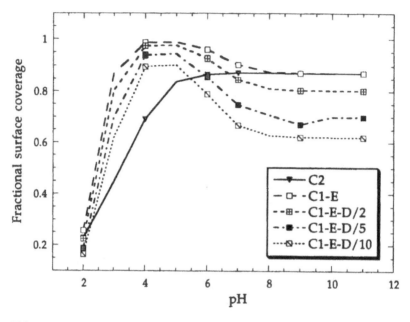

FIG. 34 Combined effects of electrostatic and dispersive interactions on changes in aniline uptake upon oxidation: C2, original (as received) carbon; C1-E, oxidized carbon (only electrostatic interaction adjusted); C1-E-D/2, oxidized carbon (dispersive attraction potential reduced by a factor of 2); C1-E-D/5, oxidized carbon (dispersive attraction potential reduced by a factor of 5); C1-E-D/10, oxidized carbon (dispersive attraction potential reduced by a factor of 10).

by a factor of 2 (C3-E-D × 2) to 10 (C3-E-D × 10). It is seen in Fig. 33 that the electrostatic effect is dominant in suppressing aniline uptake at low pH, but the dispersive effect both dampens the electrostatic effect at low pH and enhances the uptake at high pH.

Conversion of an amphoteric carbon (C2) to an "acidic" carbon (C1—E) decreases the pH_{PZC} from 6.5 to 3.0, and this is beneficial for the adsorption of anilinium cations at low pH; but it also reduces the π electron density in the graphene layers and thus decreases the dispersive potential, say, by a factor of 2 (C1—E—D/2) to 10 (C1—E—D/10). The net result is shown in Fig. 34: enhanced adsorption at low pH, made less pronounced by the detrimental dispersive effect, and suppressed adsorption at high pH.

As discussed above, in many instances the uptake of organics (especially aromatics) will be determined by a complex interplay of electrostatic and dispersion interactions. As discussed in Section III, adsorption of inorganic compounds will be dominated by the electrostatic interactions. It remains to be seen whether the parameters of this promising model will be able to accommodate these main features, as well as the many other more subtle effects in the adsorption of organics and inorganics on activated carbons.

VI. ROLE OF DISPERSION INTERACTIONS IN THE ADSORPTION OF AROMATICS: A CASE STUDY IN CITATION ANALYSIS

The complex role of surface chemistry in the adsorption of aromatics has been highlighted and (we hope) clarified in Section IV.B. The seed for the explanations offered there had been planted as early as 1968–1971 in a series of landmark papers by Coughlin and coworkers [450,331,332] and Mattson and coworkers [24,6]. In hindsight, however, it is obvious that the arguments presented in these two series of papers cannot *both* be correct. Here we quote the critical statements in the key papers and follow their fate with the help of the Science Citation Index. The statistics of their citations are summarized in Table 39.

Coughlin and Ezra [450] state,

> Possible explanations for the role of the acidic surface oxygen groups in inhibiting adsorption of phenol and nitrobenzene molecules from aqueous solution are interesting to consider . . . Oxygen chemically bound on the edges can localize electrons in surface states thereby removing them from the π electron system of the basal planes. Walker et al. [in Chem. Phys. Carbon, Vol. 1, p. 328] have explained the effect of oxygen chemisorption on the thermoelectric power of carbon by similar reasoning. Increases in thermoelectric power of carbon owing to oxygen chemisorption were attributed by these workers

TABLE 39 Comparison of Citation Histories of Publications by Coughlin and Coworkers (Refs. 331, 332, 450) and Mattson and Coworkers (Refs. 6, 24)

Citing period	Ref. 332	Ref. 450	Ref. 331	Ref. 24	Ref. 6
1970–1974	2	9	8	21	8
1975–1979	5	9	6	9	52
1980–1984	1	20	16	15	50
1985–1989	3	15	5	22	65
1990–1994	2	21	2	17	51[a]
1995–present	2	23	11	31	69[a]
Average per year	0.97	3.3	1.6	3.8	9.8

[a] Includes only citations to entire book and Chs. 6 or 7.

to a depletion in the number of electrons and an increase in the population of positive holes in the conduction band of the π system. These considerations appear consistent with the concept of dispersion forces between the phenol π electron system and the π band of the graphitic planes of the carbon as responsible for adsorption. Removal of electrons from the π band of the carbon by chemisorbed oxygen might be expected to interfere with and weaken these forces.

We shall refer to this as the "π–π interaction" argument. It has been formulated based on the following experiments: (1) adsorption isotherms of phenol, nitrobenzene, and sodium benzenesulfonate on a series of activated carbons and carbon blacks; and (2) characterization of the surface chemistry of as-received and chemically modified carbons. The authors did not report the pH; in fact, this word is not even mentioned in any of their papers on this subject [450,331,332].

In contrast, Mattson et al. [24] state,

The interaction of the aromatic ring with the surface of the active carbon must . . . be considered the major influence in the [adsorption] processes, interacting through the π electron system of the ring. There is considerable evidence in the literature for the formation of donor–acceptor complexes between phenol and several kinds of electron donors. . . It is suggested that . . . aromatic compounds adsorb on active carbon by a donor–acceptor complex mechanism involving carbonyl oxygens of the carbon surface acting as the electron donor and the aromatic ring of the solute acting as the acceptor.

Let us call this the "donor–acceptor complex" proposal, similar to that presented recently for adsorption of substituted nitrobenzenes and nitrophenols on mineral surfaces [739]. The experiments on which this proposal is based are (1) isotherms of phenol, nitrobenzene, and m- and p-nitrophenol on one commercial activated carbon at pH = 2–7 and very low solute concentrations ("<1.5% of the solubility limit of these species" [6]); and (2) detailed infrared (internal reflection) spectroscopic analysis of the surface after adsorption of p-nitrophenol. Interestingly, neither in this study, nor in any subsequent study that supports this mechanism, has a similar analysis been performed with carbons containing varying concentrations of carbonyl surface groups. Also of interest is that the authors dismiss the electrostatic explanation of the reported pH effects by assuming that the isoelectric point of the carbon (which was dried at 200°C for 12–24 h) was ca. 2.4.

In early studies, Snoeyink et al. [333], as well as Umeyama et al. [405] to some extent, paraphrase both arguments regarding the role of surface oxygen, but they fail to point out the apparent contradiction. Ward and Getzen [677] support the π–π argument; they do allow room for the role of oxygen-containing functional groups, not through a donor–acceptor mechanism but rather through their hydrogen bonding with acidic aromatics, especially at low pH (when they are not dissociated). In contrast, and perhaps not surprisingly, Epstein et al. [740] echo the donor–acceptor complex argument for *p*-nitrophenol (PNP) adsorption based on cyclic voltammetry results that showed that less PNP was available for electrochemical reduction when the surface of pyrolytic carbon was prereduced, while much more was available for reduction when it was preoxidized; by assuming the existence of quinone–hydroquinone groups on an oxidized carbon surface [41,37], which can be interconverted electrochemically, these results were interpreted as evidence for more favorable interaction of the electron-accepting PNP with the electron-donating quinone groups. Marsh and Campbell [741] also studied the adsorption of PNP, in this case on a range of polymer-derived carbons activated to different degrees of burn-off; they discussed both proposals but did not arrive at any substantive conclusion that would go beyond the useful realization that "[a]dsorption of *p*-nitrophenol is probably the most informative of adsorbates in solution."

In addition to the supporting self-citations [742,6,457,330], there was much early support for the formation of donor–acceptor complexes. Radke and Prausnitz [743] interpreted the "extensive loadings" of phenols even at very low concentrations, in comparison with lower uptakes of several aliphatic adsorbates, as evidence for "specific interaction with the activated carbon surface." Barton and Harrison [744] studied the effect of graphite outgassing temperature on the heat of immersion of benzene and attributed a shallow minimum at ca. 800°C to the effect of CO desorption, thus implicitly supporting the donor–acceptor complex proposal in terms of "a reduction in the interaction between the partial charge on the carbonyl carbon atom and the π-electron cloud of the benzene molecule."

Komori et al. [490] considered both proposals in a study of adsorption of substituted benzoic acids on two commercial activated carbons and concluded, based on regular increases in adsorptivities of *m*-substituted acids with their decreasing solubilities (in agreement with Traube's rule), that "the carbon surface does not directly interact with the functional groups of the adsorbates, but with the aromatic ring." A similar conclusion ("adsorption occurs predominantly on the purely carbon part of the adsorbent, which has no oxygen-containing groups") was reached by Glushchenko et al. [745] in their study of the effect of the nature of the carbon surface on the adsorption of nine aromatic nitro compounds from aqueous phosphate solutions. Curiously, however, these authors do not cite the work of Coughlin and coworkers. Their conclusion is based on the observation of a linear decrease in adsorption uptake with increasing degree of carbon oxidation for all adsorbates except for PNP, for which the authors invoke the argument of Mattson and coworkers by saying, rather vaguely, that its behavior "is partly determined by a specific interaction with the surface of the carbon adsorbent." Additional support for the donor–acceptor complex proposal was provided by Pershko and Glushchenko [746] who rationalized the "unusual behavior" of PNP on the surface of graphite in terms of "carbonyl groups on the adsorbent surface forming stable complexes with p-nitrophenol molecules."

The early work of Puri et al. [339], who made seminal contributions to our understanding of the surface chemistry of carbons (see Ref. 37), is significant in its rejection of the $\pi-\pi$ interaction proposal and its implicit endorsement of the donor–acceptor complex argument. These authors studied the influence of surface oxygen complexes on the adsorption of phenol from aqueous solution and concluded that the "presence of the combined oxygen as CO_2-complex imparts to the surface preference for water and thus affects adversely the adsorbability of phenol," while the "presence of oxygen as CO-complex . . . improves the adsorbability of phenol due, probably, to the interaction of quinonic oxygen of this complex with π electrons of the benzene ring, as well as the OH group of phenol." It is interesting to note, however, that the work of Mattson and coworkers is not cited either here or in a contemporary study of Puri [338], in which the author appears to be unaware of the donor–acceptor complex argument because he explains the same results in terms of the "interaction of OH groups of phenol with phenolic and quinonic oxygens associated with [the] CO-complex." In subsequent work, still ignoring the original work of Mattson and coworkers, Puri et al. [344] studied phenol adsorption from a nonaqueous (benzene) solution and reached a different conclusion: the main role of the surface oxygen is to induce the perpendicular (rather than parallel) orientation of phenol molecules, which permits the "accommodation of maximum number of phenol molecules on the surface." In neither of these studies is the importance of electrostatic effects mentioned, e.g., a decrease in repulsion between COO^- surface groups and phenolate anions upon removal of CO_2-yielding surface complexes (see Section IV.B).

A rare early endorsement of the $\pi-\pi$ interaction proposal, again implicit because the work of Coughlin and coworkers is not cited, was offered in a study of phenyl phosphonate anions on a graphitized carbon black in which the authors [747] conclude that "the π-electrons of the ring systems of adsorbent and adsorbate are involved in the adsorption of aromatic compounds." A more definitive source of support came a few years later. In the thorough reexamination of the very same issue discussed several years earlier by Puri et al. [339], Mahajan et al. [345] considered both mechanistic proposals at length but also arrived at an important practical recommendation: "phenol uptake capacity of commercial activated carbons can be maximized by heating them in N_2 up to 950°C followed by treatment with H_2 at 300°C." This is a consequence of the observation that phenol uptake is suppressed as the surface acidity increases and as the suspension pH decreases. Indeed, the authors noted that "[i]f the views of Mattson et al. are correct, then the formation of CO-evolving complex (predominantly carbonyl groups) on the carbon surface exposed to dry air should have enhanced phenol uptake.'' Intriguingly, however, they do not pursue this argument further by invoking the importance of electrostatic adsorbate–adsorbent interactions (see Section IV.B.1). Instead, they emphasize the ''importance of the nature of the carbon surface in influencing adsorption behavior of carbons'' and then switch gear to describe an elegant experiment that lends further support to the π–π interaction argument: ''the addition of boron substitutionally into the lattice of polycrystalline graphite, with an accompanying removal of π-electrons from the graphite, results in a lowering of phenol uptake from water.''

Such was then the uncertainty regarding the role of surface oxides in adsorption that a timely book by Suffet and McGuire [748] included a transcript of a very useful discussion [619] (albeit in the absence of Coughlin and with participation of two of the authors of the paper by Mattson et al. [24]), whose tone was set by the initial question posed by Suffet: ''There is some confusion in the literature regarding the major sites of adsorption on [granular activated carbon]. Is an oxidized surface adsorbing material on the oxidized functional groups, or is it adsorbing on the nonpolar carbon surface?'' Not surprisingly perhaps, even Puri acknowledged on this occasion the importance of the ''donor–acceptor complex formation with surface carbonyl groups if present and with the fused-ring system of the basal planes of the carbon if these groups are absent.'' In an accompanying paper [346], he attempted to respond to Suffet's question more explicitly, not only by reexamining the removal of phenols from different solvents but also by examining the aqueous phase adsorption of oxalic and succinic acids and the removal of stearic acid from solutions of benzene and carbon tetrachloride. The phenol adsorption results were again interpreted as evidence against the π–π interaction argument. The uptake of the aliphatic acids, like that of phenols, increased as oxygen was removed from the carbon surface by heat treatment, but Puri did not clarify whether this finding supports or casts doubt on the donor–acceptor complex argument. Curiously, the lower uptake of stearic acid from a benzene solution was attributed to the ''complexing effect of π-electrons with quinonic oxygen'' and not simply to the lower affinity of carbon for the nonaromatic solvent. In the final paper on this topic [354], Puri unambiguously embraces the donor–acceptor complex argument. Here again, however, there is no mention of electrostatic adsorbate–adsorbent interactions.

Interesting insight into the role of carbon surface chemistry was offered by Chang and coworkers [347,349]. They modified the activated carbon surface with sulfur groups and measured the heats of desorption of naphthalene as well as the uptakes of naphthalene, phenol, and nitrobenzene. They concluded [347] that ''the modification of the carbon surface with sulfur surface groups decreases the binding energy between the adsorbate and the adsorbent,'' which sounds like an endorsement of the π–π interaction argument. In more explicit statements [348,349], however, based on analysis of infrared absorption bands, the authors endorse the donor–acceptor complex proposal.

In the 1980s, the pace of research on this issue accelerated considerably, but the controversies continued. Oda and coworkers [350,410,749] investigated the role of surface acidity in both vapor-phase and liquid-phase adsorption. In the former case they arrived at a straightforward and interesting conclusion, that "polar adsorbates are obviously affected and the [chromatographic column] retention times increase with an increase in the surface acidity." In the latter case, however, their discussion is quite confusing and even inappropriate on occasions. The results are consistent with previous studies, e.g., "adsorption of phenol is retarded considerably by the acidity of the carbons" [350], but some of their citations are both inaccurate and do not give credit where credit is due. For example, they state [410], inappropriately, that "little attention has been paid to the problem of the surface acidity of activated carbons" and then attribute only to their earlier study [350], and not to the many earlier reports by other investigators, the virtue of having "demonstrated that the surface acidity . . . has a marked effect on the adsorption characteristics." A reader who now expects to learn about the important details is disappointed, however, because the authors fail to say much beyond this rather vague statement.

A longer-term effort was made by Ogino and coworkers [351,356, 357,361,546,363,371]. They cite one of the papers by Coughlin and coworkers [450] and ignore completely the work of Mattson and coworkers. For example, they confirm the by then well established finding that "the amount of phenol adsorbed decreased with increasing the surface acidity on the modified carbon blacks" but fail to give any prior credit for their statement that "[t]his may be attributed to the fact that the acidic functional groups . . . were introduced on the surface of the carbon blacks" [356]. Even more puzzling is that, in a subsequent paper [357], they "propose a model regarding the adsorption mechanism between carbon and phenol" without even mentioning the $\pi-\pi$ interaction or donor–acceptor complex arguments. Instead, they argue in favor of "competitive adsorption of phenol with water molecules" and state that "phenol adsorption due to the hydrophobic interaction on the carbon interferes with [the] selective adsorption [of water on the surface functional groups on the carbon]." In yet another example of grossly inappropriate citation, they argue in their introduction that "the effect of the surface-chemical structure of the adsorbents on adsorption is important" [546], provide nine references for this statement, all of which are to their own work, but fail to show how exactly such structure is important. Finally, in a recent review with an ambitious title [371], Ogino makes the rather puzzling introductory statement, that "the influence of oxygen compounds on the carbon surfaces regarding the adsorption of organic compounds has not been studied in detail," goes on to summarize the competitive adsorption model, and emphasizes a practically important conclusion that "carbon adsorbents treated with hydrogen gas [at 1000°C] have excellent adsorbability in the aqueous-phase adsorption."

Britton and Assubaie [750] studied the adsorption of ferrocenes on pyrolytic graphite and speculated that, since ferrocene does not contain a polar group to interact with the functional groups on the surface, "π–π (dispersion) interactions between ferrocene and graphite may be involved." A paper published one year later, by McGuire et al. [358] on the electrosorption of phenol on activated carbon, considered together with those of Müller and coworkers [523–525] (see Section IV.B), should have been an eye-opener regarding the mechanism(s) of adsorption of aromatics because, even though it did not "ascertain the mechanism of the phenol–surface interaction," it did cite the studies of both Coughlin and coworkers and Mattson and coworkers and it *also* discussed the importance of "the [pH$_{PZC}$] on the activated carbon." Nevertheless, the entire subsequent decade of relevant studies continued, for the most part, to follow the theretofore divergent or parallel paths. In Section IV.B.1 we traced the path of arguments based on electrostatic interactions; here we shall continue to trace the chronological path of arguments based on primarily dispersive interactions.

There are several notable exceptions to this trend. The poorly cited paper of Caturla et al. [414], with seven citations in eight years, discusses the relative adsorbabilities of molecules and anions as a function of pH (see Section IV.A.2). Based on uptake differences in unbuffered solution (when neutral species are dominant) and at pH > 13 (when anionic species are dominant), the authors concluded that "the oxygen surface groups of the carbon modify the interactions between the carbon and the aromatic ring of the solute" but also addressed the dispersive interactions (see Section IV.B.2) by invoking several key arguments (albeit in a somewhat incoherent manner), e.g., "phenol interacts with the carbon surface through the aromatic ring" and "[because] the aromatic ring of the solutes is increasingly deactivated by the substituents [in the order phenol, 4-chlorophenol, 2,4-dichlorophenol, 2,4-dinitrophenol]," it is then "expected that DNP would be adsorbed to a lesser extent because the nitro group is strongly deactivating." In their conclusions they are not specific enough, but they make the interesting statement that "the apparent surface area deduced from the adsorption of all phenols [is] affected by the chemical nature of the carbon because its pH increases with activation," as well as the arguably (and unknowingly!) prophetic statement (see Section V) that "it is also affected by the substituents of the aromatic ring, which modify the electron density of the aromatic ring."

In the largely ignored paper of Kastelan-Macan et al. [364], the authors discuss the surface charge of the activated carbon but also state that "it is more probable that the adsorption of phenol takes place by an interaction with the π-system of the graphite ring of the basal plane, and that the presence of acidic surface oxygen groups at edges of the basal plane serves to withdraw electrons from the π-system of the basal plane deactivating the aromatic ring system of the carbon." It is intriguing that their reference for this last statement is not a paper by Coughlin and coworkers but one by Drago et al. (*J. Amer. Chem. Soc. 86*, 1694, 1964)—

in fact the same one cited by Mattson et al. [24]—according to which "phenol forms strong donor–acceptor complexes with oxygen groups" [24].

The study of Woodard et al. [493] is even more enlightening than that of McGuire et al. [358]. These authors emphasize that "electrostatic interactions . . . between the carbon absorbent [sic] and organic cations are at least partially responsible for the adsorption of those species" and that "electrostatic repulsion prevented the adsorption of tyrosine and phenylaniline anions." They also provide an extensive discussion of the arguments of Coughlin and coworkers [450,331], Puri [619,346], and Mattson et al. [24] regarding the role of surface oxides. Even though both electrostatic and dispersion interactions are explicitly mentioned among the "contributing factors," the precise nature of the latter was not clarified. Indeed, rather than endorsing the π–π interaction argument, the authors intriguingly place emphasis on the "polar capping model," which they also attribute, quite inappropriately, to Coughlin et al. [331]. It seems rather unconvincing that the low-surface-area electrochemically oxidized carbon fibers used here (<20 m^2/g vs. >400 m^2/g powders used in the work of Coughlin et al. [331]) would exhibit significant molecular sieving and that the putative surface oxide/water complexes would thus "block access to regions deeper within the pore."

In an earlier report, which Woodard et al. [493] do not cite, Vasquez et al. [650] also studied the effect of electrochemical pretreatment of a carbon electrode on its adsorption behavior. In contrast to Woodard et al. [493], these authors endorsed the donor–acceptor complex proposal:

[A] reduction of the nucleophilic character of the nitro group for the meta derivatives [of nitrobenzene] is compatible with the Mattson et al. [op. cit.] charge transfer interaction model for adsorption of organic molecules at carbon containing quinodal and/or carbonyl-like surface species in which the adsorption process is visualized as a result of partial transference of charge from the active sites toward the low-lying acceptor orbitals of the aromatic nucleus of strong electron-withdrawing compounds.

They arrived at this conclusion by comparing the effect of electron donor or acceptor substituents in nitrobenzene on the magnitude of the potential shift at a freshly polished glassy carbon electrode vs. one that was electrochemically pretreated. In the latter case, the meta derivatives, for which "adsorption increases with increasing electron-withdrawing power of the substituent," the potential shift results were interpreted as evidence that "adsorption weakened the electronic effect exerted by electron-withdrawing substituents on the electroreactivity of their nitro group." (For a more detailed discussion of the effects of adsorption on electron transfer at carbon electrodes, see Refs. 673 and 751.)

Snoeyink and coworkers had been concerned with the technological and scientific aspects of the adsorption of aromatics for many years [333,752–754]. In what is arguably their most substantial scientific contribution to the discussion at hand, Wedeking et al. [755] confirm the results of Coughlin and Ezra [450] and Mahajan et al. [345] regarding the enhancement of PNP uptake by the removal of surface oxygen. They then argue that, if the donor–acceptor complex mechanism of adsorption were applicable, "the PNP adsorption capacity probably would have decreased after outgassing at 1000°C since the carbonyl groups would decompose to CO at this temperature. This effect was observed only for some carbons, however." They thus conclude that "[m]ore research is needed to better characterize the carbon's surface chemistry and the factors that affect it, and the effects of specific groups on adsorption."

In a well-referenced and important study, Zawadzki [415] was the next investigator to consider, in a very substantive way, the relative merits of the two proposals. He argued that "[i]f the specific interaction of the surface of carbon is with the aromatic nucleus, then it would be expected that the [infrared] vibrations that are most perturbed [upon adsorption] would be those associated with the aromatic ring of the adsorbed compound." He found no evidence of "competition between water and organic solute for the carbon surface"—albeit without mentioning the proposals of Ogino and coworkers or Woodard et al. (see refs. above)—but was less definitive otherwise. Not having found spectroscopic evidence for the formation of chemical adsorbate–adsorbent bonds, he concluded "that adsorption of the aromatic compounds is partly a specific interaction with the surface of carbon and occurs with the participation of π electrons of the adsorbed molecules." A similar FTIR study was performed later by Xing et al. [402] in which the " slight shift of the $C{=}O$ band [at 1580 cm^{-1}] to a higher wavenumber (1620 cm^{-1}) on the carbon with phenol . . . may have resulted from the formation of a putative donor–acceptor complex between the ketone or quinonic groups and phenol aromatic ring;" here the authors cite the work of Puri [346] when a reference to the work of Mattson et al. [24] would have been much more appropriate.

In a more recent study, Nelson and Yang [494] presented a "surface complexation model" to describe the effect of pH on adsorption equilibria of chlorophenols, i.e., the electrostatic effect; they also discussed the potential importance of π–π interactions and donor–acceptor complex formation but could not distinguish between the two and concluded, somewhat vaguely, that "[t]hese proposed mechanisms provide plausible explanations for the surface complexation reactions between chlorophenols (neutral or anionic forms) and the surface of activated carbon (acidic or basic sites)."

In the early 1990s a further complication in the adsorption of phenols was discovered [417,418]: that of oxidative coupling promoted in the presence of O_2 and especially at high pH when phenolate anions are dominant in solution. The role of oxygen in determining the adsorption capacity of activated carbons turned out to be not "illusive" [sic] [418], but elusive indeed. The resulting controversies contributed to the fact that the π–π interaction and donor–complex arguments became almost secondary issues for a while, mentioned only in passing, if at all, while the emphasis shifted to the phenomenological effects of both dissolved and surface oxygen. Thus, for example, Grant and King [417] cite the latter argument as "a popular hypothesis to rationalize irreversible adsorption," Chatzopoulos et al. [756] invoke the same argument for toluene, while Vidic and coworkers [418,495,463,462,757,423,424] and Abu Zeid and coworkers [425,429] cite the study of Coughlin et al. [450] only to illustrate the effect of surface oxidation on phenol uptake. More recently, though in the same context, Vidic et al. [478] addressed the role of the surface properties of activated carbons in oxidative coupling of phenolic compounds. They found "no relationship between surface functional group content as measured by the wet chemistry methods and irreversible adsorption" and reported that the "increased basic group content results in the highest oxic capacity and the greatest irreversible adsorption for [the carbon outgassed at 900°C]." Based on the reduced adsorption capacity of the carbon outgassed at 1200°C, they concluded, however, that "greater structural perfection and the increased basicity due to these structural changes are not the key to increasing adsorptive capacity." Rather, they interpreted their findings as evidence for "the importance of oxygen-containing basic surface functional groups in promoting irreversible adsorption," without clarifying whether this is an endorsement of the π–π interaction or the donor–acceptor complex mechanism of adsorption. A similar study was performed by Leng and Pinto [386], who interpreted similar results (increased concentrations of surface carboxylic groups were found to decrease physisorption capacity) as evidence for "increased water adsorption and weaker dispersion interactions with basal plane carbons," thus supporting the π–π interaction argument. Kilduff and King [384] also examined the issue at hand in some detail and concluded that their data "are consistent with several mechanisms which may contribute to the suppression of oxidative coupling by oxidized carbons." They appear to endorse, at least implicitly, the π–π interaction model, by saying that "the interaction energy between phenol and the surface may be reduced" upon oxygen chemisorption. In addition, they emphasize that "increased selectivity for water on oxidized carbons is likely and may contribute, in part, to the increased regenerability of these adsorbents."

The study by Streat et al. [376] focuses on the virtues of activated carbons derived from straw and used rubber tires, but it also alludes to the relative merits of our two competing proposals, in the context of the finding that "the sorption capacity for p-chlorophenol is greater than for phenol on each of the samples tested." The relevant statements are very confusing, however: after summarizing the two proposals and emphasizing that "the chemical properties of the surface are much more important than the porosity of the carbon," they state that "the electronegative chlorine atom attracts electrons towards the benzene ring thus enhancing the activation of the molecule'' and leave it to the reader to figure out how this argument helps to explain the higher uptake of p-chlorophenol. Polcaro and Palmas [758] studied the electrochemical oxidation of chlorinated phenols and benzoquinones, noted ''remarkable differences in [their] adsorption constants'' on a carbon felt electrode, and concluded that the ''presence of chlorine atoms in the ring . . . enhances the sorption capacity.'' They thus echoed the argument of Streat et al. [376]: after summarizing the arguments of Mattson and coworkers and Coughlin and coworkers, they simply state, without any differentiation or clarification, that ''[t]hese effects could be increased by the presence of electronegative chlorine atoms that attract electrons toward the benzene ring.''

The recent interest in the regeneration of activated carbons has also resulted in studies that address the key mechanistic aspects of desorption (and adsorption) of aromatics. Ferro-García et al. [453] studied the regeneration of AC exhausted with chlorophenols and attempted to elucidate the role that the oxygen surface complexes play in the regeneration process. Here and elsewhere [454] the donor–acceptor complex argument of Mattson et al. [24] is invoked to explain the thermal-regeneration-induced cracking of the adsorbed species interacting with the carbonyl groups on the activated carbon. Earlier Utrera-Hidalgo et al. [759] were less committing, and both Ref. 331 and Ref. 24 (together with the study by Mahajan et al. [345], which supports only Ref. 331 but not Ref. 24) were cited for the following statement: '' [A]dsorption of phenol-like molecules on carbon involves dispersive forces between the π-electrons in phenol and π-electrons in carbon, the aromatic ring of the phenol-like molecules acting as acceptor.''

Similar ambivalence persists to this day. In the already mentioned study of Nevskaia et al. [394] the same two studies are cited for the following statement: ''As has been previously established [450,6], the adsorption of phenol on the activated carbon may imply electron donor–acceptor complexes or may involve dispersion forces between π electrons.'' Not having considered the possibility that the decrease in phenol uptake upon surface oxidation was due either to more severe electrostatic repulsion or to suppressed π–π interactions, the authors were forced to speculate, apparently following the donor–acceptor argument, that ''the carbonyl groups have been added in a smaller extension than the carboxyl groups,'' without indicating whether their own experimental data support this possibility. The same ambiguities had been echoed even earlier, e.g., in the study of Amicarelli et al. [343], where the work of Coughlin and Ezra [450] is cited for the statement about ''electrostatic donor–acceptor interactions between AC surfaces and the π-electron system of aromatic rings.''

In subsequent work, Ferro-García and coworkers [760,761] stated that ''the extent of chemical regeneration . . . is a function of the strength of the adsorbent–adsorbate interactions'' [761]. In an earlier study, Rivera-Utrilla et al. [762] do make the more specific statement that ''the higher interaction of [m-chlorophenol] with the surface of the carbon may be due to the higher dipolar moment and slightly higher electronic density in the benzene ring of this compound in relation to the [o-chlorophenol]." In hindsight, reference to the work of Coughlin and coworkers should have been included here; this statement did, however, set the stage for the study of Tamon and coworkers (see below and Section V).

In contrast to these gas-phase temperature-programmed desorption studies by Moreno-Castilla and coworkers, Salvador and Merchán [432] carried out regeneration studies in the liquid phase. They do not cite the work of Coughlin and coworkers even though they report increasing stability of adsorbate–surface bonding with increasing number of (nitro and chloro) substituents on phenol. Intriguingly, and indeed inappropriately, they attribute to Mattson et al. [24]—as well as to Puri et al. [341] and Davis and Huang [400]—the statement that "[t]he hypothesis of adsorption through hydrogen bonds . . . is the most widely accepted." Based on calculated values of adsorption bonding energies, which were deemed to be in the same range as those of hydrogen bonds (26-74 kJ/mol), they embrace this argument and conclude that the adsorption energy is "directly related to the electrophilic character of the substituents . . . but not with the degree of substitution of the benzene ring."

In an intriguing comparison of the uptakes of heptanoic and benzoic acids by a mesoporous carbon adsorbent, Morgun and Khabalov [763] make an inconsequential reference to the study of Mattson [6], and when they should cite the work of Coughlin and coworkers, for the statement that there is "additional interaction between the π electrons of the benzoic ring and the conjugated π electron system of the carbon adsorbent," they fail to do so.

Tamon and coworkers undertook an extensive experimental and modeling study [764,765,81,380,733,734,766,767,735,768]. They never consider the donor–acceptor complex mechanism of adsorption, nor do they cite the study by Mattson et al. [24]. They do cite the studies by Coughlin and coworkers but do not endorse their proposal; in Section V it was argued that their results lend support to the π–π interaction argument. Thus, for example, "the difficulty of regeneration of spent carbon [was attributed] to adsorption irreversibility" [734], which was promoted by electron-donating substituents in aniline and phenols. In contrast, "[a]s for the adsorption of electron-attracting compounds on carbonaceous adsorbents, the adsorption is reversible" [733]. These findings were confirmed by semiempirical molecular orbital theory calculations according to which "irreversible adsorption appeared when the energy difference between HOMO of adsorbate and LUMO of adsorbent was small" [735].

Radovic and coworkers have been interested in this issue since the early 1990s [736,674,738]. The results of their experimental and modeling efforts were summarized in Section V. They endorse the π–π interaction proposal of Coughlin and coworkers and conclude that "[c]onsensus seems to be emerging regarding the effect of chemical surface properties of carbon adsorbents on the uptake of aromatic pollutants from aqueous effluents." For aromatic pollutants such as benzoic acid at high pH, they conclude that "[w]hile electrostatic adsorbate/adsorbent interactions are important, π–π dispersion interactions appear to be dominant" [674]. More specifically, for a weak electrolyte such as aniline, "electrostatic interactions between anilinium cations and negatively charged carbon surface" can be dominant at pH > pH_{PZC}, while for the nondissociating nitrobenzene, "adsorption results primarily from dispersive interactions between nitrobenzene molecules and graphene layers." In the latter case, the authors emphasize that "the removal of electron-withdrawing oxygen functional groups not only increases the point of zero charge [of the carbon], it also increase the dispersive adsorption potential by increasing the π-electron density" [738]. Whether or not the exclusion of potentially competitive effects of adsorbed water (especially for heavily oxidized carbons) is a reasonable simplification remains to be confirmed in future studies.

As a final illustration of the degree of progress, or lack thereof, in the last quarter of a century, it is perhaps most fitting to analyze two publications by Weber and coworkers [24,471] which appeared exactly quarter of a century apart. Mattson et al. [24] were the first to discuss the relative merits of the two proposals. They dismiss the π–π interaction argument by arguing that the results of Coughlin and Ezra [450] are actually consistent with their donor–acceptor complex argument: "As it cannot be expected that the C—C bonds broken in oxidation will rejoin upon reduction, much of the original carbonyl groups are not reformed, hence the capacity may not fully recover to its original value. Thus, it is reasonable to expect that the capacity of carbon for electron acceptors would go down upon oxidation of surface carbonyl groups to carboxylic acid groups." In hindsight, several counterarguments are now obvious, e.g., carbon oxidation does not (necessarily) involve the conversion of carbonyl groups to carboxyl groups and, contrary to what the authors affirm, Coughlin and Ezra [450] do not show this to be the case. Lynam et al. [471], with the same senior author, summarized recently their understanding of the adsorption of PNP. They reaffirm their confidence in the donor–acceptor complex argument: "It was postulated that the carbonyl groups present on the carbon surface donate electron density into the PNP ring. When all the carbonyl surface sites are exhausted adsorption continues by complexation of the PNP molecule with the carbon rings of the basal planes [24]. The presence of the nitro substituent, which is a strong electron-withdrawing group, facilitates a reduction in electron density in the pi-system of the ring." All this had been said, verbatim, in the 1969 paper. In light of the mechanistic confusions and contradictions illustrated above, an inquisitive reader might have expected that the authors would address some of them and thus provide stronger arguments for their proposal. A straightforward one would have been the agreement, or lack thereof, between the carbonyl group content and the PNP uptake at the putative high-concentration break in the isotherm. A more critical argument would be one that addresses the important point raised by Wedeking et al. [755] (see above), whose resolution appears to require not only the choice of one of the two competing dispersion interaction mechanisms but also a reconsideration of the electrostatic interactions.

Discussed above are the studies of researchers that have taken the bull by the horns, so to speak, by attempting to rationalize the importance of surface chemistry or to reconcile their own results with those of earlier investigations. It must be pointed out, however, that—as is too often the case—the majority of citing publications stay away from the main arguments, even when the topic demands otherwise, and cite the relevant papers either only in passing or for their other (minor) messages. It should be noted also that quite a few examples of rather careless referencing of prior work have been identified. This is unfortunate not only because credit ought to be given where credit is most due but also because the value of the Science Citation Index, as the most convenient and efficient tool for the evaluation of research, is greatly diminished when the authors do not do this, as is too often the case (see *C&E News*, October 28, 1996, p. 6). Some important examples that fall into this category are offered below in an attempt to set the record straight.

The citation of Coughlin et al. [331] by Khabalov et al. [769] is a typical case of confusion between facts and hypotheses. According to them, "[i]t is known that chemical oxidation of the surface of carbon adsorbents in the adsorption of organic substances from aqueous solutions leads to an increase in the adsorption of the more polar component, the water molecules [331]." What is known of course—as discussed above—is that oxidation suppresses adsorption of organics from aqueous solution. Whether this indeed leads to an increase in the adsorption of the more polar component is less certain; in fact, the study that the authors cite has certainly not shown this to be the case. In a very recent study, Juang and Swei [508] do not provide a clear distinction between dispersion and electrostatic interactions. They first inappropriately attribute to Mattson et al. [24] the statement about the existence of "many conjugated π-bonds which are nonlocalized and highly active." Then they incorrectly assign electron-acceptor character to carbonyl groups on the oxidized carbon surface (Mattson et al. postulate exactly the opposite, of course) and hypothesize that "electrostatic interaction occurs between the electron-acceptor groups on the [carbon] surface and the positive charges on basic dye compounds." Similar confusion is also apparent in a recent paper by Barton et al. [382] according to which adsorption "takes place mainly by the formation of electron donor–acceptor complexes between basic-sites [sic], principally π electron rich regions on the basal planes of the carbon, and the aromatic ring of the adsorbate." The key publications by Mattson and coworkers are not cited at all, even though a donor–acceptor complex is invoked. Here the authors should really have cited the work of Coughlin and coworkers, but they did not; instead they refer to the work of Singh [334] who, as mentioned above, cites Ref. 331 only in the context of surface geometry of phenol molecules.

In a peer-reviewed paper which mentions 13 references in the text, but provides the list of only the first 11, Tanada et al. [587] studied the aqueous solution removal of chloroform by a surface-modified activated carbon. They report a monotonic decrease in the amount adsorbed with increasing acidity of the AC; although the reported surface areas are essentially the same, they state that the "adsorption capacity . . . is dominated by pore structure" and that "hetero atoms [such as oxygen, nitrogen, and hydrogen] may affect [it]." Here they make a very vague reference to the paper by Coughlin et al. [450], do not mention the work of Mattson and coworkers, and conclude, again without an appropriate reference, that "aggregate of water molecules on activated carbon prevent [sic] the invasion of chloroform into the adsorption sites."

Finally, there are quite a few papers, some mentioned already, which should have cited our landmark references but failed to do so. For example, Nagaoka and Yoshino [652] studied the surface properties of electrochemically pretreated glassy carbon by cyclic voltammetry and adsorption of catechol (1,2-dihydroxybenzene). They do refer, in passing, to one of the papers by Mattson et al. [770]. However, they do not cite this work when they dismiss the donor–acceptor complex argument, nor do they cite any of the papers by Coughlin and coworkers when they state that "it is more likely that the interaction results from π–π interaction between catechol and the graphite-like structures." The important scanning tunneling microscopy study of quinone adsorption by McDermott and McCreery [720] is thought to offer strong evidence against the donor–acceptor mechanism, and to be essentially in favor of the π–π interaction mechanism, but neither one of the two studies are cited here. The uptakes were found to "exceed the geometric area of the graphitic edge plane by a factor of 30," which "rules out quinone adsorption to specific sites [and specific interactions with surface functional groups]." The authors concluded that "adsorption of quinones [on glassy carbon and pyrolytic graphite] depends on an electronic effect such as an electrostatic attraction between the adsorbate and partial surface charges, rather than a specific chemical effect."

In a recent authoritative treatise, Lyklema [66] has reinterpreted the results of Mattson et al. [24] and stated that "[p]robably the driving force [for adsorption of aromatics] is π-electron exchange between the aromatic ring and the surface." The author further noted that the "nitro group enhances the electronegativity; perhaps this is why phenol, which lacks such a group, falls below the others [nitrophenols]." This sounds like the argument of Mattson and coworkers, but it appears to be turned upside down, invoking an electron-donating rather than an electron-accepting character of the benzene ring.

To complete this list of ways that literature citations are treated, or mistreated, in a study coauthored by Coughlin [771] the landmark paper by Coughlin and Ezra [450] is listed among the references but is not cited in the text! (Who should be more embarrassed, the authors or the reviewers?) Finally, in reference to Table

39, it is worth pointing out that what was initially the less "popular" explanation turned out to be the more convincing one: even the most cited of Coughlin's papers [450] has been cited less frequently than the study of Mattson and coworkers. Whether these two arguments are necessarily mutually exclusive, as they appear to be, remains to be demonstrated in future work in which the basicity of the carbon surface will be quantified in a more precise way than can be done today.

VII. SUMMARY AND CONCLUSIONS

The evidence scrutinized in this review is overwhelmingly in favor of the notion that knowledge of the following is necessary *and sufficient* for understanding the adsorption of inorganic compounds on carbons:

1. The speciation diagram of the adsorbate
2. The unique amphoteric behavior of the adsorbent

Equilibrium uptakes of metallic cations, dyes, and even gold and anionic adsorbates are largely governed by electrostatic attraction or repulsion. We thus suggest that future studies of the role of the carbon surface first show whether this is indeed the case. If not, only then does the role of specific interactions (e.g., complex formation), or even nonspecific van der Waals interactions, become a reasonable alternative or complementary argument. Indeed, in the adsorption of inorganic solutes, the main fundamental challenge remains how to "activate" the entire carbon surface in order to achieve maximum removal efficiencies. The optimum aqueous solution conditions at which maximum uptake can be achieved depend on the surface chemistry of the adsorbent. In fact, once the surface chemistry of the carbon adsorbents has been thoroughly understood, primarily in terms of its unique amphoteric character, drawing analogies between the behavior of carbon and that of the well-studied oxide adsorbents [772,773] becomes both useful and striking.

In contrast, adsorption of organic compounds, and of aromatics in particular, is a decidedly more complex interplay of electrostatic and dispersive interactions. This is particularly true for phenolic compounds. The following arguments appear to have solid experimental and theoretical support:

1.   Adsorption of phenolic compounds is partly physical and partly chemical, and future work should carefully distinguish between the two, by combining adsorption studies with the increasingly important reactivation studies.

2.   The strength of $\pi-\pi$ interactions can be modified by ring substitution on both the adsorbate and the adsorbent, leading under favorable conditions (e.g., heating) even to chemical interactions that complicate the feasibility of thermal regeneration.

The following mechanistic issues remain controversial: (1) whether in the formation of EDA complexes it is the adsorbent or the adsorbate that acts as the electron donor; (2) to what extent the molecular orbital theory, and the difference between the HOMO and LUMO levels of the adsorbate and the adsorbent, can predict the degree and the direction of electron transfer that leads to chemisorption.

In contrast to the optimization of physical surface properties of activated carbons—where it has long been known *what* is needed for a particular adsorption or water treatment task (an adequate surface area and a specific pore size distribution) and *how* to achieve it—very few guidelines have been available for the optimization of the chemical surface properties, especially for water treatment applications. The progress made in our understanding of the surface chemistry of carbons, which has heretofore been applied to the design of carbon-supported catalysts and in understanding carbon gasification reactivity, has now been extended to aqueous-phase adsorption. Our contention is that we now know how to optimize the surface and/or solution chemistry for efficient removal of water pollutants. Not unexpectedly, we have also realized that compromises need to be made to meet the sometimes conflicting demands imposed by electrostatic or dispersion forces. Typical modifications of carbon surface chemistry affect the extent of both electrostatic and dispersive interactions. For example, carbon oxidation not only lowers its point of zero charge; it also reduces the dispersive adsorption potential by decreasing the π-electron density in the graphene layers. And conversely, the removal of electron-withdrawing oxygen functional groups not only increases the point of zero charge; it also increases the dispersive adsorption potential by increasing the π electron density.

A theoretical model that accounts for both electrostatic and dispersive interactions successfully described the qualitative trends observed in this study. In the particular case of aniline (a weak electrolyte), both theory and experiments indicate that maximum adsorption uptakes are attained on oxidized carbon surfaces at a solution pH near the adsorbate's pH_{PZC}. Accordingly, aniline adsorption is deemed to involve two parallel mechanisms: electrostatic interactions between anilinium cations and negatively charged carbon surface groups, as well as dispersive interactions between aniline molecules and graphene layers. On the other hand, for nondissociating nitrobenzene, maximum adsorption uptakes are found on heat-treated carbon surfaces, particularly at a solution pH $\sim pH_{PZC}$. Therefore, adsorption results primarily from dispersive interactions between nitrobenzene molecules and graphene layers. Dispersive interactions in general are promoted by conducting the experiments at solution pH values near the adsorbent's pH_{PZC}, at which the repulsive interactions between the surface functional groups and uncharged molecules are effectively minimized. The theoretical model invoked to account for all the above trends can in principle be used as a much needed predictor of the adsorbent and solution characteristics that will act in concert to maximize the adsorption uptake of aromatic compounds by activated carbons.

The analogy with the relevance of surface chemistry in gas reactions of carbon is quite striking. There [774] it is increasingly appreciated [618] that the chemistry of the graphene layer, while not as important as that of the edges, has a significant effect on carbon reactivity [775,776,719]. In liquid-phase adsorption on carbons we have argued here that the key to a more profound understanding of uptakes lies in the same interplay between the delocalized π electron system in the graphene layers and the oxygen functional groups at the edges. When this interplay is taken into account, and if it is controlled, the prophetic but premature conclusion of Coughlin et al. [331] can finally be realized: "In the manufacture of adsorptive carbon, control of chemical and physical process conditions might be harnessed to produce carbon surfaces suitable for particular adsorption applications."

ACKNOWLEDGMENTS

We are grateful to our colleagues and collaborators at the University of Granada (Grupo de Investigación en Carbones) and at Pennsylvania State University, who have helped to perform the studies and to develop some of the ideas summarized in this review.

APPENDIX: SPECIATION DIAGRAMS OF METALLIC IONS IN AQUEOUS SOLUTIONS

Hydrolysis reactions are common occurrences for the majority of metallic cations. This is because most of them form strong bonds with oxygen, and because OH^- ions are ubiquitous in water in concentrations that can vary over a wide range as a consequence of water's low dissociation constant. In a general way hydrolysis can be represented by

$$x M^{z+} + y H_2O = M_x(H_2O)_y^{(xz-y)+} + y H^+$$

It is quite complicated to determine the identity and stability of hydrolysis products [123], basically because many metallic ions form polynuclear hydroxides, i.e., these contain more than one metallic ion, and the equations that govern formation equilibria cease to be linear in concentration ranges of interest; furthermore, formation of these polynuclear species is responsible for a growth in the number of nonnegligible hydrolysis products. Also, the pH range over which the formation of soluble hydrolysis products lends itself to analysis is often limited by the precipitation of the hydroxide or oxide species. Another complication in some systems is that, due to the amphoteric character, the precipitate may redissolve as the pH is raised.

Soluble products of hydrolysis are particularly important in systems where cation concentrations are low and thus the cations can survive in a wide pH interval. Removal or recovery of metallic ions by adsorption on activated carbons is just such a system, because the ions are present in contaminated waters at trace levels. In this case, because of the low cation concentrations, the hydroxides that form are primarily mononuclear. There are exceptions, of course, such as Cr, which can exist as polynuclear species, even though their rate of formation at room temperature is quite slow [123].

Here we assume that the hydrolysis products are exclusively mononuclear and that the solution concentrations do not allow precipitation to occur (see Table A1). Speciation diagrams can thus be constructed from knowledge of global hydrolysis constants β_i, which govern the formation of mononuclear hydroxides. For example, if a metallic cation M^{z+} forms several hydroxide species $M(OH)^{(z-1)+}$, $M(OH)_2^{(z-2)+}$, $M(OH)_3^{(z-3)+}$, ... , $M(OH)_n^{(z-n)+}$, the corresponding hydrolysis equilibria are

$$M^{z+} + H_2O = M(OH)^{(z-1)+} + H^+ \qquad\qquad \beta_1$$

$$M^{z+} + 2H_2O = M(OH)_2^{(z-2)+} + 2H^+ \qquad\qquad \beta_2$$

$$M^{z+} + 3H_2O = M(OH)_3^{(z-3)+} + 3H^+ \qquad\qquad \beta_3$$

$$\vdots$$

$$M^{z+} + nH_2O = M(OH)_n^{(z-n)+} + nH^+ \qquad\qquad \beta_n$$

$$[M(OH)^{(z-1)+}] = \beta_1 \frac{[M^{z+}]}{[H^+]}$$

$$[M(OH)_2^{(z-2)+}] = \beta_2 \frac{[M^{z+}]}{[H^+]^2}$$

$$[M(OH)_3^{(z-3)+}] = \beta_3 \frac{[M^{z+}]}{[H^+]^3}$$

$$\vdots$$

$$[M(OH)_n^{(z-n)+}] = \beta_n \frac{[M^{z+}]}{[H^+]^n}$$

TABLE A1   Stability Constants (log K) for Various Lead-Hydroxo Species

Metal	Equilibrium	log K ($I = 5 \times 10^{-2}$ M)
Pb(II)	$Pb^{2+} + OH^- = Pb(OH)^+$	5.70
	$Pb^{2+} + 2OH^- = Pb(OH)_2$	9.84
	$Pb^{2+} + 3OH^- = Pb(OH)_3^-$	12.64

Source: Smith RM, Martell AE. *Critical Solubility Constants*. New York: Plenum Press, 1976.

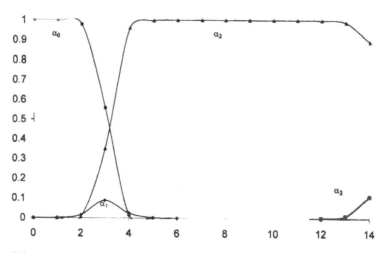

FIG. A1   Speciation diagram for Zn(II).

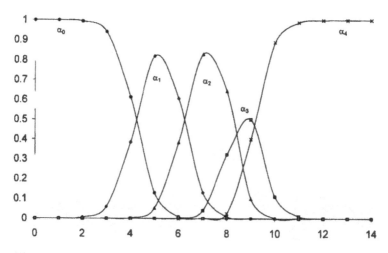

FIG. A2   Speciation diagram for Hg(II).

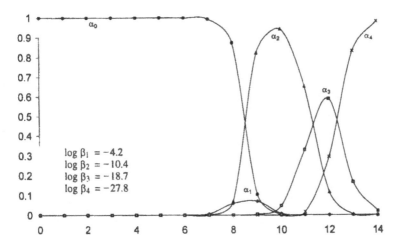

FIG. A3 Speciation diagram for Cr(III).

If C is the concentration of metal M in solution, then the material balance results in

$$C = [M^{z+}] + [M(OH)^{(z-1)+}] + [M(OH)_2^{(z-2)+}]$$

$$+ [M(OH)_3^{(z-3)+}] + \cdots + [M(OH)_n^{(z-n)+}]$$

$$C = [M] \left(1 + \frac{\beta_1}{[H^+]} + \frac{\beta_2}{[H^+]^2} + \frac{\beta_3}{[H^+]^3} + \cdots + \frac{\beta_n}{[H^+]^n} \right)$$

Therefore the fraction of metal α_i present as each one of the species can be given as a function of $[H^+]$ only:

$$\alpha_0 = \frac{[M^{z+}]}{C} = \left(1 + \frac{\beta_1}{[H^+]} + \frac{\beta_2}{[H^+]^2} + \frac{\beta_3}{[H^+]^3} + \cdots + \frac{\beta_n}{[H^+]^n} \right)^{-1}$$

$$\alpha_1 = \frac{[M(OH)^{(z-1)+}]}{C} = \frac{\beta_1 \alpha_0}{[H^+]}$$

$$\alpha_2 = \frac{[M(OH)_2^{(z-2)+}]}{C} = \frac{\beta_2 \alpha_0}{[H^+]^2}$$

$$\alpha_3 = \frac{[M(OH)_3^{(z-3)+}]}{C} = \frac{\beta_3 \alpha_0}{[H^+]^3}$$

$$\vdots$$

$$\alpha_n = \frac{[M(OH)_n^{(z-n)+}]}{C} = \frac{\beta_n \alpha_0}{[H^+]^n}$$

Thus, for example, the speciation diagrams for Zn(II), Hg(II), and Cr(III) can be obtained from their respective global hydrolysis constants [J. Kragten, *Atlas of Metal-Ligand Equilibria in Aqueous Solution*, New York: John Wiley, 1978], and this is shown in Figs. A1–A3.

Zn^{2+}	Hg^{2+}
$\log \beta_1 = -9.17$	$\log \beta_1 = -3.8$
$\log \beta_2 = -17.1$	$\log \beta_2 = -6.2$
$\log \beta_3 = -28.4$	$\log \beta_3 = -21.1$
$\log \beta_4 = -40.7$	

For polynuclear species, see Kragten's atlas.

REFERENCES

1. Deitz VR. *Bibliography of Solid Adsorbents*. Washington, DC: National Bureau of Standards, 1944.
2. Hassler JW, Goetz JW. In: Kirk RE, Othmer DF, eds. *Encyclopedia of Chemical Technology*. New York: Interscience Encyclopedia, 1948, pp. 881–899.
3. Hassler JW. *Active Carbon*. Brooklyn, NY: Chemical Publishing, 1951.
4. Deitz VR. *Bibliography of Solid Adsorbents 1943 to 1953*. Washington, DC: National Bureau of Standards, 1956.
5. Hassler JW. *Activated Carbon*. New York: Chemical Publishing, 1963.
6. Mattson JS, Mark HB Jr. *Activated Carbon: Surface Chemistry and Adsorption from Solution*. New York: Marcel Dekker, 1971.
7. Baker JR, Mihelcic JR, Luehrs DC, Hickey JP. *Water Environ. Res.* 69:136–145, 1997.
8. Gawlik BM, Sotiriou N, Feicht EA, Schulte-Hostede S, Kettrup A. *Chemosphere* 34(12):2525–2551, 1997.
9. Parfitt GD, Rochester CH, eds. *Adsorption from Solution at the Solid/Liquid Interface*. London: Academic Press, 1983.
10. Radeke KH, Lohse U, Struve K, Weiss E, Schröder H. *Zeolites* 13(1):69–70, 1993.
11. Blum DJW, Suffet IH, Duguet JP. *Water Res.* 28:687–699, 1994.
12. Jackson PT, Schure MR, Weber TP, Carr PW. *Anal. Chem.* 69:416–425, 1997.
13. Matsumoto MR, Jensen JN, Reed BE. *Water Environ. Res.* (June 1995):419.
14. Matsumoto MR, Jensen JN, Reed BE, Lin W. *Water Environ. Res.* 68(4):431, 1996.
15. Reed BE, Li W, Matsumoto MR, Jensen JN. *Water Environ. Res.* 69(4):444, 1997.
16. Reed BE, Matsumoto MR, Jensen J, Viadero R, Lin W. *Water Environ. Res.* 70(4):449–473, 1998.
17. Reed BE, Matsumoto MR, Viadero R, Segar R, Vaughan R, Masciola D. *Water Environ. Res.* 71(5):584–618, 1999.

18. Zettlemoyer AC, Narayan KS. In: Walker PL Jr., ed. *Chemistry and Physics of Carbon*, Vol. 2. New York: Marcel Dekker, 1966, pp. 197–224.
19. Weber WJ Jr. In: Faust SD, Hunter JV, eds. *Principles and Applications of Water Chemistry*. New York: John Wiley, 1967, pp. 89–126.
20. Weber WJ Jr., Van Vliet BM. In: Suffet IH, McGuire MJ, eds. *Activated Carbon Adsorption of Organics from the Aqueous Phase*, Vol. 1. Ann Arbor, MI: Ann Arbor Science, 1980, pp. 15–41.
21. Weber WJ Jr., DiGiano FA. *Process Dynamics in Environmental Systems*. New York: John Wiley, 1996.
22. Radovic LR, Rodríguez-Reinoso F. In: Thrower PA, ed. *Chemistry and Physics of Carbon*, Vol. 25. New York: Marcel Dekker, 1997, pp. 243–358.
23. Cookson JT. In: Cheremisinoff PN, Ellerbusch F, eds. *Carbon Adsorption Handbook*. Ann Arbor, MI: Ann Arbor Science, 1978, pp. 241–279.
24. Mattson JS, Mark HB Jr., Malbin MD, Weber WJ Jr., Crittenden JC. *J. Colloid Interf. Sci.* 31(1):116–130, 1969.
25. Cini R, Pantani F, Sorace G. In: Suffet IH, McGuire MJ, eds. *Activated Carbon Adsorption of Organics from the Aqueous Phase*, Vol. 1. Ann Arbor, MI: Ann Arbor Science, 1980, pp. 425–446.
26. Perrich JR. *Activated Carbon Adsorption for Wastewater Treatment*. Boca Raton, FL: CRC Press, 1981.
27. McDougall GJ, Fleming CA. In: Streat M, Naden D, eds. *Ion Exchange and Sorption Processes in Hydrometallurgy*. Chichester: John Wiley, 1987, pp. 56–126.
28. Derylo-Marczewska A, Jaroniec M. *Surf. Colloid Sci.* 14:301–377, 1987.
29. Najm IN, Snoeyink VL, Lykins BW Jr. Adams JQ. *J. AWWA* (January 1991):65–76.
30. Haist-Gulde B, Baldauf G, Brauch H-J. In: Hrubec J, ed. *Water Pollution. Quality and Treatment of Drinking Water*. New York: Springer-Verlag, 1995, pp. 129–138.
31. Haist-Gulde B, Baldauf G, Brauch H-J. In: Hrubec J, ed. *Water Pollution. Quality and Treatment of Drinking Water*. New York: Springer-Verlag, 1995, pp. 103–128.
32. Faust SD, Aly OM. *Chemistry of Water Treatment*. Chelsea, MI: Ann Arbor Press, 1998.
33. Faust SD, Aly OM. *Adsorption Processes for Water Treatment*. Boston: Butterworths, 1987.
34. Jankowska H, Swiatkowski A, Choma J. *Active Carbon*. New York: Ellis Horwood, 1991.
35. Bansal RC, Donnet J-B, Stoeckli F. *Active Carbon*. New York: Marcel Dekker, 1988.
36. Kinoshita K. *Carbon: Electrochemical and Physicochemical Properties*. New York: Wiley-Interscience, 1988.
37. Leon y Leon CA, Radovic LR. In: Thrower PA, ed. *Chemistry and Physics of Carbon*, Vol. 24. New York: Marcel Dekker, 1994, pp. 213–310.
38. Boehm HP. *Carbon* 32(5):759–769, 1994.
39. Rodríguez-Reinoso F, Linares-Solano A. In: Thrower PA, ed. *Chemistry and Physics of Carbon*, Vol. 21. New York: Marcel Dekker, 1989, pp. 1–146.
40. Patrick JW, ed. *Porosity in Carbons*. New York: Halsted Press, 1995.

41. Garten VA, Weiss DE. *Rev. Pure Appl. Chem.* 7:69–122, 1957.
42. James AM. In: Good RJ, Stromberg RR, eds. *Surface and Colloid Science.* New York: Plenum Press, 1979, pp. 121–185.
43. Huang CP, Ostovic FB. *J. Environ. Eng.* 104:863–878, 1978.
44. Huang CP, Smith EH. In: Cooper WJ, ed. *Chemistry in Water Reuse.* Ann Arbor, MI: Ann Arbor Science, 1981, pp. 355–400.
45. Jevtitch MM, Bhattacharyya D. *Chem. Eng. Commun.* 23:191–213, 1983.
46. Huang CP, Fu PLK. *J. WPCF 56*:233–242, 1984.
47. Dobrowolski R, Jaroniec M, Kosmulski M. *Carbon 24*:15–20, 1986.
48. Huang CP, Vane LM. *J. WPCF 61*:1596–1603, 1989.
49. Abotsi GMK, Scaroni AW. *Carbon 28*(1):79–84, 1990.
50. Solar JM, Leon y Leon CA, Osseo-Asare K, Radovic LR. *Carbon 28*:369–375, 1990.
51. López-Ramón MV, Moreno-Castilla C, Rivera-Utrilla J, Hidalgo-Alvarez R. *Carbon 31*(5):815–819, 1993.
52. Menéndez JA, Illán-Gómez MJ, Leon y Leon CA, Radovic LR. *Carbon 33*:1655–1657, 1995.
53. Polovina, MJ, Babic BM, Kaludjerovic BV, Dekanski, A. *Carbon 35*:1047–1052, 1997.
54. García AB, Cuesta A, Montés-Morán MA, Martínez-Alonso A, Tascón JMD. *J. Colloid Interf. Sci.* 192:363–367, 1997.
55. Frampton VL, Gortner RA. *J. Phys. Chem.* 41:567–582, 1937.
56. Kratohvil S, Matijevic E. *Colloids Surfaces 5*:179–186, 1982.
57. Fuerstenau DW. *Pure Appl. Chem.* 24(1):135–164, 1970.
58. Brunelle JP. *Pure Appl. Chem.* 50:1211–1229, 1978.
59. Schultz M, Grimm S, Burckhardt W. *Solid State Ionics 63–65*:18–24, 1993.
60. Snoeyink VL, Weber WJ Jr. In: Danielli JF, Rosenberg MD, Cadenhead DA, eds. *Progress in Surface and Membrane Science*, Vol. 5. New York: Academic Press, 1972, pp. 63–115.
61. James RO, Parks GA. In: Matijevic E, ed. *Surface and Colloid Science*, Vol. 12. New York: Plenum Press, 1982, pp. 119–216.
62. Akratopulu KC, Kordulis C, Lycourghiotis A. *J. Chem. Soc. Faraday Trans.* 86: 3437–3440, 1990.
63. Mieth JA, Schwarz JA, Huang Y-J, Fung SC. *J. Catal.* 122:202–205, 1990.
64. Zalac S, Kallay N. *J. Colloid Interf. Sci.* 149(1):233–240, 1992.
65. Jayaweera P, Hettiarachchi S, Ocken H. *Colloids Surfaces A 85*:19–27, 1994.
66. Lyklema J. *Fundamentals of Interface and Colloid Science.* Volume II. *Solid–Liquid Interfaces.* London: Academic Press, 1995.
67. Leon y Leon CA, Solar JM, Calemma V, Radovic LR. *Carbon 30*:797–811, 1992.
68. Dixon S, Cho EH, Pitt CH. *AIChE Symp. Ser.* 74(173):75–83, 1978.
69. Moreno-Castilla C, Rivera-Utrilla J, López-Ramón MV, Carrasco-Marín F. *Carbon 33*:845–851, 1995.
70. Lau AC, Furlong DN, Healy TW, Grieser F. *Colloids Surfaces 18*:93–104, 1986.
71. Fabish TJ, Schleifer DE. *Carbon 22*:19–38, 1984.
72. Fabish TJ, Schleifer DE. In: Sarangapani S, Akridge JR, Schumm B, eds. *Proceed-*

ings of the Workshop on the Electrochemistry of Carbon. Pennington, NJ: Electrochemical Society, 1984, pp. 79–109.

73. Papirer E, Li S, Donnet J-B. Carbon 25(2):243–247, 1987.
74. Hunter CA, Sanders JKM. J. Amer. Chem. Soc. 112:5525–5534, 1990.
75. Hunter CA, Singh J, Thornton JM. J. Mol. Biol. 218:837–846, 1991.
76. Hunter CA, Lu X-Y, Kapteijn GM, van Koten G. J. Chem. Soc. Faraday Trans. 91:2009–2015, 1995.
77. Chipot C, Jaffe R, Maigret B, Pearlman DA, Kollman PA. J. Amer. Chem. Soc. 118:11217–11224, 1996.
78. Topsom RD. In: Taft RW, ed. Progress in Physical Organic Chemistry, Vol. 12. New York: Wiley-Interscience, 1976, pp. 1–20.
79. Grunwald E. CHEMTECH (November 1984):698–705.
80. Taft RW, Topsom RD. In: Taft RW, ed. Progress in Physical Organic Chemistry, Vol. 16. New York: Wiley-Interscience, 1987, pp. 1–83.
81. Tamon H, Aburai K, Abe M, Okazaki M. J. Chem. Eng. Japan 28:823–829, 1995.
82. Menendez JA, Phillips J, Xia B, Radovic LR. Langmuir 12:4404–4410, 1996.
83. Menéndez JA, Radovic LR, Xia B, Phillips J. J. Phys. Chem. 100:17243–17248, 1996.
84. Menéndez JA, Xia B, Phillips J, Radovic LR. Langmuir 13:3414–3421, 1997.
85. Radovic LR. In:Schwarz JA, Contescu CI, eds. Surfaces of Nanoparticles and Porous Materials. New York: Marcel Dekker, 1999, pp. 529–565.
86. Brédas JL, Street GB. J. Chem. Phys. 90:7291–7299, 1989.
87. Suzuki S, Green PG, Bumgarner RE, Dasgupta S, Goddard WA III, Blake GA. Science 257:942–945, 1992.
88. Dougherty DA. Science 271:163–168, 1996.
89. Montes-Morán MA, Menéndez JA, Fuente E, Suárez D. J. Phys. Chem. B 102: 5595–5601, 1998.
90. Padin J, Yang RT. Ind. Eng. Chem. Res. 36:4224–4230, 1997.
91. Snoeyink VL, Jenkins D. Water Chemistry. New York: John Wiley, 1980.
92. Walker PL Jr., Janov J. J. Colloid Interf. Sci. 28:449–458, 1968.
93. Kaneko Y, Ohbu K, Uekawa N, Fujie K, Kaneko K. Langmuir 11:708–710, 1995.
94. Müller EA, Rull LF, Vega LF, Gubbins KE. J. Phys. Chem. 100:1189–1196, 1996.
95. Farmer RW, Dussert BW, Kovacic SL. Amer. Chem. Soc. Div. Fuel Chem. Preprints 41(1):456–460, 1996.
96. Greinke RA, Bretz RI, U.S. Patent 5,582,811 (UCAR Carbon Technology Corporation), December 10, 1996.
97. Huang CP. In: Cheremisinoff PN, Ellerbusch F, eds. Carbon Adsorption Handbook. Ann Arbor, MI: Ann Arbor Science, 1978, pp. 281–329.
98. Peters RW, Ku Y. Reactive Polymers 5:93–104, 1987.
99. Bagreev AA, Tarasenko YA. Russ. J. Phys. Chem. 67(6):1111–1113, 1993.
100. Bao ML, Griffini O, Santianni D, Barbieri K, Burrini D, Pantani F. Water Res. 33(13):2959–2970, 1999.
101. Adams MD. Miner. Eng. 7:1165–1177, 1994.
102. Ilic MR, Jovanic PB, Radosevic PB, Rajakovic LV. Sep. Sci. Technol. 30:2707–2729, 1995.
103. Camara S, Wang Z, Ozeki S, Kaneko K. J. Colloid Interf. Sci. 162:520–522, 1994.

104. Gonce E, Voudrias E. *Water Res. 28*:1059, 1994.
105. Siddiqi M, Zhai W, Amy G, Mysore C. *Water Res. 30*:1651, 1996.
106. Rajakovic LV, Ristic MD. *Carbon 34*:769–774, 1996.
107. Govindarao VMH, Gopalakrishna KV. *Ind. Eng. Chem. Res. 34*:2258–2271, 1995.
108. Pizzio LR, Cáceres CV, Blanco MN. *J. Colloid Interf. Sci. 190*:318–326, 1997.
109. Tarasenko YA, Antonov SP, Bagreev AA, Mardanenko VK, Reznik GV. *Ukr. Khim. Zh. 57*(4):385–389, 1991.
110. Seron A, Benaddi H, Beguin F, Frackowiak E, Bretelle JL, Thiry MC, Bandosz TJ, Jagiello J, Schwarz JA. *Carbon 34*:481–487, 1996.
111. Frackowiak E. *Fuel 77*(6):571–575, 1998.
112. Green-Pedersen H, Korshin G. *Environ. Sci. Technol. 33*(15):2633–2637, 1999.
113. Qadeer R, Hanif J, Saleem M, Afzal M. *Collect. Czech. Chem. Commun. 57*:1–8, 1992.
114. Onganer Y, Temur C. *J. Colloid Interf. Sci. 205*:241–244, 1998.
115. Pakula M, Biniak S, Swiatkowski A. *Langmuir 14*:3082–3089, 1998.
116. Yang OB, Kim JC, Lee JS, Kim YG. *Ind. Eng. Chem. Res. 32*:1692–1697, 1993.
117. Qadeer R, Hanif J. *Carbon 33*(2):215–220, 1995.
118. Puri BR. In: Walker PL Jr., ed. *Chemistry and Physics of Carbon*, Vol. 6. New York: Marcel Dekker, 1970, pp. 191–282.
119. Cheremisinoff PN, Ellerbusch F, eds. *Carbon Adsorption Handbook*. Ann Arbor, MI: Ann Arbor Science, 1978.
120. Mahajan OP, Youssef A, Walker PL Jr. *Sep. Sci. Technol. 13*(6):487–499, 1978.
121. Huguenin D, Stoeckli F. *J. Chem. Soc. Faraday Trans. 89*(6):939–941, 1993.
122. Richard FC, Bourg ACM. *Water Res. 25*:807, 1991.
123. Baes Jr. CF, Mesmer RE. *The Hydrolysis of Cations*. New York: John Wiley, 1976.
124. Leyva-Ramos R, Juarez-Martinez A, Guerrero-Coronado RM. *Water Sci. Technol. 30*:191–197, 1994.
125. Leyva-Ramos R, Fuentes-Rubio L, Guerrero-Coronado RM, Mendoza-Barron J. *J. Chem. Technol. Biotechnol. 62*:64–67, 1995.
126. Huang CP, Wu M-H. *J. Water Poll. Contr. Fed. 47*:2437–2446, 1975.
127. Huang CP, Wu MH. *Water Res. 11*:673–679, 1977.
128. Bowers AR, Huang CP. *Prog. Water Technol. 12*:629–650, 1980.
129. Hart O. *Prog. Water Technol. 12*: 1308–1312, 1980.
130. Hayashi K, Hirashima T, Kitagawa M, Manabe O. *Nippon Kagaku Kaishi* 1981: 1951–1956.
131. Alaerts GJ, Jitjaturunt V, Kelderman P. *Water Sci. Technol. 21*:1701–1704, 1989.
132. Golub D, Oren Y. *J. Appl. Electrochem. 19*:311–316, 1989.
133. Lee SE, Shin HS, Paik BC. *Water Res. 23*:67–72, 1989.
134. Srivastava SK, Tyagi R, Pant N. *Water Res. 23*:1161–1165, 1989.
135. D'Avila JS, Matos CM, Cavalcanti MR. *Water Sci. Technol. 26*:2309–2312, 1992.
136. Gajghate DG, Saxena ER, Aggarwal AL. *Water, Air, and Soil Pollution 65*:329–337, 1992.
137. Jayson GG, Sangster JA, Thompson G, Wilkinson MC. *Carbon 31*(3):487–492, 1993.
138. Sharma DC, Forster CF. *Water Res. 27*:1201–1208, 1993.
139. Bautista-Toledo I, Rivera-Utrilla J, Ferro-García MA, Moreno-Castilla C. *Carbon 32*(1):93–100, 1994.

140. Moreno-Castilla C, Ferro-García MA, Rivera-Utrilla J, Joly JP. *Energy Fuels 8*: 1233–1237, 1994.
141. Sharma DC, Forster CF. *Biores. Technol. 47*:257–264, 1994.
142. Sharma DC, Forster CF. *Biores. Technol. 49*:31–40, 1994.
143. Zouboulis AI, Lazaridis NK, Zamboulis D.*Sep. Sci. Technol. 29*:385–400, 1994.
144. Namasivayam C, Yamuna RT. *Chemosphere 30*:561–578, 1995.
145. Perez-Candela M, Martin-Martinez JM, Torregrosa-Macia R. *Water Res. 29*:2174–2180, 1995.
146. Sharma DC, Forster CF. *Process Biochem. 31*:213–218, 1996.
147. Sharma DC, Forster CF. *Water SA 22*(2):153–160, 1996.
148. Tereshkova SG. *Russ. J. Phys. Chem. 70*: 1020–1025, 1996.
149. Bulewicz EM, Kozak A, Kowalski Z. *Ind. Eng. Chem. Res. 36*:4381–4384, 1997.
150. Farmer JC, Bahowick SM, Harrar JE, Fix DV, Martinelli RE, Vu AK, Carroll KL. *Energy Fuels 11*:337–347, 1997.
151. Ouki SK, Neufeld RD. *J. Chem. Technol. Biotechnol. 70*:3–8, 1997.
152. Ferro-García MA, Rivera-Utrilla J, Bautista-Toledo I, Moreno-Castilla C. *Langmuir 14*:1880–1886, 1998.
153. Lalvani SB, Wiltowski T, Hübner A, Weston A, Mandich N. *Carbon 36*(7–8): 1219–1226, 1998.
154. Aggarwal D, Goyal M, Bansal RC. *Carbon 37*:1989–1997, 1999.
155. Bello G, Cid R, García R, Arriagada R. *J. Chem. Technol. Biotechnol. 74*(9):904–910, 1999.
156. Selomulya C, Meeyoo V, Amal R. *J. Chem. Technol. Biotechnol. 74*(2):111–122, 1999.
157. Park S-J, Park B-J, Ryu S-K. *Carbon 37*:1223–1226, 1999.
158. El-Bayoumy S, El-Kolaly M. *J. Radioanal. Chem. 68*:7–13, 1982.
159. Cruywagen JJ, de Wet HF. *Polyhedron 7*:547, 1988.
160. Dun JW, Gulari E, Ng KYS. *Appl. Catal. 15*:247–263, 1985.
161. Rondon S, Proctor A, Houalla M, Hercules DM. *J. Phys. Chem. 99*:327–331, 1995.
162. Rondon S, Wilkinson WR, Proctor A, Houalla M, Hercules DM. *J. Phys. Chem. 99*:16709–16713, 1995.
163. Rivera-Utrilla J, Ferro-Garcia MA, Mata-Arjona A, Gonzalez-Gomez C. *J. Chem. Technol. Biotechnol. 34A*:243–250, 1984.
164. Rivera-Utrilla J, Ferro-García MA. *Ads. Sci. Technol. 3*:293–302, 1986.
165. Rivera-Utrilla J, Ferro-García MA. *Carbon 25*:645–652, 1987.
166. Netzer A, Hughes DE. *Water Res. 18*:927–933, 1984.
167. Huang CP, Tsang MW, Hsieh YS. In: Peters RW, Kim BM, eds. *Separation of Heavy Metals and Other Trace Contaminants*. New York: American Institute of Chemical Engineers, 1985, pp. 85–98.
168. Paajanen A, Lehto J, Santapakka T, Morneau JP. *Sep. Sci. Technol. 32*:813–826, 1997.
169. Teker M, Saltabas O, Imamoglu M. *J. Environ. Sci. Health A 32*(8):2077–2086, 1997.
170. Takiyama LR, Huang CP. Surface acidity of activated carbon and its relation to metal ion adsorption. 23rd *Biennial Conference on Carbon*. State College, PA: American Carbon Society, 1997, 56–57.
171. Corapcioglu MO, Huang CP. *Water Res. 21*(9):1031–1044, 1987.

172. Seco A, Marzal P, Gabaldón C, Ferrer J. *J. Chem. Technol. Biotechnol. 68*:23–30, 1997.
173. Bhattacharyya D, Cheng CYR. *Environ. Prog. 6*:110–118, 1987.
174. Reed BE, Nonavinakere SK. *Sep. Sci. Technol. 27*(14):1985–2000, 1992.
175. Brennsteiner A, Zondlo JW, Stiller AH, Stansberry PG, Tian D, Xu Y. *Energy Fuels 11*:348–353, 1997.
176. Deorkar NV, Tavlarides L. *Environ. Prog. 17*(2):120–125, 1998.
177. Mohapatra SP, Siebel M, Alaerts G. *J. Environ. Sci. Health A 28*(3):615–629, 1993.
178. Khalfaoui B, Meniai AH, Borja R. *J. Chem. Technol. Biotechnol. 64*:153–156, 1995.
179. Periasamy K, Namasivayam C. *Chemosphere 32*(4):769–789, 1996.
180. Wilkins E, Yang Q. *J. Environ. Sci. Health A 31*(9): 2111–2128, 1996.
181. Karthikeyan KG, Elliott HA, Cannon FS. *Environ. Sci. Technol. 31*:2721–2725, 1997.
182. Gupta VK. *Ind. Eng. Chem. Res. 37*:192–202, 1998.
183. Devi PR, Naidu GRK. *Analyst 115*:1469–1471, 1990.
184. Chang C, Ku Y. *Sep. Sci. Technol. 80*:899–915, 1995.
185. Chang C, Ku Y. *J. Chin. Inst. Eng. 20*(6):651–659, 1997.
186. Chang C, Ku Y. *Sep. Sci. Technol. 33*(4):483–501, 1998.
187. Ferro-Garcia MA, Rivera-Utrilla J, Rodriguez-Gordillo J, Bautista-Toledo J. *Carbon 26*:363–373, 1988.
188. Kahn MA, Khattak YI. *Carbon 30*:957–960, 1992.
189. Petrov N, Budinova T, Khavesov I. *Carbon 30*:135–139, 1992.
190. Budinova TK, Gergova KM, Petrov NV, Minkova VN. *J. Chem. Technol. Biotechnol. 60*:177–182, 1994.
191. Khattak Y, Kian M. *J. Chem. Soc. Pakistan 17*(4):194–199, 1995.
192. Chen J, Yiacoumi S, Blaydes T. *Separations Technology 6*(2):133–146, 1996.
193. Namasivayam C, Kadirvelu K. *Chemosphere 34*(2):377–399, 1997.
194. Biniak S, Pakula M, Szymanski G, Swiatkowski A. *Langmuir 15*:6117–6122, 1999.
195. Seco A, Gabaldón C, Marzal P, Aucejo A. *J. Chem. Technol. Biotechnol. 74*(9): 911–918, 1999.
196. Allen SJ, Brown PA. *J. Chem. Technol. Biotechnol. 62*:17–24, 1995.
197. Allen SJ, Whitten LJ, Murray M, Duggan O, Brown P. *J. Chem. Technol. Biotechnol. 68*:442–452, 1997.
198. Johns MM, Marshall W, Toles C. *J. Chem. Technol. Biotechnol. 71*(2):131–140, 1998.
199. Mishra SP, Chaudhury GR. *J. Chem. Technol. Biotechnol. 59*:359–364, 1994.
200. Gabaldon C, Marzal P, Ferrer J, Seco A. *Water Res. 30*(12):3050–3060, 1996.
201. Marzal P, Seco A, Gabaldón C, Ferrer J. *J. Chem. Technol. Biotechnol. 66*:279–285, 1996.
202. Carrott P, Ribero Carrott M, Nabias J, Prates Ramalho J. *Carbon 35*(3):403–410, 1997.
203. Seco A, Marzal P, Gabaldón C, Ferrer J. *Sep. Sci. Technol. 34*(8):1577–1593, 1999.
204. Reed BE, Matsumoto MR. *Sep. Sci. Technol. 28*:2179–2195, 1993.
205. Polovina M, Surbek A, Lausevic M, Kaludjerovic B. *J. Serb. Chem. Soc. 60*:43–49, 1995.

206. Rubin AJ, Mercer DL. In: Anderson MA, Rubin AJ, eds. *Adsorption of Inorganics at Solid–Liquid Interfaces*. Ann Arbor, MI: Ann Arbor Science, 1981, pp. 295–325.
207. Periasamy K, Namasivayam C. *Ind. Eng. Chem. Res. 33*:317–320, 1994.
208. Macías-García A, Valenzuela-Calahorro C, Espinosa-Mansilla A, Gómez-Serrano V. *Anal. Quím. 91*(7–8):547–552, 1995.
209. Huang CP, Wirth PK. *J. Environ. Eng. 108*:1280–1299, 1982.
210. Mostafa MR. *Ads. Sci. Technol. 15*(8):551–557, 1997.
211. Jia Y, Thomas K. *Langmuir 16*:1114–1122, 2000.
212. Namasivayam C, Periasamy K. *Water Res. 27*:1663–1668, 1993.
213. Thiem L, Badorek D, O'Connor JT. *J. AWWA* (August 1976): 447–451.
214. Yoshida H, Kamegawa K, Arita S. *Nippon Kagaku Kaishi 1976*: 1596–1601.
215. Yoshida H, Kamegawa K, Arita S. *Nippon Kagaku Kaishi 1976*: 808–813.
216. Humenick Jr. MJ, Schnoor JL. *J. Environ. Eng. 100*:1249–1262, 1974.
217. Rivin D. *Rubber Chem. Technol. 44*:307–343, 1971.
218. López-González JD, Moreno-Castilla C, Guerrero-Ruiz A, Rodriguez-Reinoso F. *J. Chem. Technol. Biotechnol. 32*:575–579, 1982.
219. Sinha RK, Walker PL Jr. *Carbon 10*:754–756, 1972.
220. Adams MD. *Hydrometallurgy 26*:201–210, 1991.
221. Huang CP, Blankenship DW. *Water Res. 18*:37–46, 1984.
222. Jayson GG, Sangster JA, Thompson G, Wilkinson MC. *Carbon 25*:523–531, 1987.
223. Gómez-Serrano V, Macías-García A, Espinosa-Mansilla A, Valenzuela-Calahorro C. *Water Res. 32*:1–4, 1998.
224. Pearson RG. *Chemical Hardness*. Weinheim, Germany: Wiley-VCH, 1997.
225. Kaneko K. *Carbon 26*(6):903–905, 1988.
226. Youssef AM, Khalil LB, El-Nabarawy T. *Afinidad 49*(441):339–341, 1992.
227. Polcaro AM, Slavik E, Palmas S. *Annali di Chimica 85*:583–599, 1995.
228. Ma X, Subramanian KS, Chakrabarti CL, Guo R, Cheng J, Lu Y, Pickering WF. *J. Environ. Sci. Health A27*:1389–1404, 1992.
229. Peräniemi S, Hannonen S, Mustalahti H, Ahlgrén M. *Fresenius J. Anal. Chem. 349*(7):510–515, 1994.
230. Shirakashi T, Tanaka K, Tamura T, Yosihara S. *Nippon Kagaku Kaishi* 1999:137–143.
231. Carrott P, Ribero Carrott M, Nabias J. *Carbon 36*(1–2):11–17, 1998.
232. Greenwood NN, Earnshaw A. *Chemistry of the Elements*. Oxford, UK: Pergamon Press, 1984.
233. Kuennen RW, Taylor RM, Van Dyke K, Groenevelt K. *J. AWWA* (February 1992): 91–101.
234. Cheng J, Subramanian K, Chakrabarti C, Guo R, Ma X, Lu Y, Pickering WF. *J. Environ. Sci. Health A 28*(1):51–71, 1993.
235. Reed BE, Arunachalam S, Thomas B. *Environ. Progress 13*(1):60–64, 1994.
236. Periasamy K, Namasivayam C. *Sep. Sci. Technol. 30*:2223–2237, 1995.
237. Reed BE. *Sep. Sci. Technol. 30*(1):101–116, 1995.
238. Reed BE, Jamil M, Thomas B. *Water Environ. Res. 68*:877, 1996.
239. Raji C, Manju G, Anirudhan T. *Ind. J. Eng. Mater. Sci. 4*(6):254–260, 1997.
240. Lee M, Shin H, Lee S, Park J, Yang J. *Sep. Sci. Technol. 33*(7):1043–1056, 1998.

241. Taylor RM, Kuennen R. *Environ. Prog. 13*(1):65–71, 1994.
242. Carriere P, Mohaghegh S, Gaskari R, Reed B, Jamil M. *Sep. Sci. Technol. 31*:965–985, 1996.
243. Rinkus K, Reed BE, Lin W. *Sep. Sci. Technol. 32*:2367–2384, 1997.
244. Saleem M, Afzal M, Qadeer R, Hanif J. *Sep. Sci. Technol. 27*:239–253, 1992.
245. Abbasi WA, Streat M. *Sep. Sci. Technol. 29*(9):1217–1230, 1994.
246. Qadeer R, Saleem M. *Ads. Sci. Technol. 15*(5):373–376, 1997.
247. Park GI, Park H, Woo S. *Sep. Sci. Technol. 34*(5):833–854, 1999.
248. Taskaev E, Apostolov D. *J. Radioanal. Chem. 45*:65–71, 1978.
249. van Deventer JSJ, van der Merwe PF. *Metall. Trans. B. 24*(3):433–440, 1993.
250. Liebenberg SP, van Deventer JSJ. *Sep. Sci. Technol. 32*:1787–1804, 1997.
251. Liebenberg SP, van Deventer JSJ. *Miner. Eng. 11*(6):551–562, 1998.
252. McDougall GJ, Hancock RD. *Gold Bull. 14*(4):138–153, 1981.
253. Ibrado AS, Fuerstenau DW. *Min. Metall. Process.* (February 1989):23–28.
254. Ladeira ACQ, Figueira M, Ciminelli V. *Miner. Eng. 6*(6):585–596, 1993.
255. Hughes HC, Linge HG. *Hydrometallurgy 22*:57–65, 1989.
256. Hiskey JB, Qi PH. *Min. Metall. Process.* (August 1997):13–17.
257. Pesic B, Storhok VC. *Metall. Trans. B 23*:557–566, 1992.
258. Davidson RJ. *J. South Afr. Inst. Min. Metall.* (November 1974):67–76.
259. Grabovskii AI, Ivanova LS, Korostyshevskii NB, Shirshov VM, Storozhuk RK, Matskevich ES, Arkadakskaya NA. *Russ. J. Appl. Chem. 49*:1405–1408, 1976.
260. Grabovskii AI, Grabchak SL, Ivanova LS, Storozhuk RK, Shirshov VM. *Russ. J. Appl. Chem. 50*:504–506, 1977.
261. Grabovskii AI, Ivanova LS, Matskevich ES, Storozhuk RK, Shirshov VM. *Russ. J. Appl. Chem. 51*:1440–1443, 1978.
262. Cho EH, Dixon SN, Pitt CH. *Metall. Trans. B. 10*:185–189, 1979.
263. Cho EH, Pitt CH. *Metall. Trans. B 10*:165–169, 1979.
264. Cho EH, Pitt CH. *Metall. Trans. B 10*:159–164, 1979.
265. McDougall GJ, Hancock RD, Nicol MJ, Wellington OL, Copperthwaite RG. *J. South Afr. Inst. Min. Metall.* (September 1980):344–356.
266. Adams MD, McDougall GJ, Hancock RD. *Hydrometallurgy 19*:95–115, 1987.
267. Fleming CA, Nicol MJ. *J. South Afr. Inst. Min. Metall.* (April 1984):85–93.
268. Qadeer R, Hanif J, Saleem M, Afzal M. *J. Chem. Soc. Pakistan 17*(2):82–86, 1995.
269. Raji C, Anirudhan T. *J. Sci. Ind. Res. 57*:10–15, 1998.
270. van der Merwe PF, van Deventer JSJ. *Chem. Eng. Commun. 65*:121–138, 1988.
271. Tsuchida N, Muir DM. *Metal. Trans. B 17B*:529–533, 1986.
272. Petersen FW, van Deventer JSJ. *Chem. Eng. Sci. 46*(12):3053–3065, 1991.
273. Vegter NM, Sandenbergh RF, Prinsloo LC, Heyns AM. *Miner. Eng. 11*(6):545–550, 1998.
274. Adams MD. *Hydrometallurgy 25*:171–184, 1990.
275. Adams MD, Fleming CA. *Metall. Trans. B 20*:315–325, 1989.
276. Adams MD, Friedl J, Wagner F. *Hydrometallurgy 31*(3):265–275, 1992.
277. Adams MD. *Hydrometallurgy 31*:121–138, 1992.
278. Adams MD, Friedl J, Wagner FE. *Hydrometallurgy 37*:33–45, 1995.
279. Klauber C. *Surf. Sci. 203*:118–128, 1988.
280. Cook R, Crathorne EA, Monhemius AJ, Perry DL. *Hydrometallurgy 22*:171–182, 1989.

281. Cook R, Crathorne EA, Monhemius AJ, Perry DL. *Hydrometallurgy* 25:394–396, 1990.
282. Klauber C, Vernon CF. *Hydrometallurgy* 25:387–393, 1990.
283. Klauber C. *Langmuir* 7:2153–2159, 1991.
284. Groenewold GS, Ingram JC, Appelhans AD, Delmore JE, Pesic B. *Anal. Chem.* 67:1987–1991, 1995.
285. Cashion JD, Cookson DJ, Brown LJ, Howard DG. In: Long GJ, Stevens JG, eds. *Industrial Applications of the Mössbauer Effect.* New York: Plenum Press, 1986, pp. 595–608.
286. Cashion JD, McGrath AC, Volz P, Hall JS. *Trans. Inst. Min. Metall.* 97C:129–133, 1988.
287. Groszek AJ, Partyka S, Cot D. *Carbon* 29:821–829, 1991.
288. Lagerge S, Zajac J, Partyka S, Groszek AJ, Chesneau M. *Langmuir* 13:4683–4692, 1997.
289. Lagerge S, Zajac J, Partyka S, Groszek AJ. *Langmuir* 15:4803–4811, 1999.
290. Ibrado AS, Fuerstenau DW. *Hydrometallurgy* 30:243–256, 1992.
291. Ibrado AS, Fuerstenau DW. *Miner. Eng.* 8(4/5):441–458, 1995.
292. Papirer E, Polania-Leon A, Donnet J-B, Montagnon P. *Carbon* 33:1331–1337, 1995.
293. Kongolo K, Kinabo C, Bahr A. *Hydrometallurgy* 44:191–202, 1997.
294. Jia YF, Steele C, Hayward I, Thomas K. *Carbon* 36:1299–1308, 1998.
295. Teirlinck PAM, Petersen FW. *Sep. Sci. Technol.* 30:3129–3142, 1995.
296. Arriagada R, García R. *Hydrometallurgy* 46:171–180, 1997.
297. Jeong HS, Sohn H-J. *Min. Metall. Process.* (February 1988):21–25.
298. Soto AM, Machuca RA. *J. Chem. Technol. Biotechnol.* 44:219–223, 1989.
299. Gupta SK, Chen KY. *J. WPCF* (March 1978):493–506.
300. Kamegawa K, Yoshida H, Arita S. *Nippon Kagaku Kaishi* 1979:1365–1370.
301. Rajakovic LV. *Sep. Sci. Technol.* 27(11):1423–1433, 1992.
302. Lorenzen L, van Deventer J, Landi W. *Miner. Eng.* 8(4/5):557–569, 1995.
303. Diamadopoulos E, Samaras P, Sakellaropoulos GP. *Water Sci. Technol.* 25(1):153–160, 1992.
304. Metcalf R. In: White-Stevens R, ed. *Pesticides in the Environment*, Vol. 1. New York: Marcel Dekker, 1971, pp. 93 ff.
305. Cox J, Ramsey D. *Chemical Reviews* 64:317, 1964.
306. Fukoto T. In: Narchashi T, ed. *Neurotoxicology of Insecticides and Pheromones.* New York: Plenum Press, 1979.
307. Jayson GG, Lawless TA, Fairhurst D. *J. Colloid Interf. Sci.* 86(2):397–410, 1982.
308. Ferro-Garcia MA, Carrasco-Marín F, Rivera-Utrilla J, Utrera-Hidalgo E, Moreno-Castilla C. *Carbon* 28(1):91–95, 1990.
309. Nightingale E. *J. Phys. Chem.* 63:1381, 1959.
310. Wei C-H, Cao X-L. *Carbon* 31(8):1319–1324, 1993.
311. Hingston FJ. In: Anderson MA, Rubin AJ, eds. *Adsorption of Inorganics at Solid–Liquid Interfaces.* Ann Arbor, MI: Ann Arbor Science, 1981, pp. 51–90.
312. Carrasco-Marín F, Solar JM, Radovic LR. On the importance of electrokinetic properties of carbon for the gasification reactivity of carbon-supported palladium catalysts. International Carbon Conference (Paris), 1990, pp. 672–673.

313. Hatori H, Radovic LR. Manuscript in preparation.
314. Yajnik NA, Rana TC. *J. Phys. Chem.* 28:267–278, 1924.
315. Israelachvili JN. *Intermolecular and Surface Forces.* 2d ed. London: Academic Press, 1992.
316. Evans DF, Wennerström H. *The Colloidal Domain: Where Physics, Chemistry, Biology and Technology Meet.* New York: VCH Publishers, 1994.
317. Ferro-Garcia MA, Rivera-Utrilla J, Bautista-Toledo I, Mingorance MD. *Carbon* 28:545–552, 1990.
318. Verwey EJW, Overbeek JTG. *Theory of the Stability of Lyophobic Colloids.* New York: Elsevier, 1948.
319. Lewis KE, Parfitt GD. *Trans. Faraday Soc.* 62:1652–1661, 1966.
320. Lyklema J, van Leeuwen H, Minor M. *Adv. Colloid Interf. Sci.* 83:33–69, 1999.
321. Cookson Jr. JT, North WJ. *Environ. Sci. Technol.* 1(1):46–52, 1967.
322. Gerba CP, Sobsey MD, Wallis C, Melnick JL. *Environ. Sci. Technol.* 9(8):727–731, 1975.
323. George N, Davies J. *J. Chem. Technol. Biotechnol.* 43(2):117–129, 1988.
324. George N, Davies J. *J. Chem. Technol. Biotechnol.* 43(3):173–186, 1988.
325. van Loosdrecht MCM, Lyklema J, Norde W, Zehnder AJB. *Microb. Ecol.* 17:1–15, 1989.
326. van Loosdrecht MCM, Norde W, Zehnder AJB. *J. Biomater. Appl.* 5(October):91–106, 1990.
327. Jiang ZP, Yang ZH, Yang JX, Zhu WP, Wang ZS. In: Mallevialle J, Suffet IH, Chan US, eds. *Influence and Removal of Organics in Drinking Water.* Boca Raton, FL: Lewis, 1992, pp. 79–95.
328. Juang R-S, Tseng R-L, Wu F-C, Lee S-H. *Sep. Sci. Technol.* 31:1915–1931, 1996.
329. Juang R-S, Wu F-C, Tseng R-L. *J. Chem. Eng. Data* 41:487–492, 1996.
330. Mattson JS, Mark HB Jr., Weber WJ Jr. *J. Colloid Interf. Sci.* 48:181, 1974.
331. Coughlin RW, Ezra FS, Tan RN. *J. Colloid Interf. Sci.* 28(3/4):386–396, 1968.
332. Coughlin RW, Tan RN. Role of functional groups in adsorption of organic pollutants on carbon. Water-1968: AIChE, 1968, pp. 207–214.
333. Snoeyink VL, Weber Jr. WJ, Mark Jr. HB. *Environ. Sci. Technol.* 3(10):918–926, 1969.
334. Singh DD. *Ind. J. Chem.* 9:1369–1371, 1971.
335. Vaccaro RF. *Environ. Sci. Technol.* 5:134–138, 1971.
336. Pahl RH, Mayhan KG, Bertrand GL. *Water Res.* 7:1309, 1973.
337. Ganho R, Gibert H, Angelino H. *Chem. Eng. Sci.* 30:1231, 1975.
338. Puri BR. In: Mittal KL, ed. *Adsorption at Interfaces*, Vol. 8. Washington, DC: American Chemical Society, 1975, pp. 212–224.
339. Puri BR, Bhardwaj SS, Kumar V, Mahajan OP. *J. Indian Chem. Soc.* 52:26–29, 1975.
340. Economy J, Lin RY. *Appl. Polym. Symp.* 29:199–211, 1976.
341. Puri BR, Bhardwaj SS, Gupta U. *J. Indian Chem. Soc.* 53:1095–1098, 1976.
342. Dondi F, Betti A, Blo G, Bighi C. *Annali Chim.* 68:293–301, 1978.
343. Amicarelii V, Baldassare G, Liberti L. *Thermochim. Acta* 30:247–253, 1979.
344. Puri BR, Singh DD, Gupta U. *J. Indian Chem. Soc.* 56:1193–1195, 1979.

345. Mahajan OP, Moreno-Castilla C, Walker PL Jr. *Sep. Sci. Technol.* 15(10):1733–1752, 1980.
346. Puri BR. In: Suffet IH, McGuire MJ, eds. *Activated Carbon Adsorption of Organics from the Aqueous Phase*, Vol. 1. Ann Arbor, MI: Ann Arbor Science, 1980, pp. 353–378.
347. Chang CH, Kleppa OJ. *Carbon* 19:187–192, 1981.
348. Chang CH, Savage DW. *Environ. Sci. Technol.* 15(1):201–206, 1981.
349. Chang CH, Savage DW, Longo JM. *J. Colloid Interf. Sci.* 79(1):178–191, 1981.
350. Oda H, Kishida M, Yokokawa C. *Carbon* 19:243–248, 1981.
351. Ogino K, Tsukamoto H, Yamabe K, Takahashi H. *Nippon Kagaku Kaishi 1981*(3): 321–325.
352. Cha T-H, Glasgow LA. *Ind. Eng. Chem. Process Des. Dev.* 22:198–202, 1983.
353. Ehrburger P, Wozniak E, Lahaye J. *J. Chim. Phys.* 80:685–690, 1983.
354. Puri BR. In: McGuire MJ, Suffet IH, eds. *Treatment of Water by Granular Activated Carbon* (Advances in Chemistry Series Vol. 202). Washington, DC: American Chemical Society, 1983, pp. 77–93.
355. Sutikno T, Himmelstein KJ. *Ind. Eng. Chem. Fundam.* 22:420–425, 1983.
356. Asakawa T, Ogino K. *J. Colloid Interf. Sci.* 102(2):348–355, 1984.
357. Asakawa T, Ogino K, Yamabe K. *Bull. Chem. Soc. Jpn.* 58:2009–2014, 1985.
358. McGuire J, Dwiggins CF, Fedkiw PS. *J. Appl. Electrochem.* 15:53–62, 1985.
359. Kim BR, Chian ESK, Cross WH, Cheng S-S. *J. Water Poll. Contr. Fed.* 58(January):35–40, 1986.
360. Magne P, Walker PL Jr. *Carbon* 24(2):101–107, 1986.
361. Ogino K, Kaneko Y, Asakawa T. *Nippon Kagaku Kaishi 1986*:839–843.
362. Le Cloirec P, Martin G, Gallier J. *Carbon* 26:275–282, 1988.
363. Kaneko Y, Abe M, Ogino K. *Colloids Surfaces* 37:211–222, 1989.
364. Kastelan-Macan M, Cerjan-Stefanovic S, Petrovic M. *Chromatog.* 27:297–300, 1989.
365. Korczak M, Kurbiel J. *Water Res.* 23:937–946, 1989.
366. Sheveleva IV, Khabalov VV, Pavlova GS. *Russ. J. Phys. Chem.* 64:86–88, 1990.
367. Abuzeid N, Harrazin L. *J. Environ. Sci. Health A* 26(2):257–271, 1991.
368. Pollard SJT, Sollars CJ, Perry R. *J. Chem. Technol. Biotechnol.* 50:277–292, 1991.
369. Pollard SJT, Sollars CJ, Perry R. *J. Chem. Technol. Biotechnol.* 50:265–275, 1991.
370. Bushmakin L. Gallyamova E. *Colloid J. Russ. Acad. Sci.* 54(4):633–635, 1992.
371. Ogino K. In: Suzuki M, ed. *Fundamentals of Adsorption*. Tokyo: Kodansha, 1993, pp. 491–498.
372. Yenkie MKN, Natarajan GS. *Sep. Sci. Technol.* 28(5):1177–1190, 1993.
373. Fleig M, Knoll H, Quitzsch K. *Z. Physik, Chem.* 190:83–98, 1995.
374. Halhouli KA, Darwish NA, Al-Dhoon NM. *Sep. Sci. Technol.* 30:3313–3324, 1995.
375. Rajeshwari K, Ghosh P. *J. Microbial Biotechnol.* 10(2):59–68, 1995.
376. Streat M, Patrick JW, Camporro-Perez MJ. *Water Res.* 29:467–472, 1995.
377. Király Z, Dékány I, Klumpp E, Lewandowski H, Narres HD, Schwuger MJ. *Langmuir* 12:423–430, 1996.
378. Leng C-C, Pinto NG. *Ind. Eng. Chem. Res.* 35:2024–2031, 1996.
379. Loughlin KF, Hassan MM, Ekhator EO, Nakhla GF. In: LeVan MD, ed. *Fundamentals of Adsorption*. Kluwer Academic, 1996, pp. 537–544.

380. Tamon H, Abe M, Okazaki M. *J. Chem. Eng. Japan* 29:384–386, 1996.
381. Tukac V, Hanika J. *Collect. Czech. Chem. Commun.* 61:1010–1017, 1996.
382. Barton SS, Evans MJB, MacDonald JAF. *Polish J. Chem.* 71:651–656, 1997.
383. Halhouli KA, Darwish NA, Al-Jahmany YY. *Sep. Sci. Technol.* 32:3027–3036, 1997.
384. Kilduff JE, King CJ. *Ind. Eng. Chem. Res.* 36:1603–1613, 1997.
385. Lee HW, Kim KJ, Fane AG. *Sep. Sci. Technol.* 32:1835–1849, 1997.
386. Leng C-C, Pinto NG. *Carbon* 35:1375–1385, 1997.
387. Liu X, Pinto NG. *Modeling the Adsorption of Phenol and Aniline on Active Carbon. 23rd Biennial Conference on Carbon.* State College, PA: American Carbon Society, 1997.
388. Mehta MP, Flora JRV. *Water Res.* 31:2171, 1997.
389. Michot LJ, Didier F, Villiéras F, Cases JM. *Polish J. Chem.* 71:665–678, 1997.
390. Ivancev-Tumbas I, Dalmacija B, Tamas Z, Karlovic E. *Water Res.* 32(4):1085–1094, 1998.
391. Teng H, Hsieh C-T. *Ind. Eng. Chem. Res.* 37:3618–3624, 1998.
392. Arafat HA, Franz M, Pinto NG. *Langmuir* 15:5997–6003, 1999.
393. Leyva-Ramos R, Soto-Zúñiga J, Mendoza-Barron J, Guerrero Coronado R. *Ads. Sci. Technol.* 17(7):533–543, 1999.
394. Nevskaia DM, Santianes A, Muñoz V, Guerrero-Ruíz A. *Carbon* 37:1065–1074, 1999.
395. Wu F, Tseng R, Juang R. *J. Environ. Sci. Health A* 34(9):1753–1775, 1999.
396. Pollio FX, Kunin R. *Environ. Sci. Technol.* 1:160–163, 1967.
397. Boyd SA. *Soil Sci.* 134:337–343, 1982.
398. Yost EC, Anderson MA. *Environ. Sci. Technol.* 18:101–106, 1984.
399. Goto M, Hayashi N, Goto S. *Environ. Sci. Technol.* 20(5):463–467, 1986.
400. Davis AP, Huang CP. *Langmuir* 6:857–862, 1990.
401. Liu JC, Huang CP. *J. Colloid Interf. Sci.* 153(1):167–176, 1992.
402. Xing B, McGill WB, Dudas MJ, Maham Y, Hepler L. *Environ. Sci. Technol.* 28:466–473, 1994.
403. Cleveland TG, Garg S, Rixey WG. *J. Environ. Eng.* 122(3):235–238, 1996.
404. Aksu Z, Yener J. *J. Environ. Sci. Health A* 34(9):1777–1796, 1999.
405. Umeyama H, Nagai T, Nogami H. *Chem. Pharm. Bull.* 19:1714–1721, 1971.
406. Kitagawa H. *Nippon Kagaku Kaishi* 1975:1631–1634.
407. Peel RG, Benedek A. *Environ. Sci. Technol.* 14:66–71, 1980.
408. Urano K, Koichi Y, Nakazawa Y. *J. Colloid Interf. Sci.* 81:477–485, 1981.
409. Fuerstenau DW, Pradip [sic]. *Colloids Surfaces* 4:229–243, 1982.
410. Oda H, Yokokawa C. *Carbon* 21:485–489, 1983.
411. Itaya A, Kato N, Yamamoto J, Okamoto K-I. *J. Chem. Eng. Japan* 17:389–395, 1984.
412. Yonge DR, Keinath TM, Poznanska K, Jiang ZP. *Environ. Sci. Technol.* 19:690–694, 1985.
413. Thakkar S, Manes M. *Environ. Sci. Technol.* 21:546–549, 1987.
414. Caturla F, Martin-Martinez JM, Molina-Sabio M, Rodriguez-Reinoso F, Torregrosa R. *J. Colloid Interf. Sci.* 124:528–534, 1988.
415. Zawadzki J. *Carbon* 26:603–611, 1988.
416. Mostafa MR, Samra SE, Youssef AM. *Ind. J. Chem.* 28A:946–948, 1989.

417. Grant TM, King CJ. *Ind. Eng. Chem. Res. 29*:264–271, 1990.
418. Vidic RD, Suidan MT, Traegner UK, Nakhla GF. *Water Res. 24*:1187–1195, 1990.
419. Baudu M, Le Cloirec P, Martin G. *Water Sci. Technol. 23*:1659–1666, 1991.
420. Sheveleva IV, Zryanina NV, Voit AV. *Russ. J. Phys. Chem. 65*(4):596–599, 1991.
421. Salvador F, Merchán MD. *Langmuir 8*:1226–1229, 1992.
422. Shirgaonkar IZ, Joglekar HS, Mundale VD, Joshi JB. *J. Chem. Eng. Data 37*:175–179, 1992.
423. Vidic RD, Suidan MT, Brenner RC. *Environ. Sci. Technol 27*(10):2079–2085, 1993.
424. Vidic RD, Suidan MT, Sorial GA, Brenner RC. *Water Environ. Res. 65*(2):156–161, 1993.
425. Abuzaid NS, Nakhla GF. *Environ. Sci. Technol. 28*:216–221, 1994.
426. Cooney DO, Xi Z. *AIChE J. 40*:361–364, 1994.
427. Nakhla G, Abuzaid N, Farooq S. *Water Environ. Res. 66*:842–850, 1994.
428. Singh BK, Rawat NS. *J. Chem. Technol. Biotechnol. 61*:307–317, 1994.
429. Abu Zeid N, Nakhla G, Farooq S, Osei-Twum E. *Water Res. 29*:653–660, 1995.
430. Abuzaid NS, Nakhla GF. *J. Hazard. Mater. 49*:217–230, 1996.
431. Dargaville TR, Guerzoni FN, Looney MG, Solomon DH. *J. Colloid Interf. Sci. 182*:17–25, 1996.
432. Salvador F, Merchán MD. *Carbon 34*:1543–1551, 1996.
433. Deng X, Yue Y, Gao Z. *J. Colloid Interf. Sci. 192*:475–480, 1997.
434. Khan AR, Al-Bahri TA, Al-Haddad A. *Water Res. 31*(8):2102–2112, 1997.
435. Khan AR, Ataullah R, Al-Haddad A. *J. Colloid Interf. Sci. 194*:154–165, 1997.
436. Le Cloirec P, Brasquet C, Subrenat E. *Energy Fuels 11*:331–336, 1997.
437. Liu X, Pinto NG. *Carbon 35*:1387–1397, 1997.
438. Matatov-Meytal YI, Sheintuch M. *Ind. Eng. Chem. Res. 36*:4374–4380, 1997.
439. Wang R-C, Kuo C-C, Shyu C-C. *J. Chem. Technol. Biotechnol. 68*:187–194, 1997.
440. Daifullah AAM, Girgis BS. *Water Res. 32*(4):1169–1177, 1998.
441. Edgehill R, Lu G. *J. Chem. Technol. Biotechnol. 71*:27–34, 1998.
442. Gee I, Sollars C, Fowler G, Ouki S, Perry R. *J. Chem. Technol. Biotechnol. 72*:329–338, 1998.
443. Haghseresht F, Lu GQ. *Energy Fuels 12*:1100–1107, 1998.
444. Ravi V, Jasra R, Bhat T. *J. Chem. Technol. Biotechnol. 71*:173–179, 1998.
445. Zhou ML, Martin G, Taha S, Sant'Anna F. *Water Res. 32*(4):1109–1118, 1998.
446. Zogorski JS, Faust SD, Haas JH Jr. *J. Colloid Interf. Sci. 65*:329–341, 1976.
447. Friedrich M, Seidel A, Gelbin D. *Chem. Eng. Process. 24*(1):33–38, 1988.
448. Bhatia SK, Kalam A, Joglekar HS, Joshi JB. *Chem. Eng. Comm. 98*:139–154, 1990.
449. Sankaran N, Anirudhan T. *Ind. J. Eng. Mater. Sci. 6*(4)229–236, 1999.
450. Coughlin RW, Ezra FS. *Environ. Sci. Technol. 2*(4):291–297, 1968.
451. Graham D. *J. Phys. Chem. 59*(9):896–900, 1955.
452. Farrier DS, Hines AL, Wang SW. *J. Colloid Interf. Sci. 69*:233–237, 1979.
453. Ferro-García MA, Utrera-Hidalgo E, Rivera-Utrilla J, Moreno-Castilla C. *Carbon 31*(6):857–863, 1993.
454. Moreno-Castilla C, Rivera-Utrilla J, Joly JP, López-Ramón MV, Ferro-Garcia MA, Carrasco-Marín F. *Carbon 33*:1417–1423, 1995.
455. Teng H, Hsieh C-T. *J. Chem. Technol. Biotechnol. 74*(2):123–130, 1999.

456. Freundlich H, Heller W. *J. Amer. Chem. Soc. 61*:2228–2230, 1939.
457. Mattson JS. *Ind. Eng. Chem. Prod. Res. Develop. 12*:312–317, 1973.
458. Chiou CCT, Manes M. *J. Phys. Chem. 78*:622, 1974.
459. Saperstein DD. *J. Phys. Chem. 90*:3883–3885, 1986.
460. Gürses A, Bayrakçeken S, Gülaboglu MS. *Colloids Surfaces 64*:7–13, 1992.
461. Miyahara M, Okazaki M. *J. Chem. Eng. Japan 25*:408–414, 1992.
462. Vidic RD, Suidan MT. *Water Sci. Technol. 26*:1185–1193, 1992.
463. Vidic RD, Suidan MT. *J. Amer. Water Works Assoc.* (March 1992):101–109.
464. El-Shahawi MS. *Chromatogr. 36*:318–322, 1993.
465. Miyahara M, Okazaki M. *J. Chem. Eng. Japan 26*:510–516, 1993.
466. Sorial GA, Suidan MT, Vidic RD, Brenner RC. *Water Environ. Res. 65*(1):53–57, 1993.
467. Chatzopoulos D, Varma A, Irvine RL. *Environ. Prog. 13*(1):21–25, 1994.
468. Vidic RD, Suidan MT, Sorial GA, Brenner RC. *J. Hazard. Mater. 38*:373, 1994.
469. Allen SJ, Balasundaram V, Armenante PM, Thom L, Kafkewitz D. *J. Chem. Technol. Biotechnol. 64*:261–268, 1995.
470. Chatzopoulos D, Varma A. *Chem. Eng. Sci. 50*:127–141, 1995.
471. Lynam MM, Kilduff JE, Weber J. WJ. *J. Chem. Educ. 72*:80–84, 1995.
472. Srivastava SK, Tyagi R. *Water Res. 29*:483–488, 1995.
473. Klecka GM, McDaniel SG, Wilson PS, Carpenter CL, Clark JE, Thomas A, Spain JC. *Environ. Prog. 15*(2):93–107, 1996.
474. Mollah AH, Robinson CW. *Water Res. 30*:2901–2906, 1996.
475. Mollah AH, Robinson CW. *Water Res. 30*:2907, 1996.
476. Derylo-Marczewska A, Marczewski AW. *Polish J. Chem. 71*:618–629, 1997.
477. Karimi-Jashni A, Narbaitz RM. *Water Res. 31*:3039–3044, 1997.
478. Vidic RD, Tessmer CH, Uranowski LJ. *Carbon 35*:1349–1359, 1997.
479. Uranowski LJ, Tessmer CH, Vidic RD. *Water Res. 32*(6):1841–1851, 1998.
480. Arriagada R, García R. *J. Chem. Technol. Beotechnol. 74*:870–876, 1999.
481. Wheeler OH, Levy EM. *Can J. Chem. 37*:1235–1240, 1959.
482. Holmes HN, McKelvey JB. *J. Phys. Chem. 32*:1522, 1928.
483. Kipling JJ, Wright EHM. *J. Chem. Soc. 1965*:4340–4348.
484. Edsall JT. *Proc. Amer. Philos. Soc. 129*:371, 1985.
485. Kamegawa K, Yoshida H. *Bull. Chem. Soc. Jpn. 63*:3683–3685, 1990.
486. González-Martin ML, Valenzuela-Calahorro C, Gómez-Serrano V. *Langmuir 7*: 1296–1298, 1991.
487. Gómez-Serrano V, Beltrán FJ, Durán-Segovia A. *Chem. Eng. Technol. 15*:124–130, 1992.
488. González-Martín ML, Valenzuela-Calahorro C, Gómez-Serrano V. *Langmuir 10*: 844–854, 1994.
489. Hobday MD, Li PHY, Crewdson DM, Bhargava SK. *Fuel 73*(12):1848–1854, 1994.
490. Komori M, Urano K, Nakai T. *Nippon Kagaku Kaishi 1974*:1795–1799.
491. El-Dib MA, Badawy MI. *Water Res. 13*:255–258, 1979.
492. Martin RJ. *Ind. Eng. Chem. Prod. Res. Dev. 19*:435–441, 1980.
493. Woodard FE, McMackins DE, Jansson REW. *J. Electroanal. Chem. 214*:303–330, 1986.

494. Nelson PO, Yang M. *Water Environ. Res. 67*:892–898, 1995.
495. Vidic RD, Suidan MT. *Environ. Sci. Technol. 25*:1612–1618, 1991.
496. Tessmer CH, Uranowski LJ, Vidic RD. *Environ. Sci. Technol. 31*(7):1872–1878, 1997.
497. Zaror CA. *J. Chem. Technol. Biotechnol. 70*:21–28, 1997.
498. Adamson AW. *Physical Chemistry of Surfaces.* 5th ed. New York: John Wiley, 1990.
499. Nandi SP, Walker PL Jr. *Fuel 50*:345–366, 1971.
500. McKay G. *J. Chem. Technol. Biotechnol. 32*:759–772, 1982.
501. Perineau F, Molinier J, Gaset A. *J. Chem. Technol. Biotechnol. 32*:749–758, 1982.
502. McKay G. *Chem. Eng. Res. Des. 61*:29–36, 1983.
503. Potgieter JH. *Colloids Surfaces 50*:393–399, 1990.
504. Lin SH. *J. Chem. Technol. Biotechnol. 57*:387–391, 1993.
505. Bousher A, Shen X, Edyvean RJ. *Water Res. 31*:2084–2092, 1997.
506. Chern J-M, Huang S-N. *Ind. Eng. Chem. Res. 37*:253–257, 1997.
507. Gupta VK, Srivastava SK, Mohan D. *Ind. Eng. Chem. Res. 36*:2207–2218, 1997.
508. Juang R-S, Swei S-L. *Sep. Sci. Technol. 31*:2143–2158, 1997.
509. Shmidt JL, Pimenov AV, Lieberman AI, Chen HY. *Sep. Sci. Technol. 32*:2105–2114, 1997.
510. Walker GM, Weatherley LR. *Water Res. 31*:2093–2101, 1997.
511. Moreira RFPM, Kuhnen NC, Peruch MG. *Latin Amer. Appl. Res. 28*:37–41, 1998.
512. Krupa NE, Cannon FS. *J. AWWA 88*(June):94–108, 1996.
513. Giles CH, D'Silva AP. *Trans. Faraday Soc. 65*:1943–1951, 1969.
514. Giles CH, D'Silva AP, Trivedi AS. In: Everett DH, Ottewill RH, eds. *Surface Area Determination.* London: Butterworths, 1970, pp. 317–329.
515. Giles CH. Adsorption of dyes. In: Parfitt GD, Rochester CH, eds. *Adsorption from Solution at the Solid/Liquid Interface.* London: Academic Press, 1983, pp. 321–376.
516. Linares-Solano A, Rodriguez-Reinoso F, Molina-Sabio M, Lopez-Gonzalez JD. *Ads. Sci. Technol. 1*:223–234, 1984.
517. Brina R, De Battisti A. *J. Chem. Educ. 64*:175–176, 1987.
518. Duff DG, Ross SMC, Vaughan DH. *J. Chem. Educ. 65*(9):815–816, 1988.
519. Kasaoka S, Sakata Y, Tanaka E, Naitoh R. *Intern. Chem. Eng. 29*:734–742, 1989.
520. Potgieter JH. *J. Chem. Educ. 68*:349–350, 1991.
521. Dai M. *J. Colloid Interf. Sci. 164*:223–228, 1994.
522. Dai M. *J. Colloid Interf. Sci. 198*:6–10, 1998.
523. Müller G, Radke CJ, Prausnitz JM. *J. Phys. Chem. 84*:369–376, 1980.
524. Müller G, Radke CJ, Prausnitz JM. *J. Colloid Interf. Sci. 103*(2):466–483, 1985.
525. Müller G, Radke CJ, Prausnitz JM. *J. Colloid Interf. Sci. 103*(2):484–492, 1985.
526. (See Ref. 532.)
527. Sasaki M, Tamai H, Yoshida T, Yasuda H. *Tanso 1998*(183):151–155.
528. Tamai H, Sasaki M, Yasuda H. *Tanso 1998*(184):219–224.
529. Tamai H, Yoshida T, Sasaki M, Yasuda H. *Carbon 37*:983–989, 1999.
530. Khalil L, Girgis B. *Ads. Sci. Technol. 16*(5):405–414, 1998.
531. Krupa NE. Characterization of the pore structure of thermally regenerated activated carbon using adsorbates of varying molecular dimensions. M.S. thesis, Pennsylvania State University, 1994.

532. Al-Degs Y, Khraisheh M, Allen S, Ahmad M. *Adv. Environ. Res. 3*(2):132–138, 1999.
533. Donati C, Drikas M, Hayes R, Newcombe G. *Water Res. 28*:1735–1742, 1994.
534. Huang C, Van Benschoten JE, Jensen JN. *J. AWWA 1996*:116–128.
535. Chen G, Dussert BW, Suffet IH. *Water Res. 31*:1155, 1997.
536. Newcombe G, Drikas M, Hayes R. *Water Res. 31*:1065–1073, 1997.
537. Gillogly TET, Snoeyink VL, Elarde JR, Wilson CM, Royal EP. *J. AWWA 90*(1): 98–108, 1998.
538. Gillogly TET, Snoeyink VL, Holthouse A, Wilson CM, Royal EP. *J. AWWA 90*(2): 107–114, 1998.
539. McCreary JJ, Snoeyink VL. *Water Res. 14*:151–160, 1980.
540. Weber Jr. WJ, Pirbazari M, Long JB, Barton DA. In: Suffet IH, McGuire MJ, eds. *Activated Carbon Adsorption of Organics from the Aqueous Phase*, Vol. 1. Ann Arbor, MI: Ann Arbor Science, 1980, pp. 317–336.
541. Youssefi M, Faust SD. In: Suffet IH, McGuire MJ, eds. *Activated Carbon Adsorption of Organics from the Aqueous Phase*, Vol. 1. Ann Arbor, MI: Ann Arbor Science, 1980, pp. 133–143.
542. Lee MC, Snoeyink VL, Crittenden JC. *J. Amer. Water Works Assoc.* (August 1981): 440–446.
543. Randtke SJ, Jepsen CP. *J. Amer. Water Works Assoc.* (August 1981):411–418.
544. Randtke SJ, Jepsen CP. *J. Amer. Water Works Assoc.* (February 1982):84–93.
545. Weber WJ Jr., Voice TC, Jodellah A. *J. Amer. Water Works Assoc.* (December 1983):612–619.
546. Ogino K, Kaneko Y, Minoura T, Agui W, Abe M. *J. Colloid Interf. Sci. 121*(1): 161–169, 1988.
547. Summers RS, Roberts PV. *J. Colloid Interf. Sci. 122*(2):382–397, 1988.
548. Summers RS, Roberts PV. *J. Colloid Interf. Sci. 122*(2):367, 1988.
549. Lafrance P, Mazet M. *J. Amer. Water Works Assoc.* (April 1989):155–162.
550. Summers RS, Haist B, Koehler J, Ritz J, Zimmer G, Sontheimer H. *J. Amer. Water Works Assoc.* (May 1989):66–74.
551. Najm IN, Snoeyink VL, Richard Y. *J. Amer. Water Works Assoc.* (August 1991): 57–63.
552. Carter MC, Weber WJ Jr., Olmstead KP. *J. AWWA* (August 1992):81–91.
553. Morris G, Newcombe G. *J. Colloid Interf. Sci. 159*:413–420, 1993.
554. Newcombe G, Hayes R, Drikas M. *Colloids Surfaces 78*:65–71, 1993.
555. Newcombe G. *J. Colloid Interf. Sci. 164*:452–462, 1994.
556. Annesini MC, Gironi F, Lamberti L. *Annali Chimica 85*:613–620, 1995.
557. Cerminara PJ, Sorial GA, Papadimas SP, Suidan MT, Moteleb MA, Speth TF. *Water Res. 29*:409–419, 1995.
558. Jacangelo JG, DeMarco J, Owen DM, Randtke SJ. *J. AWWA* (January 1995):64–77.
559. Owen DM, Amy GL, Chowdhury ZK, Paode R, McCoy G, Viscosil K. *J. AWWA* (January 1995):46–63.
560. Warta CL, Papadimas SP, Sorial GA, Suidan MT, Speth TF. *Water Res. 29*:551–562, 1995.
561. Karanfil T, Schlautman MA, Kilduff JE, Weber WJ Jr. *Environ. Sci. Technol. 30*: 2195–2201, 1996.

562. Karanfil T, Schlautman MA, Kilduff JE, Weber WJ Jr. *Environ. Sci. Technol.* 30: 2187–2194, 1996.
563. Kilduff JE, Karanfil T, Chin Y-P, Weber WJ Jr. *Environ. Sci. Technol.* 30:1336–1343, 1996.
564. Kilduff JE, Karanfil T, Weber WJ Jr. *Environ. Sci. Technol.* 30:1344–1351, 1996.
565. Fukushima M, Oba K, Tanaka S, Nakayasu K, Nakamura H, Hasebe K. *Environ. Sci. Technol.* 31:2218–2222, 1997.
566. Newcombe G, Drikas M. *Carbon* 35:1239–1250, 1997.
567. Karanfil T, Kilduff JE, Schlautman MA, Weber WJ Jr. *Water Res.* 32:154–164, 1998.
568. Karanfil T, Kitis M, Kilduff JE, Wigton A. *Environ. Sci. Technol.* 33:3225–3233, 1999.
569. Radovic LR, Cannon FC, Moore B, Mazyck D. In: Radovic LR, ed. *Chem. Phys. Carbon.* Vol. 28. New York: Marcel Dekker, 2001 (in preparation).
570. Amy GL, Sierka RA, Bedessem J, Price D, Tan L. *J. AWWA* (June 1992):67–75.
571. Philip JC, Jarman J. *J. Phys. Chem.* 28:346–350, 1924.
572. Garner W, McKie D, Knight B. *J. Phys. Chem.* 31:641–648, 1927.
573. Linner ER, Gortner RA. *J. Phys. Chem.* 39:35–67, 1935.
574. Badorek DL, Thiem LT, O'Connor JT. In: McGuire MJ, Suffet IH, eds. *Activated Carbon Adsorption of Organics from the Aqueous Phase*, Vol. 2. Ann Arbor, MI: Ann Arbor Science, 1980, pp. 71–84.
575. Mullins RL Jr., Zogorski JS, Hubbs SA, Allgeier GD. In: Suffet IH, McGuire MJ, eds. *Activated Carbon Adsorption of Organics from the Aqueous Phase*, Vol. 1. Ann Arbor, MI: Ann Arbor Science, 1980, pp. 273–307.
576. Neely JW. In: McGuire MJ, Suffet IH, eds. *Activated Carbon Adsorption of Organics from the Aqueous Phase*, Vol. 2. Ann Arbor, MI: Ann Arbor Science, 1980, pp. 417–424.
577. Alben KT, Shpirt E, Kaczmarczyk JH. *Environ. Sci. Technol.* 22:406, 1988.
578. Takeuchi Y, Suzuki Y, Koizumi A, Soeda N. *Water Sci. Technol.* 23:1687–1694, 1991.
579. Razvigorova M, Budinova T, Petrov N, Minkova V. *Water Res.* 32(7):2135–2139, 1998.
580. Kim BR, Snoeyink VL. In: Suffet IH, McGuire MJ, eds. *Activated Carbon Adsorption of Organics from the Aqueous Phase*, Vol. 1. Ann Arbor, MI: Ann Arbor Science 1980, pp. 463–481.
581. Chakma A, Meisen A. *Carbon* 27:573–584, 1989.
582. Kuo Y, Munson CL, Rixey WG, Garcia AA, Frierman M, King CJ. *Sep. Purif. Methods* 16(1):31–64, 1987.
583. González-Pradas E, Villafranca-Sánchez M, Socias-Viciana M, del Rey-Bueno F, García-Rodríguez A. *J. Chem. Technol. Biotechnol.* 39:19–27, 1987.
584. Urano K, Yamamoto E, Tonegawa M, Fujie K. *Water Res.* 25:1459–1464, 1991.
585. Carter MC, Weber WJ Jr. *Environ. Sci. Technol.* 28:614–623, 1994.
586. Cho SY, Lee YY. *Ind. Eng. Chem. Res.* 34:2468–2472, 1995.
587. Tanada S, Uchida M, Nakamura T, Kawasaki N, Doi H, Takebe Y. *J. Environ. Sci. Health* A32:1451–1458, 1997.
588. Kilduff JE, Karanfil T, Weber WJ Jr. *J. Colloid Interf. Sci.* 205:271–279, 1998.

589. Yun J-H, Choi D-K, Kim S-H. *Ind. Eng. Chem. Res. 37*:1422, 1998.
590. Karanfil T, Kilduff JE. *Environ. Sci. Technol. 33*:3217–3224, 1999.
591. Kilduff JE, Wigton A. *Environ. Sci. Technol. 33*:250–256, 1999.
592. Hansen RS, Craig RP. *J. Phys. Chem. 58*:211–215, 1954.
593. Naono H, Hakuman M, Shimoda M, Nakai K, Kondo S. *J. Colloid Interf. Sci. 182*: 230–238, 1996.
594. Urano K, Koichi Y, Yamamoto E. *J. Colloid Interf. Sci. 86*:43–50, 1982.
595. Nagy M. *Langmuir 7*:344–349, 1991.
596. Groszek AJ, Partyka S. *Langmuir 9*:2721–2725, 1993.
597. Isirikyan AA, Polyakov NS, Tatarinova LI. *Colloid Journal 56*(4):519–520, 1994.
598. Wright EHM. *J. Colloid Interf. Sci. 24*:180–184, 1967.
599. Urano K, Kano H, Tabata T. *Bull. Chem. Soc. Japan 57*:2307–2308, 1984.
600. Rybolt TR, Burrell DE, Shults JM, Kelley AK. *J. Chem. Educ. 65*:1009–1010, 1988.
601. Dusart O, Bouabane H, Mazet M. *J. Chim. Phys. 88*:259–270, 1991.
602. Rochester CH, Strachan A. *J. Colloid Interf. Sci. 177*:339–342, 1996.
603. Akolekar DB, Hind AR, Bhargava SK. *J. Colloid Interf. Sci. 199*:92–98, 1998.
604. Shaw J, Harris RK, Norman PR. *Langmuir 14*:6716, 1998.
605. Weber WJ Jr., Morris JC. *J. San. Eng. Div.* (Proc. Amer. Soc. Civil Eng.) *90(SA3)*: 79–107, 1964.
606. Corkill JM, Goodman JF, Tate JR. *Trans. Faraday Soc. 63*(537):2264–2269, 1967.
607. Wang LK, Leonard RP, Wang MH, Goupil DW. *J. Appl. Chem. Biotechnol. 25*: 491–502, 1975.
608. Chobanu MM, Ropot VM. *Zh. Prikl. Khim. 56*(2):271–274, 1983.
609. Stenby EH, Birdi KS. *Prog. Colloid Polym. Sci. 70*:89–91, 1985.
610. Asakawa T, Ogino K. *Colloid Polym. Sci. 264*:1085–1089, 1986.
611. Mazet M, Yaacoubi A, Lafrance P. *Water Res. 22*:1321–1329, 1988.
612. Kamegawa K, Yoshida H. *Nippon Kagaku Kaishi 63*:789–794, 1989.
613. Leyva-Ramos R. *J. Chem. Technol. Biotechnol. 45*:231–240, 1989.
614. Allali-Hassani M, Dusart O, Mazet M. *Water Res. 24*:699–708, 1990.
615. Krishnakumar S, Somasundaran P. *Colloids Surfaces A 117*:227–233, 1996.
616. Rudzinski W, Dabrowski A, Narkiewicz-Michalek J, Podkoscielny P, Partyka S. *Polish J. Chem. 70*:231–252, 1996.
617. Kameya T, Urano K, Momonoi K. *Adsorption*, submitted for publication.
618. Radovic LR. in *Encyclopedia of Materials Science*, 2001, in press.
619. Anonymous. Discussion IV: Surface chemistry and physical properties of carbon affecting adsorption. In: Suffet IH, McGuire MJ, eds. *Activated Carbon Adsorption of Organics from the Aqueous Phase*, Vol. 1. Ann Arbor, MI: Ann Arbor Science, 1980, pp. 483–492.
620. Deshiikan SR, Papadopoulos KD. *J. Colloid Interf. Sci 174*:302–312, 1995.
621. Deshiikan SR, Papadopoulos KD. *J. Colloid Interf. Sci 174*:313–318, 1995.
622. Rosene MR, Manes M. *J. Phys. Chem. 81*(17):1651–1657, 1977.
623. Ha K-S, Hinago H, Sakoda A, Suzuki M. In: Suzuki M, ed. *Fundamentals of Adsorption*. Tokyo: Kodansha, 1993, pp. 251–258.
624. Longchamp S, Randriamahazaka HN, Nigretto J-M. *J. Colloid Interf. Sci. 166*:444–450, 1994.

625. Miller EJ. *J. Phys. Chem. 36*:2967–2980, 1932.
626. Snoeyink VL, Weber WJ. In: Weber WJ, Matijevic E, eds. *Adsorption from Aqueous Solution*. Washington, DC: American Chemical Society, 1968, pp. 112–134.
627. Tanabe K. *Solid Acids and Bases*. New York: Academic Press, 1970.
628. Tomita A, Tamai Y. *J. Phys. Chem. 75*:649–654, 1971.
629. Murata T, Matsuda Y. *Electrochim. Acta 27*(6):795–798, 1982.
630. Tobias H, Soffer A. *J. Electroanal. Chem. Interf. Electrochem. 148*:221–232, 1983.
631. Bard AJ. *J. Chem. Educ. 60*:302–304, 1983.
632. van der Linden WE, Dieker JW. *Anal. Chim. Acta 119*:1–24, 1980.
633. Beilby AL, Carlsson A. *J. Electronal. Chem. 248*:283–304, 1988.
634. Panzer RE, Elving PJ. *J. Electrochem. Soc. Electrochem. Sci. Technol. 119*(7):864–874, 1972.
635. Tarasevich MR, Khrushcheva EI. In: Conway BE, Bockris JO'M, White RE, eds. *Modern Aspects of Electrochemistry*, Vol. 19. New York: Plenum Press, 1989, pp. 295–358.
636. Pocard NL, Alsmeyer DC, McCreery RL, Neenan TX, Callstrom MR. *J. Mater. Chem. 2*:771–784, 1992.
637. Broderick PA. *Electroanal. 2*:241–251, 1990.
638. Rajeshwar K, Ibanez JG. *Environmental Electrochemistry*. San Diego: Academic Press, 1997.
639. Panzer RE, Elving PJ. *Electrochim. Acta 20*:635–647, 1975.
640. Randin J-P, Yeager E. *Electroanal. Chem. Interfac. Electrochem. 58*:313–322, 1975.
641. Fabish TJ, Hair ML. *J. Colloid Interf. Sci. 62*(1):16–23, 1977.
642. Oren Y, Tobias H, Soffer A. *J. Electroanal. Chem. 162*:87–99, 1984.
643. Oren Y, Soffer A. *J. Electroanal. Chem. 186*:63–77, 1985.
644. Oren Y, Soffer A. *J. Electroanal. Chem. 206*:101–114, 1986.
645. Cohen H, Soffer A, Oren Y. *J. Colloid Interf. Sci. 120*(1):272–280, 1987.
646. Golub D, Oren Y, Soffer A. *Carbon 25*:109–117, 1987.
647. Deakin MR, Stutts KJ, Wightman RM. *J. Electroanal. Chem. 182*:113–122, 1985.
648. Kamau GN, Willis WS, Rusling JF. *Anal. Chem. 57*:545–551, 1985.
649. Rubinstein I. *J. Electroanal. Chem. 183*:379–386, 1985.
650. Vasquez RE, Hono M, Kitani A, Sasaki K. *J. Electroanal. Chem. 196*:397–415, 1985.
651. Deakin MR, Kovach PM, Stutts KJ, Wightman RM. *Anal. Chem. 58*:1474–1480, 1986.
652. Nagaoka T, Yoshino T. *Anal. Chem. 58*:1037–1042, 1986.
653. Hance GW, Kuwana T. *Anal. Chem. 59*:131–134, 1987.
654. Micheal AC, Justice JB Jr. *Anal. Chem. 59*:405–410, 1987.
655. Bodalbhai L, Brajter-Toth A. *Anal. Chem. 60*:2557–2561, 1988.
656. Childers-Peterson TE, Brajter-Toth A. *J. Electroanal. Chem. 239*:161–173, 1988.
657. Saraceno RA, Ewing AG. *Anal. Chem. 60*:2016–2020, 1988.
658. Anjo DM, Kahr M, Khodabaksh MM, Nowinski S, Wanger M. *Anal. Chem. 61*:2603–2608, 1989.
659. Goyal RN. *Bull. Soc. Chim. France 1989*:343–347.
660. Stutts KJ, Scortichini CL, Repucci CM. *J. Org. Chem. 54*:3740–3744, 1989.

661. Sujaritvanichpong S, Aoki K. *Electroanal. 1*:397–403, 1989.
662. Bai Z-P, Nakamura T, Izutsu K. *Anal. Sci. 6*:443–447, 1990.
663. Bianco P, Haladjian J, Draoui K. *J. Electroanal. Chem. 279*:305–314, 1990.
664. Wang J, Martinez T, Yaniv DR, McCormick LD. *J. Electroanal. Chem. 313*:129–140, 1991.
665. Allred CD, McCreery RL. *Anal. Chem. 64*:444–448, 1992.
666. Zhang X, Wang C, Zhou X. *Anal. Chim. Acta 265*:27–34, 1992.
667. Downard AJ, Roddick AD. *Electroanalysis 6*:409–414, 1994.
668. Goyal RN, Jain AK, Jain N. *J. Chem. Soc. Perkin Trans. 2*:1055–1061, 1995.
669. Goyal RN, Jain AK, Jain N. *Bull. Chem. Soc. Jpn. 69*:1987–1995, 1996.
670. Eisinger RS, Keller GE. *Environ. Prog. 9*:235–244, 1990.
671. McCreery RL, Cline KK, McDermott CA, McDermott MT. *Colloids Surfaces 93*:211–219, 1994.
672. Surmann P, Peter B. *Electroanal. 8*:692–697, 1996.
673. McCreery RL. In: Bard AL, ed. *Electroanalytical Chemistry*, Vol. 17. New York: Marcel Dekker, 1991, pp. 221–374.
674. Radovic LR, Ume JI, Scaroni AW. In: LeVan MD, ed. *Fundamentals of Adsorption*. Norwell, MA: Kluwer Academic, 1996, pp. 749–756.
675. Snoeyink VL, Weber WJ Jr. *Environ. Sci. Technol. 1*(3):228–234, 1967.
676. Getzen FW, Ward TM. *J. Colloid Interf. Sci. 31*(4):441–453, 1969.
677. Ward TM, Getzen FW. *Environ. Sci. Technol. 4*(1):64–67, 1970.
678. Jossens L, Prausnitz JM, Fritz W, Schlünder EU. *Chem. Eng. Sci. 33*:1097–1106, 1978.
679. Myers AL, Zolandz RR. In: Suffet IH, McGuire MJ, eds. *Activated Carbon Adsorption of Organics from the Aqueous Phase*, Vol. 1. Ann Arbor, MI: Ann Arbor Science, 1980, pp. 243–250.
680. Al-Bahrani KS, Martin RJ. *Water Res. 10*:731–736, 1976.
681. Martin RJ, Al-Bahrani KS. *Water Res. 11*:991–999, 1977.
682. Martin RJ, Al-Bahrani KS. *Water Res. 12*:879–888, 1978.
683. Martin RJ, Al-Bahrani KS. *Water Res. 13*:1301–1304, 1979.
684. Martin RJ, Iwugo KO. *Water Res. 16*:73–82, 1982.
685. Abe I, Hayashi K, Kitagawa M. *Nippon Kagaku Kaishi 1977*:1905–1910.
686. Abe I, Hayashi K, Kitagawa M, Urahata T. *Bull. Chem. Soc. Japan 52*:1899–1904, 1979.
687. Abe I, Hayashi K, Kitagawa M, Urahata T. *Bull. Chem. Soc. Japan 53*:1199–1205, 1980.
688. Abe I, Hayashi K, Kitagawa M. *Bull. Chem. Soc. Japan 54*:2819–2820, 1981.
689. Abe I, Hayashi K, Kitagawa M. Bull. *Chem. Soc. Japan 55*:687–689, 1982.
690. Abe I, Hayashi K, Tatsumoto H, Kitagawa M, Hirashima T. *Water Res. 19*:1191–1193, 1985.
691. Tanada S, Kawasaki N, Nakamura T, Abe I. *J. Colloid Interf. Sci. 177*:329–334, 1996.
692. Chiou CT, Peters LJ, Freed VH. *Science 206*:831–832, 1979.
693. Belfort G. *Environ. Sci. Technol. 13*:939–946, 1979.
694. Miller S. *Environ. Sci. Technol. 14*:1037–1049, 1980.
695. Kamlet MJ, Doherty RM, Abraham MH, Taft RW. *Carbon 23*(5):549–554, 1985.

696. Nirmalakhandan N, Speece RE. *Environ. Sci. Technol.* 22:606–615, 1988.
697. Nirmalakhandan N, Speece RE. *Environ. Sci. Technol.* 24:575–580, 1990.
698. Blum DJW, Suffet IH. In: Mallevialle J, Suffet IH, Chan US, eds. *Influence and Removal of Organics in Drinking Water.* Boca Raton, FL: Lewis, 1992; pp. 67–78.
699. Blum DJW, Suffet IH, Duguet JP. *Crit. Rev. Environ. Sci. Technol.* 23(2):121–136, 1993.
700. Luehrs DC, Hickey JP, Nilsen PE, Godbole KA, Rogers TN. *Environ. Sci. Technol.* 30:143–152, 1996.
701. Brasquet C, Bourges B, Le Cloirec P. *Environ. Sci. Technol.* 33:4226–4231, 1999.
702. Crittenden JC, Sonongraj S, Bulloch JL, Hand DW, Rogers TN, Speth TF, Ulmer M. *Environ. Sci. Technol.* 33:2926–2933, 1999.
703. Urano K, Sonai M, Nakayama R, Kobayashi Y. *Nippon Kagaku Kaishi 1976*:1773–1778.
704. Urano K, Kano H. *Bull. Chem. Soc. Japan* 57:2051–2054, 1984.
705. Altshuler G, Belfort G. In: McGuire MJ, Suffet IH, eds. *Treatment of Water by Granular Activated Carbon* (Advances in Chemistry Series Vol. 202). Washington, DC: American Chemical Society, 1983, pp. 29–61.
706. Anonymous. Discussion I: Theoretical approaches. In: McGuire MJ, Suffet IH, eds. *Treatment of Water by Granular Activated Carbon* (Advances in Chemistry Series Vol. 202). Washington, DC: American Chemical Society, 1983, pp. 107–118.
707. Semmens MJ, Norgaard GE, Hohenstein G, Staples AB. *J. Amer. Water Works Assoc.* (May 1986):89–93.
708. van Driel J. The chemical state of the surface of activated carbon and its relevance for gas and liquid phase adsorption. In: Liapis AI, ed. *Fundamentals of Adsorption.* Amer. Inst. Chem. Eng., 1987.
709. Helmy AK, Ferreiro EA, De Bussetti SG. *Ads. Sci. Technol.* 4:211–216, 1987.
710. Cooney DO, Wijaya J. In: Liapis AI, ed. *Fundamentals of Adsorption.* Amer. Inst. Chem. Eng., 1987, pp. 185–194.
711. Mazet M, Farkhani B, Baudu M. *Water Res.* 28:1609–1617, 1994.
712. Tung LA, King CJ. *Ind. Eng. Chem. Res.* 33:3217–3223, 1994.
713. Newcombe G, Drikas M. *Water Res.* 27(1):161–165, 1993.
714. Chipot C, Maigret B, Pearlman DA, Kollman PA. *J. Amer. Chem. Soc.* 118:2998–3005, 1996.
715. Mecozzi S, West AP Jr., Dougherty DA. *J. Amer. Chem. Soc.* 118:2307–2308, 1996.
716. Heard GL, Boyd RJ. *J. Phys. Chem. A* 101:5374, 1997.
717. Ma JC, Dougherty DA. *Chem. Rev.* 97:1303–1324, 1997.
718. Williams VE, Lemieux RP. *J. Amer. Chem. Soc.* 120:11311–11315, 1998.
719. Miyake M, Yasuda K, Kashihara T, Teranishi T. *Chem. Lett. 1999*:1037–1038.
720. McDermott MT, McCreery RL. *Langmuir* 10:4307–4314, 1994.
721. Hirschfelder JO, Curtiss CF, Bird RB. *Molecular Theory of Gases and Liquids.* New York: John Wiley, 1954.
722. Bruch LW, Cole MW, Zaremba E. *Physical Adsorption: Forces and Phenomena.* Oxford: Clarendon Press, 1997.
723. Kaliszan R, Osmialowski K, Bassler BJ, Hartwick RA. *J. Chromat.* 499:333–344, 1990.

724. Abraham MH, Doherty RM, Kamlet MJ, Taft RW. *Chem. Britain* (June 1986): 551–554.
725. Kamlet MJ, Doherty RM, Abboud JLM, Abraham MH, Taft RW. *CHEMTECH* 23(September):566–576, 1986.
726. Abraham MH, Buist GJ, Grellier PL, McGill RA, Doherty RM, Kamlet MJ, Taft RW, Maroldo SG. *J. Chromat.* 409:15–27, 1987.
727. Abraham MH, Andonian-Haftvan J, Whiting GS, Leo A, Taft RW. *J. Chem. Soc. Perkin Trans* 2:1777–1791, 1994.
728. McGuire MJ, Suffet IH. In: Suffet IH, McGuire MJ, eds. *Activated Carbon Adsorption of Organics from the Aqueous Phase*, Vol. 1. Ann Arbor, MI: Ann Arbor Science, 1980, pp. 91–115.
729. Manes M. In: Suffet IH, McGuire MJ, eds. *Activated Carbon Adsorption of Organics from the Aqueous Phase*, Vol. 1. Ann Arbor, MI: Ann Arbor Science, 1980, pp. 43–64.
730. Klopman G. *J. Amer. Chem. Soc.* 90(2):223–234, 1968.
731. Jensen WB. *The Lewis Acid-Base Concepts*. New York: Wiley-Interscience, 1980.
732. Jean Y, Volatron F. *An Introduction to Molecular Orbitals*. New York: Oxford University Press, 1993.
733. Tamon H, Atsushi M, Okazaki M. *J. Colloid Interf. Sci.* 177:384–390, 1996.
734. Tamon H, Atsushi M, Okazaki M. In: LeVan MD, ed. *Fundamentals of Adsorption*. Norwell, MA: Kluwer Academic, 1996, pp. 961–968.
735. Tamon H, Okazaki M. *J. Colloid Interf. Sci.* 179:181–187, 1996.
736. Ume JI, Scaroni AW, Radovic LR. Effect of surface chemical properties of activated carbons on the adsorption of carboxyl anions. *21st Biennial Carbon Conference*. Buffalo, NY, 1993, pp. 468–469.
737. Silva IF, Ume JI, Scaroni AW, Radovic LR. *ACS Preprints* (Div. Fuel Chem.) 41(1):461–465, 1996.
738. Radovic LR, Silva IF, Ume JI, Menéndez JA, Leon y Leon CA, Scaroni AW. *Carbon* 35:1339–1348, 1997.
739. Haderlein S, Schwarzenbach RP. *Environ. Sci. Technol.* 27:316–326, 1993.
740. Epstein B, Dalle-Molle E, Mattson JS. *Carbon* 9:609–615, 1971.
741. Marsh H, Campbell HG. *Carbon* 9:489–498, 1971.
742. Mattson JS, Lee L, Mark HB Jr., Weber WJ Jr. *J. Colloid Interf. Sci.* 33(2):284–293, 1970.
743. Radke CJ, Prausnitz JM. *AIChE J.* 18(4):761–768, 1972.
744. Barton SS, Harrison BH. *Carbon* 10:245–251, 1972.
745. Glushchenko VY, Levagina TG, Pershko AA. *Russ. Colloid J.* (Koll. Zh.) 37:111–113, 1975.
746. Pershko AA, Glushchenko VY. *Russ. J. Phys. Chem.* 50:1247–1249, 1976.
747. Balzer D, Lange H. *Colloid Polym. Sci.* 255:140–152, 1977.
748. Suffet IH, McGuire MJ, eds. *Activated Carbon Adsorption of Organics from the Aqueous Phase*, Vol. 1. Ann Arbor, MI: Ann Arbor Science, 1980.
749. Oda H, Yokokawa C. *Carbon* 21:303–309, 1983.
750. Britton WE, Assubaie F. *J. Electroanal. Chem.* 178:153–163, 1984.
751. McCreery RL. In: Wieckowski A, ed. *Interfacial Electrochemistry: Theory, Experiment, and Applications*. New York: Marcel Dekker, 1999, pp. 631–647.

752. Kim BR. *J. Water Poll. Contr. Fed. 48*(1):120–133, 1976.
753. Chen ASC, Larson RA, Snoeyink VL. *Environ. Sci. Technol. 16*:268–273, 1982.
754. Chudyk WA, Snoeyink VL. *Environ. Sci. Technol. 18*:1–5, 1983.
755. Wedeking CA, Snoeyink VL, Larson RA, Dung J. *Water Res. 21*:929–937, 1987.
756. Chatzopoulos D, Varma A, Irvine RL. *AIChE J. 39*:2027–2041, 1993.
757. Vidic RD. *Catalytic Properties of Activated Carbon for Oxidative Coupling of Organic Compounds*, unpublished report, University of Pittsburgh, 1993.
758. Polcaro AM, Palmas S. *Ind. Eng. Chem. Res. 36*:1791–1798, 1997.
759. Utrera-Hidalgo E, Moreno-Castilla C, Rivera-Utrilla J, Ferro-García MA, Carrasco-Marín F. *Carbon 30*:107–111, 1992.
760. Ferro-Garcia MA, Joly JP, Rivera-Utrilla J, Moreno-Castilla C. *Langmuir 11*: 2648–2651, 1995.
761. Ferro-Garcia MA, Rivera-Utrilla J, Bautista-Toledo I, Moreno-Castilla C. *J. Chem. Technol. Biotechnol. 67*:183–189, 1996.
762. Rivera-Utrilla J, Utrera-Hidalgo E, Ferro-García MA, Moreno-Castilla C. *Carbon 29*:613–619, 1991.
763. Morgun NP, Khabalov VV. *Russ. J. Phys. Chem. 70*:815–818, 1996.
764. Tamon H, Saito T, Kishimura M, Okazaki M, Toei R. *J. Chem. Eng. Japan 23*(4): 426–432, 1990.
765. Tamon H, Okazaki M. In: Suzuki M, ed. *Fundamentals of Adsorption*. Tokyo: Kodansha, 1993, pp. 663–669.
766. Tamon H, Kitamura K, Okazaki M. *AIChE J. 42*(2):422–430, 1996.
767. Tamon H, Okazaki M. *Carbon 34*:741–746, 1996.
768. Tamon H, Okazaki M. *J. Colloid Interf. Sci. 196*:120–122, 1997.
769. Khabalov VV, Pershko AA, Gorchakova NK, Gluschchenko VY. *Izv. Akad. Nauk SSSR Ser. Khim 1984*:263–265.
770. Mattson JS, Mark HB Jr. *J. Colloid Interf. Sci. 31*(1):131–144, 1969.
771. Ditl P, Coughlin RW, Jere EH. *J. Colloid Interf. Sci. 63*:410–420, 1978.
772. Contescu C, Vass MI. *Appl. Catal. 33*:259–271, 1987.
773. Fuerstenau DW, Oseo-Asare K. *J. Colloid Interf. Sci. 118*(2):524–542, 1987.
774. Walker PL Jr., Taylor RL, Ranish JM. *Carbon 29*(3):411–421, 1991.
775. Chen SG, Yang RT, Kapteijn F, Moulijn JA. *Ind. Eng. Chem. Res. 32*:2835–2840, 1993.
776. Skokova K, Radovic LR. In: *American Chemical Society Div. Fuel Chem. Preprints*. New Orleans, 1996, pp. 143–147.
777. Yopps JA, Fuerstenau DW. *J. Colloid Sci. 19*:61–71, 1964.

Index

407